NATURAL ATTENUATION OF HAZARDOUS WASTES

SPONSORED BY
Natural Attenuation Task Committee
Environmental and Multi-Media Council
Environmental and Water Resources Institute (EWRI) of the American
Society of Civil Engineers

EDITED BY
Rao Surampalli
Say Kee Ong
Eric Seagren
Julio Nuno
Shankha Banerji

Published by the American Society of Civil Engineers

Cataloging-in-Publication Data on File with the Library of Congress

Published by American Society of Civil Engineers
1801 Alexander Bell Drive
Reston, Virginia 20191
www.pubs.asce.org

Any statements expressed in these materials are those of the individual authors and do not necessarily represent the views of ASCE, which takes no responsibility for any statement made herein. No reference made in this publication to any specific method, product, process or service constitutes or implies an endorsement, recommendation, or warranty thereof by ASCE. The materials are for general information only and do not represent a standard of ASCE, nor are they intended as a reference in purchase specifications, contracts, regulations, statutes, or any other legal document. ASCE makes no representation or warranty of any kind, whether express or implied, concerning the accuracy, completeness, suitability, or utility of any information, apparatus, product, or process discussed in this publication, and assumes no liability therefore. This information should not be used without first securing competent advice with respect to its suitability for any general or specific application. Anyone utilizing this information assumes all liability arising from such use, including but not limited to infringement of any patent or patents.

ASCE and American Society of Civil Engineers—Registered in U.S. Patent and Trademark Office.

Photocopies: Authorization to photocopy material for internal or personal use under circumstances not falling within the fair use provisions of the Copyright Act is granted by ASCE to libraries and other users registered with the Copyright Clearance Center (CCC) Transactional Reporting Service, provided that the base fee of $18.00 per article is paid directly to CCC, 222 Rosewood Drive, Danvers, MA 01923. The identification for ASCE Books is 0-7844-0740-1/04/ $18.00. Requests for special permission or bulk copying should be addressed to Permissions & Copyright Dept., ASCE.

Copyright © 2004 by the American Society of Civil Engineers.
All Rights Reserved.
ISBN 0-7844-0740-1
Manufactured in the United States of America.

Table of Contents

Contributing Authors v

Chapter 1   Introduction 1
    1.1   Background 1
    1.2   Definition 2
    1.3   Government Policies and Rules 2
    1.4   General Evaluation Strategy 4
    1.5   Advantages and Disadvantages 6
    1.6   Organization of book 6

Chapter 2   Site Characterization 8
    2.1   Project Planning 8
    2.2   Evaluation of Historical Data 12
    2.3   Site Characterization Methods 14
    2.4   Site Characterization Report 25
    2.5   Development of a Site Conceptual Model 25

Chapter 3   Natural Attenuation Processes 29
    3.1   Physical Processes 29
    3.2   Chemical Processes 58
    3.3   Biological Processes 63
    3.4   Modeling of Natural Attenuation 73

Chapter 4   Long Term Monitoring 113
    4.1   Introduction 113
    4.2   Monitoring Wells 114
    4.3   Sampling and Analysis 115
    4.4   Contingency Plan 116

Chapter 5   Natural Attenuation in Soils, Sediments and Groundwater 120
    5.1   Petroleum Hydrocarbons and MTBE 120
    5.2   Chlorinated Solvents 167
    5.3   Polycyclic Aromatic Hydrocarbons 182
    5.4   Metals 200
    5.5   Radioactive Contaminants 209

Contributors

Contributing Authors

Eric Seagren (Chapter 3.1 and 5.1)
University of Maryland, College Park, Maryland

Say Kee Ong (Chapter 3.2 and 5.3)
Iowa State University, Ames, Iowa

Shane Rogers (Chapter 5.3)
Iowa State University, Ames, Iowa

Rao Y. Surampalli (Chapter 1 and 4)
U.S. Environmental Protection Agency, Kansas City, Kansas

Julio Nuno (Chapter 2)
SCS Engineers, Long Beach, California

Shanka Benerji (Chapter 1 and 4)
University of Missouri, Columbia, Missouri

Brian Wrenn (Chapter 3.3 and 5.2)
Washington University, St. Louis, Missouri

Jennifer Becker (Chapter 3.1 and 5.1)
University of Maryland, College Park, Maryland

Bruce Jacobs (Chapter 3.4)
Jacobs Consulting Services, Brookline, Massachusetts

Warren Brady (Chapter 5.5)
IT Group, Baton Rouge, Louisiana

Jeff Marshall (Chapter 5.4)
SCS Engineers, Reston, Virginia

Robert H. Fitzgerald, P.E. (Chapter 3.4)
Camp Dresser & McKee, Inc.

Tom Buechler (Chapter 2)
Black and Veatch, Kansas City, Missouri

Fernando Miralles-Wilhelm (Chapter 3.4)
Northeastern University, Boston, Massachusetts

Nelson Mix (Chapter 1)
U.S. Environmental Protection Agency, Washington, D.C.

Patrick V. Brady (Chapter 5.5)
Sandia National Laboratories, Albuquerque, NM

Matthew J. Eick (Chapter 5.5)
Virginia Poly Technic Institute and University

Paul R. Grossl (Chapter 5.5)
Utah State University

Mark DiSteffano (Chapter 5.5)
IT Corporation, Las Vegas, NV

Reviewers

Raveendra Damera, P.E.
General Physics Corporation, Clarksville, MD

Bijay K. Panigrahi, Ph.D., P.E., P.G.
BP Consulting Group. Inc., Orlando, Florida

Joseph Devinny
University of Southern California, Los Angeles, CA

CHAPTER 1

Introduction

1.1 Background

In the United States and other industrial countries, many groundwater and valuable land resources have been contaminated with petroleum hydrocarbons, chlorinated solvents and other chemicals through leaky tanks, lagoons, spills, improper disposal, landfills and industrial discharges. The remediation of these sites involves expensive site investigations and engineered solutions. In addition, it was found that some conventional processes (such as pump and treat) for remediation of such sites were not capable of achieving desired goals of contaminant concentrations approaching drinking water standards within reasonable time periods. The 1994 report of the National Research Council clearly indicated that for many sites there were few technical remedies available to restore them to safe conditions (National Research Council, 1994). Within this context, natural attenuation (NA), a process that relies upon the natural assimilative capacity of the site to reduce or stabilize contaminants to desirable levels appears to be an attractive alternative. The natural attenuation process was thought to be a "no action" alternative by environmental groups as very often not much human intervention was needed. But presently, it is a viable alternative for site clean up on a case-by-case basis.

For managing the pollutant loads (BOD, nutrients, etc.) from point sources discharged into surface waters, the natural assimilative capacity (or natural attenuation) of the receiving water has been considered to determine the treatment necessary to maintain the desired water quality goals in the particular stream section. Thus, the concept of natural attenuation is not new to pollution control strategies. The application of this concept to groundwater remediation is perhaps new, but natural attenuation of underground contaminants occurs regardless of our perception about the process. These natural processes occur when contaminants are introduced in the environment. Physical, chemical, and biochemical interactions of the contaminants with soil, water and gases in the media occur that may result in a decline in concentration or toxicity or stabilization of the contaminants. This process possibly occurs at all sites to a varying extent depending on the types and concentration of the contaminants present and the environment of the size.

The use of natural attenuation as a remediation method in the Superfund program climbed from 6 % in early 1990s to about 25% in later 1990s (MacDonald, 2000). The increase of its use has been even more dramatic for remediation of underground storage tank sites where it is the leading remediation method at the present time.

1.2 Definition

The process of natural attenuation of contaminants, also referred to as intrinsic or passive remediation, or natural recovery, or natural assimilation, occurs at contaminated sites (groundwater and soils) resulting in a decrease in observed concentration of the contaminants with time. For situations that use NA as a site remedial tool, US Environmental Protection Agency (US EPA) requires that monitoring be a key feature of the process and has renamed the process as monitored natural attenuation (MNA). EPA definition of MNA is as follows (US EPA, 1997): the term *monitored natural attenuation* refers to the reliance on natural attenuation processes (within the context of a carefully controlled and monitored site cleanup approach) to achieve site-specific remedial objectives within a time frame that is reasonable compared to that offered by other more active methods. The "natural attenuation processes" that are at work in such a remediation approach include a variety of physical, chemical, or biological processes that, under favorable conditions, act without human intervention to reduce the mass, toxicity, mobility, volume, or concentration of contaminants in soil and groundwater. These in-situ processes include biodegradation, dispersion, dilution, sorption, volatilization and chemical or biological stabilization, transformation or destruction of contaminants, complexation and other unknown processes without human intervention. Natural attenuation includes many different processes that reduce or stabilize the contaminants underground while the key process involved in intrinsic bioremediation is biodegradation of the contaminants.

1.3 Government Policies and Rules

Soon after it was realized that at many sites natural attenuation was reducing the hazards of disposed contaminants, government regulations and policies were formulated to properly address the use of MNA such that environmental contamination was reduced. The EPA memorandum (US EPA, 1997) was issued to clarify the use of MNA at various Superfund, underground storage tank and Resources Conservation and Recovery Act (RCRA) sites. This memorandum indicated that it was only one component of the total remedy at the site and that great care should be used if it were the sole remedy at a site. The process was to be used with other remediation objectives such as source control and restoration of contaminated groundwater, and the selection of this alternative had to be supported by detailed, site specific information demonstrating its efficacy. It also indicated that MNA was not to be considered as a default or presumptive remedy at any contaminated site, but as an appropriate remediation method as long as it was capable of achieving site-specific remediation objectives within the time frame specified.

The latest EPA directive (US EPA,1999) reiterates some of these points and clarifies others. It is expected that MNA will be appropriate for sites that have a low potential for contaminant migration. It is possible that for sites that have mixtures of contaminants, some of the contaminants may be amenable to MNA while others may not. A good example of this could be a site contaminated with gasoline containing benzene, toluene, ethylbenzene and xylene (BTEX) and methyl tertiary-butyl ether (MTBE). The BTEX compounds could attenuate well with time but MTBE may not. In such cases, care must be exercised to investigate the threat from these other compounds. There may be other examples where the natural attenuation of the compounds of interest may lead to transformations in the soil, causing the formation of toxic and/or mobile products (e.g., formation of vinyl chloride from trichloroethylene biodegradation). In some situations the compound of interest may, after natural attenuation, transform into a different phase (i.e., solid phase to liquid phase or liquid phase to gaseous phase), which could make the products more mobile and hazardous. In all such cases, evaluation must be made to see if MNA is an appropriate technology to reduce the long-term risk from the site. For petroleum-based contaminants, MNA can reduce the concentrations of BTEX compounds present by biodegradation, form non-toxic end products and stop the spread of the contaminant plume at the site. However, the heavier components of the petroleum product may still remain, causing potential harm to human health. Thus, application of MNA may require supplementary activities such as source and institutional controls to remediate the site.

The EPA regulations and policies for remediating site vary with types of contaminants and locations, but they share some key principles listed below:
- Source control measures should use treatment to address "principal threat" wastes whenever possible and engineering controls such as containment for waste that pose a relatively low long-term threat, or where treatment is impractical. Principal threat wastes are highly toxic or mobile materials that cannot be reliably contained or would present a significant risk to human health or the environment should exposure occurs.
- Contaminated groundwaters should be returned to "their beneficial uses whenever practicable, within a time frame that is reasonable given the particular circumstances of the site". When restoration of the groundwater is not practicable, further migration of the plume should be prevented.
- Contaminated soil should be remediated to achieve an acceptable level of risk to human beings and the environmental receptors and to prevent any transfer of contaminants to other media that would result in an unacceptable risk or exceed required cleanup levels.
- Remedial actions in general should include opportunities for public involvement that serve to educate interested parties and to solicit their feedback concerning the decision making process.

The selection of MNA as a remediation process at a site does not change the remedy selection principles mentioned above. It also does not reduce the responsibility of the regulated party to achieve long-term site remediation objectives. The effectiveness of MNA should be demonstrated to EPA for both short-term and long-term time frames. The efficacy of MNA must be demonstrated by obtaining site-specific data that provide historical groundwater and/or soil information showing declining contaminant concentrations, hydrogeologic and geochemical data along with groundwater modeling showing that MNA processes are at work at reasonable rates. Some field or microcosm studies may be required to substantiate the success of the process.

The MNA remediation process is appropriate at sites where it achieves remedial objectives within a reasonable time frame and meets the applicable remedy selection criteria for the particular program. It may be used in conjunction with other active remediation methods, or as a follow-up method. It should not be used at sites where such an approach may result in significant contaminant migration or unacceptable impacts to receptors.

The reasonable time frame mentioned above is site-specific, but it should not be excessive compared with other options. Groundwater modeling may be used in the estimation of the reasonable time frame. Some of the factors that impact the decision about reasonableness of the time frame are: current and potential future uses of affected groundwater; the time when the aquifer might be needed for water supply; public acceptance of the extended time for remediation; reliability of monitoring and institutional controls, and funding availability over the project duration.

1.4 General Evaluation Strategy

Demonstration of MNA at a site has been evaluated on the basis of three types of lines of evidence (National Research Council,1994, Wiedemeier, 1995). These lines of evidence are:
1. Historical data showing stabilization of plume and/or contaminant mass reduction over time
2. Chemical and geochemical data that include depletion of electron acceptors or donors; increase in metabolic by-product concentrations; decreasing concentration of contaminant compound; increase in concentration of daughter compounds
3. Microbiological data supporting the occurrence of biodegradation and providing estimates of biodegradation rates.

These lines of evidence or "footprint" can be found in locations where MNA is actively taking place (Rittman, 2000). Example of chemical or geochemical changes due to MNA would be the consumption of dissolved oxygen and production of inorganic carbon (bicarbonates and carbonates) at the contaminated site when aerobic biodegradation of petroleum hydrocarbon occurs. Many organizations, including the federal government, state government, technical societies and private companies have prepared protocols for evaluating MNA at specific sites (National Research Council, 2000). US Air Force, in collaboration with EPA and Parsons Engineering Science, Inc., developed one of the first such protocols to evaluate MNA of fuel contamination of groundwater (Wiedemeier et al., 1995). The protocol described the procedures for evaluating a site, the extent of the nonaqueous phase zones, groundwater chemical processes, aquifer parameters, and biological activities. It provided methods for calculating a mass balance of electron donors and acceptors for confirming MNA activities.

Later, US Air Force in collaborative efforts with EPA, USGS and Parsons Engineering Science, Inc. also prepared a protocol for evaluating MNA for chlorinated solvent sites (Wiedemeier et al., 1996). Other notable protocols for evaluating MNA were by EPA (US EPA. 1998), DOE (Brady et al., 1998) and ASTM (ASTM, 1998). The EPA protocol was prepared in collaboration with the US Air Force and dealt with chlorinated solvents. Most of the protocol address contamination problems from fuel hydrocarbons and chlorinated solvents as there is considerable amount of MNA data available on these contaminants. The DOE protocol does address other contaminants, such as inorganic compounds.

The National Research Council report on MNA for groundwater remediation (National Research Council, 2000) included a section discussing various protocols and listed attributes for assessing the adequacy of these protocols. These attributes included such things as: community involvement, institutional controls and long-term monitoring, contingency plans, establishment of cause and effect relationship, site condition assessment, sustainability of the process, peer review of the protocol and implementation details.

The decision to use MNA as a remediation method at a site depends upon a detailed site investigation to evaluate the local conditions, potential for natural attenuation and any site specific constraints that may be present. This monitoring effort must be continued over time to confirm if contaminant concentrations are being reduced at reasonable rates to ensure protection of human health and the environment. The data should clearly indicate if the soil and groundwater concentrations of contaminant are being reduced without any active remedial steps. If this is not the case, MNA remedial process needs to be replaced by more aggressive treatment options.

In the MNA process, generally the contaminant plume migrates farther than in cases where active remedial measures are in use. It is important to determine whether human beings or sensitive environmental areas might be affected by this contaminant migration either through direct contact or inhalation of vapors. In case of use of the groundwater for drinking purposes by future residential communities down gradient from the site, the potential future risks should be evaluated. If direct contact with contaminated soil is possible, methods to control such contact or other alternative remedial methods should be implemented.

In summary, the key components of the remediation of a contaminated site by MNA process are (ASTM,1998):
1. Site characterization and establishment of remedial goals
2. Evaluation of the plume status
3. Comparison of the MNA process performance to remedial goals
4. Comparing MNA process to other remedial options, and
5. Development and implementation of an appropriate monitoring program.

1.5 Advantages and Disadvantages

Like most processes, MNA has some distinct advantages and disadvantages. These should be carefully considered during site characterization and evaluation of remediation alternatives (US EPA, 1999).

Advantages:

- It being an in-situ process, generates lesser volume of remediation wastes, lesser chances of cross-media transfer of contaminants as might happen for ex-situ processes, has reduced risk of human exposure to contaminants, and less disturbance to ecological receptors.
- It has lower costs than most active remediation alternatives.
- It has less intrusion and minimal disturbance to the site operations.
- It can be used with, or as a follow-up process to other active remedial measures.
- In some natural attenuation processes in-situ destruction of contaminant will occur.
- It can be applied to all or part of a given site depending on the site conditions and remediation goals.

Disadvantages:

- It requires more time to achieve remediation objectives than other active processes.
- The site characterization will be more complex and expensive.
- It requires more extensive and prolonged long-term monitoring.
- The transformation products of remediation in some cases may be more toxic and /or more mobile than the original compound.
- There is a possibility of continued contaminant migration and/or cross-media transfer during the remediation phase.
- It may require institutional control and the site may not be available for use until contaminant levels are acceptable.
- The hydrologic and geochemical conditions amenable to natural attenuation may change over time which may result in renewed mobility of previously stabilized contaminants that may adversely impact the remedial effectiveness.
- It may require more extensive education and outreach efforts to gain acceptance by the public.

1.6 Organization of Book

This book has five chapters. The chapter one introduces the subject matter with definitions, government policies, evaluation strategies, and advantages and disadvantages of the process. Site characterization is the subject for Chapter 2. In Chapter 3, all processes (physical, chemical, geochemical, and biological) that effect natural attenuation are discussed in detail. Modeling of the MNA process is also included in this chapter. Long-term monitoring is described in Chapter 4. In Chapter 5, natural attenuation in soils, sediments and groundwater for specific contaminants are discussed. The contaminants discussed are: petroleum hydrocarbons, MTBE, chlorinated solvents/hydrocarbons, polyaromatic hydrocarbons, metals, and radioactive materials.

References

American Society for Testing and Materials, (1998). Standard Guide for Remediation of Groundwater by Natural Attenuation at Petroleum Release Sites, ASTM E – 1943-98, ASTM, Philadelphia, PA.

Brady, P. V., Spalding, B. P., Krupa, K. M., Waters, R. D., Zhang, P., Borns, D. J., and Brady, W. D., (1998). Site screening and technical guidance for monitoring natural attenuation at DOE Site, Draft Report, Sandia National Laboratory, Albuquerque, NM.

Macdonald, J.A., (2000). "Evaluating natural attenuation for groundwater cleanup" Environmental Sci. Technol., 33,(15), 346A-353A.

National Research Council, (1994). Alternatives for Groundwater Cleanup, National Academy Press, Washington, D.C.

National Research Council, (2000). Natural Attenuation for Groundwater Remediation, National Academy Press, Washington, D.C.

Rittman, B., (2000). Natural attenuation's promise and application, Water 21, International Water Association, pp.20-22.

U.S. EPA, (1997). Use of Monitored Natural Attenuation at Superfund, RCRA Corrective Action and Underground Storage Tank Sites, Office of Solid Waste and Emergency Response Directive 9200.4-17P, US EPA, Washington , D.C.

U.S. EPA, (1998). Technical Protocol for Evaluating Natural Attenuation of Chlorinated Solvents in Groundwater, Office of Research and Development, US EPA, Cincinnati, OH.

U.S. EPA, (1999). Use of Monitored natural Attenuation at Superfund, RCRA Corrective Action and Underground Storage Tank Sites, Office of Solid Waste and Emergency Response Directive 9200.4-17P, USEPA, Washington, D.C.

Wiedemeier, T.H., Wilson, J. T., Kambell, D.H., Miller, R. N., and Hansen, J. E., (1995). Technical Protocol for Implementing Intrinsic Remediation with Long-Term Monitoring for Natural Attenuation of Fuel Contamination Dissolved in Groundwater, U.S. Air Force Center for Environmental Excellence, Brooks Air Force Base, San Antonio, TX.

Wiedemeier, T. H., Swanson, M. A., Moutoux, D. E., Gordon, E., K., Wilson, J. T., Wilson, B. H., Kampbell, D. H., Hansen, J. E., Haas, P., and Chapelle, F. H., (1996), Technical Protocol for Evaluating Natural Attenuation of Chlorinated Solvents in Groundwater, U.S. Air Force Center for Environmental Excellence, Brooks Air Force Base, San Antonio, TX.

CHAPTER 2

Site Characterization

2.1 Project Planning

Project plan development is an important aspect of any site characterization. Planning detail depends on the complexity of the characterization investigation. Planning is important because it ensures that the scope of work is well thought out and possible contingencies are addressed (Lewis and Wilson, 1995). Proper planning minimizes mid-course modifications to the project and also ensures that budgets can be met.

The planning process should be formalized in writing, either through a work plan submitted to a client or regulator or through preparation of a detailed proposal. Most planning steps become part of the effort expended in preparing the work plan and proposal. Initial work plans may be general, becoming more detailed after consultant selection and in-house evaluations.

Site characterization investigations are usually conducted in phases which may be iterative (Figure 2-1). Information obtained in one phase is used to determine the course of the next. Although this approach can be repetitive, it is often cost-effective because the investigation is further focused and refined at each phase. For example, contaminants whose presence was initially suspected may be eliminated from subsequent analysis plans if data from a previous phase shows they are not present. Similarly, areas within the site can be eliminated from consideration during later phases if a round of data collection shows them to be clean. By the same token, additional analytical parameters or areas of investigation may be added during later phases if initial investigations suggest that this is warranted.

Typically, a site characterization or investigation project is conducted in response to a suspected or confirmed spill or other release of hazardous substance; a recommendation for additional investigation in a Phase I Preliminary Site Assessment; closure of a former industrial operation or facility; a regulatory agency inspection or order; a consent order; or a regulatory directive or requirement. During the planning process, the planning professional (including both owner's representative and engineers) should be aware of the reason that the investigation is required and the ultimate objectives so that the characterization can be directed accordingly. If possible, the final use of the resulting data or report should also be identified so that any report is written to address this anticipated use and audience. Data collection methods and detection limits should be closely scrutinized during the planning step if there is a possibility that the data will be used for risk assessment purposes. Similarly, close scrutiny will be required if litigation is involved or contemplated.

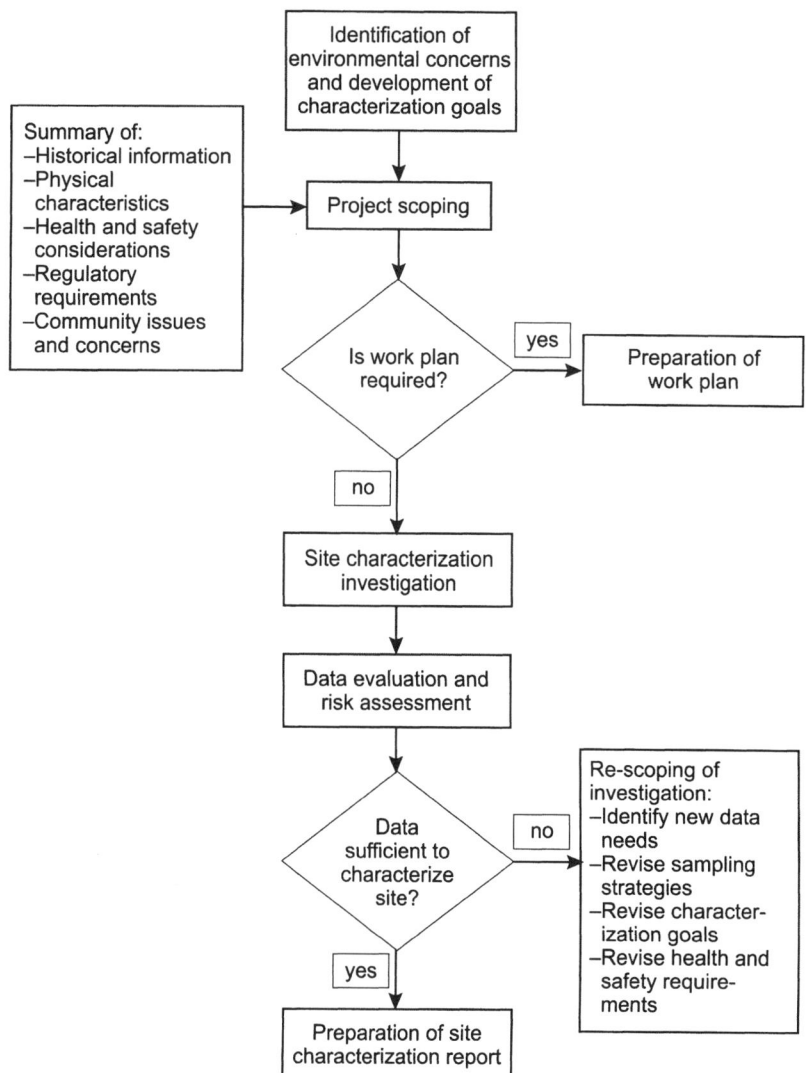

Figure 2-1. Site characterization process

Table 2-1. Seven steps of the data quality objectives (DQO) process

Step 1   **Problem statement**
Concisely describe the problem to be studied. Review prior studies and existing information to gain a sufficient understanding to define the problem.

Step 2   **Identify the decision that addresses the problem**
Identify what questions the study will attempt to resolve and what actions may result.

Step 3   **Inputs affecting decision**
Identify the information that must be obtained and the measurements that need to be taken to implement the decision statement.

Step 4   **Define the boundaries of the study**
Specify the time periods and spatial areas to which decisions will apply. Determine when and where data should be collected.

Step 5   **Decision rules**
Define the statistical parameter of interest (if applicable), specify the action level, and integrate the previous DQO inputs into a single statement that describes the logical basis for choosing among the alternative actions.

Step 6   **Limits on uncertainty**
Define the decision maker's tolerable decision error rates considering the consequences of making an incorrect decision.

Step 7   **Optimize the design**
Design the field investigation, giving adequate consideration to the results of Steps 4, 5, and 6 and the available resources. This step is often done in preliminary or outline form within the DQOs and is addressed fully in the alternatives evaluation plan.

EPA's seven-step data quality objectives (DQO) process has often been a valuable tool for defining specific goals (US EPA 1994). It is not necessary to apply every step to every project, but the analytical and decision-making process will be valuable.

The DQO process is a strategic planning approach developed by the EPA quality assurance management staff to facilitate the planning of data collection activities (Table 2-1). The process is designed to help planners achieve program goals by focusing the purpose of the investigation and directing attention to the potential uses of the data. It requires that likely response actions to potential investigation results be identified before the investigation begins.

The first step in planning a project is to collect pertinent background information. The experienced professional will determine what information exists and decide what is relevant. A good source of background information is a Phase I

Environmental Site Assessment, if one has been conducted (ASTM, 2000). In the absence of a Phase I Assessment, a history of previous site use obtained through verbal communications and review of chain-of-title, historical aerial photographs, and/or topographical maps, will be useful to establish past uses of the site. Available reports, even though not directly applicable to an environmental investigation, may also provide useful site information. For example, a geotechnical report prepared for the purpose of designing a building foundation will identify soil types and may indicate depth to groundwater or the presence of perched water on the site. Real estate reports may contain plot plans and site layouts which will help to accurately plot site features and identify locations for soil borings or groundwater wells. These reports may also identify features or operations which have contributed to suspected contamination. However, data from documents prepared by others should be verified.

The planning phase must also include collection of information regarding: site access or access requirements; surface materials (unpaved or type and thickness of pavement); depth to groundwater; location and use of nearest wells; site topography; surface water drainage; presence of overhead or subsurface utilities or other obstructions; regional land use; and, location and demographics of residents in the vicinity of the site. The decision to obtain information will be dictated by its intended use, the cost of obtaining it, its anticipated use, the particular phase of the investigation, and available budget. Before proceeding with the investigation, the planning professional should identify additional information that may be required. Following the identification of data gaps, the means for obtaining this information and its possible sources can be identified. Other factors to consider as part of the planning process may include availability of staff or subcontractors; availability of equipment needed for the investigation; presence of equipment storage and staging areas; possible encroachment onto adjacent properties or public right-of-ways; and storage and disposal of wastes (soil cuttings, purge water, protective equipment) generated during the investigation.

Budget often is the limiting factor when planning a site characterization investigation and is inevitably a significant factor in determining the scope of an investigation. As such, the project planner can be placed in a position where the thoroughness of the investigation is sacrificed to fit a budget. The skill of the planner is challenged to determine the least expensive characterization method that will address the goals of the investigation and maintain project quality. Variables to be considered in determining the most cost-effective approach include investigation methods, regulatory requirements, number of samples collected, analytical parameters selected, site screening techniques and report presentation.

Laboratory analyses may be done in an established commercial laboratory (a fixed or stationary lab), or with analytical instruments taken to the site in a transportable vehicle such as a trailer, van, or converted recreational vehicle (a mobile lab). Both of these alternatives have advantages and disadvantages. A mobile lab will provide rapid and possibly less expensive results, particularly for large investigations projects. The use of a mobile lab can also facilitate field decisions. However, the fixed lab can be used for the analysis of a more comprehensive suite of analyses and results could be more reliable. Often a mobile lab is used for the bulk of the analyses,

Table 2-2. Questions addressed as part of a site characterization plan

Why is the project being carried out?
What are the regulatory agency requirements?
What is the objective of the characterization investigation?
How will the information obtained during this investigation be used?
What data and information are currently available and what do they indicate?
What additional data and information should be obtained?
What site constraints are present?
Will the data support the conclusions?
What is the available budget?

with occasional duplicate samples being sent to a fixed lab for confirmation. The throughput of the laboratory (i.e. number of samples which can be analyzed per unit time) should also be considered when evaluating mobile versus fixed laboratories for a project.

As with any environmental project, compliance with rules, regulations, and guidelines of local, state and federal regulatory agencies is a key issue. It is important that the planning professional understand and comply with these rules. Some regulatory agency involvement will be required on most projects and should be considered early in the process, particularly if the investigation report is to be submitted to the agency for evaluation. Under some conditions, a work plan which outlines the scope of the investigation must be submitted to the lead regulatory agency for approval in advance. A pre-approved work plan is often desirable as a means to avoid conflicts and prevent project delays. In addition, regulators may require specific data for their evaluation and may specify the format of submittals. Required plans may include a sampling and analysis plan (including the field sampling plan and the quality assurance project plan), a data management plan, a health and safety plan, and a management plan for investigation-derived waste. Several state, counties and local authorities require permits to install monitoring wells or dig permits for any intrusive subsurface sampling. In summary, several key questions should be answered before the site characterization plan is developed (Table 2-2).

2.2 Evaluation of Historical Data

Historical data play a key role in establishing the scope of a site characterization investigation. Historical information can identify time periods during which potentially polluting operations were conducted and may serve to limit areas that

Table 2-3. Sources of historical information

Fire insurance maps
Aerial photographs
Historical directories
Building Department records/blueprints
Local Planning offices
Historical topographic maps
Site maps
Sanborn Maps
Regulatory Files
Conversations with personnel having a long on-site history
Chain-of-title records

warrant investigation. In addition, it can be used to decide which analyses and tests are appropriate for site characterization.

As an example, a site investigation was conducted for a former plating shop at which very little historical information was available. The investigation included an extensive sampling program, with samples analyzed for several parameters, including heavy metals, pH, and volatile organics. At another similar site, detailed historical information revealed a specific location for a vapor degreaser which remained unchanged throughout the history of operations. This detailed historical information also identified specific hazardous materials used at the facility together with hazardous materials storage areas. The investigation at the latter site required less sampling because specific areas were targeted and fewer parameters were included, even as the goals of both investigations were met.

No single source of historical information should be relied on exclusively. Collectively, data sources may provide sufficient information to formulate a comprehensive picture of site conditions (a partial list of data sources is contained in Table 2-3).

Where available, fire insurance maps may provide important information. These maps often identify past site use and can also record important historical site features such as the types and locations of flammable storage (including above ground and underground tanks).

Aerial photographs are extremely helpful in determining past site use. However, their use is limited to identification of surface structure locations and types, and identification of some activities conducted outside of buildings, such as on-site disposal. An experienced reviewer can obtain considerable information from aerial photographs, particularly those having good coverage spaced over a number of years.

Departments of transportation and county roadway departments are inexpensive sources for aerial photographs.

Larger industrial facilities or military operations may have an archive of historical plot maps. Information from past on-site operations can be obtained through the review of building names identified on the maps. Although they are not as valuable, historical topographic maps may provide some of the information that can be obtained from site-specific maps and can fill data gaps. Historical topographic maps can also be useful in identifying previous landforms which may have been later altered or covered, such as streams or swampy areas.

Conversations with personnel who know the history of the site can also provide useful data. However, the information should be verified through one or more sources to ensure the reliability of the data. Of limited use is review of chain-of-title records. Although some important information can be obtained through these records, information in chain-of-title records should not be relied upon exclusively.

Sources of historical information and its review are contained in ASCE Manuals and Reports on Engineering Practice No. 83 titled "Environmental Site Investigation Guidance Manual". This manual should be consulted for additional information.

2.3 Site Characterization Methods

No two site characterizations are exactly the same. Each has its unique set of associated circumstances, such as contaminants of concern, volume and extent of contaminants, media affected, and site access. As part of the planning process, a professional will review the characteristics of the site and define an approach which addresses the goals of the project and controls costs. Because there are discretionary steps, two equally proficient investigators may devise different approaches for a particular set of conditions. This is further complicated because site characterization projects are typically completed in phases, and the results from one phase influence the approach in the following phase.

The objective of a site characterization is to not only identify and delineate the contaminants and concentrations of concern, but to develop a broad understanding of the site. A site characterization should include physical characteristics of the site, characteristics of the contaminant source, nature and extent of contamination, and fate and transport mechanisms, and their environmental impact.

The concept of the completed exposure pathways is intrinsic to the decision-logic for any site characterization or remediation. A completed exposure pathway consists of the following five elements: a source of contamination, an environmental medium, a point of exposure, route(s) of exposure, and a receptor population. Receptor populations include community residents and any relevant worker populations. All five elements must be present for a pathway to be considered as complete and for the potential for exposure to exist. The completed exposure pathway, with all five elements present, is the foundation upon which the case is built for determining which populations are being exposed to hazardous substances, the relative hazards to human health posed by a site, and what remedial actions should be considered.

2.3.1 Physical Characteristics

Physical characteristics determine the environmental setting of a site, and include factors such as soils, geology, hydrology, meteorology and ecology. An analysis of these characteristics for a site should emphasize elements that are important in determining the fate and transport of contaminants in the exposure pathways of concern. For example, if migration to groundwater is feared, physical characteristics evaluated would include soil types in the unsaturated zone, depth to groundwater, precipitation, etc. A different set of physical characteristics would be examined if there is potential exposure to site workers through contaminant inhalation.

2.3.2 Contamination Source Characteristics

The characteristics of the contaminant(s) are a factor in determining potential exposure scenarios. The contaminants may have been generated at one time in the past, or the release may be current and continuing. The number and types of contaminants released vary, along with their rates of release. Compounds may be entering the environment as gases, liquids or solids. For example, fate and transport scenarios associated with a landfill are different than those associated with a leaking underground tank. The contaminants of concern are different as is the timing of releases. A release from an underground storage tank can be stopped relatively easily, while releases from a landfill may continue for many years. In addition, the magnitude and chemical nature of releases from each of these sources are likely to be different.

2.3.3 Environmental Data for Public Health Decisions

In general, the information collected for engineering decisions at hazardous waste sites is similar to that needed by public health agencies to make decisions regarding the hazards posed by the site. However, often there are insufficient environmental data to determine whether more rigorous health investigations should be conducted. Environmental data are critical to the public health algorithm for assessing sites, but only to the extent that such data can be used in a manner that contributes to and facilitates good public health practice. Generally, public health professionals will need additional information in the following categories:

1. Contaminant concentrations in all off-site media to which the public may be exposed.
2. An appropriate detection limit and level of quality assurance/quality control in samples to ensure resulting data are adequate for assessing possible human exposures.
3. Discrete samples that reflect the potential range of exposure of the public.
4. Shallow surface soil and sediment samples (not deeper than 3-inches).
5. Extensive biota studies and analyses of edible portions only.
6. Ambient and indoor air samples.
7. Lists of physical hazards and barrier to site access.

In order to ensure the data needs of the public health professional are addressed, it is recommended that the engineering consultants discuss the plans for the site characterization with local, state, and/or federal public agencies. Local and state agencies can provide guidance on what data are needed for site-specific public health determinations. In addition, the Agency for Toxic Substances and Disease Registry's "Environmental Data Needed for Public Health Assessments – A Guidance Manual" (June 1994) provides guidance on the data needs of the environmental professional.

2.3.4 Nature and Extent of Contamination

Sampling is conducted to define the nature and extent of contamination at a site. Pollutants may affect soil, groundwater, surface water, sediments and air. Analyses of samples are required to establish concentrations within the affected media so that the lateral and vertical extent of contamination can be determined. This information forms the basis upon which potential of the release can be assessed, the public health implications determined, and remedial alternatives can be evaluated.

2.3.4.1 Soil Sampling

The medium which is most commonly analyzed during site characterization is soil. There are numerous methods for soil characterization. They include geophysical techniques, grab bulk soil sampling, hand augering, backhoe excavation, soil vapor surveys, direct push sampling, and rotary auger or cable tool drilling (Devinny et al., 1990).

Each of these methods has inherent strengths and weaknesses. The choice of method depends upon the specific circumstances and goals of the investigation as well as site-specific constraints. It is not uncommon for one characterization investigation to utilize many of these techniques.

2.3.4.2 Surface Geophysical Techniques

Surface geophysical techniques are non-invasive and are used as a preliminary step to provide subsurface information for a site. Geophysical methods include magnetics, electromagnetics, ground penetrating radar, resistivity sounding, and seismic refraction. These methods can survey large areas of a site at relatively low cost.

Because of the non-invasive nature of surface geophysical techniques, they can be effective in identifying features on a site that warrants further investigation without disturbance of the site. For example, surface geophysical techniques can determine the physical features of a site by locating the groundwater surface, the base of refuse or other fill, fractures, or other geological inhomogeneities. These techniques can also identify subsurface structures such as buried pipelines, underground tanks, or buried wastes. It is particularly valuable to locate and avoid possible subsurface structures such as utilities.

The limitations of each method must be thoroughly evaluated before a selection is made. For example, the value of many surface geophysical techniques depends on the skills of the investigator as well as on the limitations of the methodologies. The experience of the surveyor may determine which data are taken and how they are interpreted. Results of a surface geophysical survey, though, cannot be confirmed without sampling suspicious materials.

2.3.4.3 Grab Soil Samples

Grab sampling is employed frequently to assess contamination of surface soil (e.g. upper 12 inches to 30 inches of soil). Grab samples, for example, could be used as a preliminary step to assess releases to surface soil in an area that had been used to store drums of hazardous materials. Surface soil samples are particularly important to determine human health risks. People, and especially children, are far more likely to be exposed to surface contaminants through routine work and play than to subsurface materials.

The basic approach for collecting grab soil samples is simple. A hand trowel, shovel, or similar sampling instrument is used to put soil into a sampling container. This method is easily implemented and can be used in places where access is limited. This method, however, may not be suitable for collection of samples for volatile organic compound (VOC) analysis since the sample would have been disturbed. Although this method is often used in the collection of samples for VOC analysis, caution should be exercised in the collection of the samples and in the subsequent interpretation of VOC data.

One way to collect relatively undisturbed soil samples is to use a coring device driven or pushed into the soil surface. Undisturbed samples can be collected with a slide hammer sampler with 2 inch diameter brass tubes. Additional subsampling in accordance with EPA guidelines (EPA, 1986) will be necessary to obtain samples acceptable for measuring concentrations of VOCs.

Samples collected as part of a site characterization can be either discrete or composite. Discrete samples are collected from soil at a specific location and depth on a site. Composite samples are a mix of a number of discrete samples which have been combined in the field or in the analytical laboratory to form one sample in which an average concentration can be determined. A composite sample can represent conditions at a single depth across a site or at various depths at one sampling location. Composite samples are usually collected as part of preliminary assessments to evaluate whether an area warrants further investigation. The disadvantage of composite samples is that re-analysis of discrete samples is often required if composite sample results indicate the presence of contaminants. Composite samples are not generally acceptable for analysis of VOCs because vapors may be lost.

2.3.4.4 Hand Augering Methods

A modification of the grab soil sampling method is the use of a hand auger coupled with a bulk soil sampler. Commonly, a two- to four-inch diameter auger is used to bore to the desired sampling depth. A bulk sampler, comprised of a sample barrel lined with a stainless steel or brass sleeve, is hand driven into soil using a weighted slide hammer. A relatively undisturbed core sample is collected. Various power assisted devices are available to facilitate sample collection.

Like grab sampling, hand augering methods can be used when access by larger equipment is not possible. It can also be used during preliminary phases of investigation since samples can be collected relatively inexpensively. However, because this method relies on hand power to advance to the desired depth, it is limited by site soils. Borings can be made to as much as 20 feet in firm but cuttable soils (such as silty sand), but only a foot or two may be possible in hard clay. The approach generally fails in rocky soils. This method does not work successfully in gravelly soils, loose sands, or high groundwater conditions where the soil collapses into the hole as it is being bored.

2.3.4.5 Trenching

Conventional excavating equipment such as a backhoe can also be used in sample collection. Trench (or pothole) excavation is particularly useful in conducting investigations for those contaminants that impart color to soil, such as oil. Inspection of the side walls of the excavation allows visual assessment of the vertical extent of contamination. A backhoe can excavate a trench rapidly. Disturbed samples can be collected directly from the bucket of the backhoe while a bulk density sampler is used to collect undisturbed soils from the trench. Using a backhoe, large numbers of samples may be collected over a short period of time for various parameters, including samples for geotechnical testing that may be necessary to evaluate remedial alternatives.

In addition to depth limitations, a disadvantage of backhoe excavation is that it is extremely disruptive to the site under investigation. Because a large amount of soil is disturbed and exposed to the atmosphere, there can be troublesome releases of toxic or odorous gases during the operation which may cause severe unintended consequences. Therefore, this method is not recommended for sites where highly toxic materials are anticipated. Through backhoe trenching, a large volume of excavated material is generated which must be properly stored and managed. Health and safety measures for workers and other receptors should also be considered when evaluating this method for sample collection.

2.3.4.6 Soil Vapor Surveys

Soil vapor surveys are used extensively for assessing contamination by VOCs and other gases (Dorrance et al., 1995). The method collects samples of the gas trapped between soil particles by advancing a sealed probe attached to tubing into the soil to the desired depth. The probe is then opened and evacuated. The soil vapor is collected for direct analysis in a mobile laboratory or held in gas sampling canisters or bags for transport to a fixed laboratory (Ullom, 1995). Samples are sometimes analyzed directly using field instruments.

This method can be effective in identifying the lateral and vertical extent of soil contamination associated with releases of volatile materials such as solvents and gasoline. It is also a useful tool for assessing potential groundwater contamination. Although soil vapor results are not always directly related to bulk soil results, the data provide a method for determining suitable locations for soil sampling. Under certain geological conditions, soil vapor data can be more representative of the extent of contamination than data from bulk soil samples, particularly in course-grained sediments. Indeed, regulatory agencies in California are specifying vapor surveys as the preferred method over bulk soil sampling for investigations of VOCs in coarse-grained alluvial soils.

Soil vapor surveys are rapid. Ten to twenty points can be investigated in a normal work day, although the actual number is highly dependent upon soil quality and sampling depth.

Soil vapor investigation does not provide lithological data. In addition, cobbles or other obstructions can impede the investigation. Finally, data from fine grained sediments may not be representative of actual site conditions because good vapor recovery may not be achieved.

2.3.4.7 Direct Push Sampling Methods

Direct push methods utilize hydraulics, sometimes coupled with vibration or driving action, to advance a sampling device mounted at the end of a relatively narrow (normally one inch) steel rod through soil to the desired depth. Direct push methods can be used to collect bulk soil, soil vapor or groundwater samples.

One reason for the increased popularity of direct push methods is recent advances in equipment and technology. Some equipment requires that the entire tool string be retrieved from the hole each time a sample is collected, which may result in caving of the hole. However, some equipment now utilizes a dual wall system which allows the sampling device to be removed while maintaining the integrity of the hole.

A significant advantage to this method is that there is little production of soil cuttings, which may require controlled disposal. The equipment can be less expensive than conventional drilling rigs. The disadvantage is that in many cases, a groundwater well cannot be constructed in a hole created using a direct push method because the pressure of driving the hole tends to seal the side walls, particularly in finer grained soils. In addition, direct push methods are generally limited to a maximum depth of approximately 100 feet or less depending upon soil lithology and the specific equipment used to advance or retrieve the probe.

2.3.4.8 Drilling

As with other methods, the drilling method selected for each investigation is dependent upon project goals, site constraints, and geological conditions. Typical drilling methods include hollow stem auger, air rotary, solid stem auger, cable tool, and mud rotary. Each method has applications in which it is most effective, and their suitability depends on geological conditions at the site. Emphasis should be placed on the ability to collect samples that fulfill the goals of the investigation.

The hollow stem auger is the most common tool utilized in collecting soil samples for site characterization investigations at shallow to moderate depths. As the name implies, the auger is hollow, which enables a sampling device, typically a split barrel sampler, to be passed through the auger for collecting soil samples. Undisturbed soil samples can be easily collected at desired depths. Groundwater wells can also be constructed within the hollow stem of the auger. As the bit is advanced, soil is removed from the boring through helical flights on the outside of the auger. Hollow stem augers are effective for drilling through unconsolidated materials, but are generally ineffective in consolidated soils, rock, or in soils containing a large proportion of cobbles or boulders.

The commonly used split barrel sampler is a cylinder cut in half along its longitudinal axis. Threads on the outside edge of both ends allow the sampler to remain intact when a drive head coupler and drive shoe are threaded onto each end. A relatively undisturbed sample is collected by driving the split barrel sampler into the soil after drilling has reached the desired depth. A sleeve placed inside the sampler contains collected soil. After retrieval, the sampler is disassembled and the sleeve is removed from the sampler. Samplers must be decontaminated between sampling intervals to prevent cross contamination.

Air rotary drilling is typically used for environmental applications where hollow stem techniques cannot be used. A rotary bit is used to bore a hole and pressurized air is used to remove drill cuttings. Because of the high pressure air typically associated with this type of drilling, this drilling method should not be selected for use in investigations of VOCs since the VOCs will be flushed from the soil which is to be sampled.

Solid stem, cable tool and mud rotary drilling techniques generally have limited application in site investigations. As the name implies, a solid stem auger has a narrow steel stem and wide flights. A solid stem auger is effective for drilling shallow boreholes, but must be removed for collection of undisturbed soil samples. Disturbed samples may be collected from the flights of the auger. In cable tool drilling, the bit cuts by repeated lifting and dropping of the drill string. In unconsolidated material, casing is driven into the formation following the bit. Cable tool drilling is particularly effective in loose materials and is capable of drilling deep holes in hard rock, though slowly. Mud rotary drilling is a rapid and effective method for drilling holes, but is not generally acceptable for environmental purposes because samples will be contaminated by drilling fluids. Additional details regarding drilling methods are provided in Devinny et al. (1990).

2.3.4.9 Surface Waters

Ponds, lakes, or flowing streams may be part of a contaminated site. Because stratification can occur in standing water, the number and location of samples depend on factors such as size, depth and configuration. Contaminants tend to mix throughout the cross section of a stream, but longitudinal variation is great. Upstream and downstream measurements are important.

Samples from surface waters can either be grab or composite samples. Common grab sampling techniques use pond samplers, weighted bottle samplers, peristaltic pumps, and Van Dorn, Nannsen, or Kemmerer depth samplers. Composite samples are collected to develop a general characterization of a water body. Composite samples provide an average concentration and cannot be used to assess peak or minimum concentrations. Waste or water streams can be sampled using an automatic composite sampler. If the sampler is used with a flow measuring device such as a weir with a water level recorder, flow proportioned composites can be collected.

Excavations used for sampling or in the remediation process are sometimes flooded with groundwater, and sampling may be appropriate. The appropriate methods are similar to those for ponds and lakes.

2.3.4.10 Sediment Sampling

Sediments located near or beneath water bodies can be impacted by various contaminants, through contaminated surface water contact, deposition, or direct discharge of contaminants. Most commonly, sediments are impacted by halogenated hydrocarbons (PCBs, dioxins, pesticides, etc.), polycyclic aromatic hydrocarbons, and heavy metals. These contaminants are denser than water, have an affinity for adsorbing to particles, and/or form precipitates which settle out of water.

The characteristics of sediments vary widely depending upon the overlying water body. Some sediments can be extremely fine-grained, having characteristics which are unlike those of water, while others are compacted and dense. Therefore, sampling of these media is dependent not only upon the specific analyses to be completed on the samples, but also upon the characteristics of the sediment.

Sampling methods identified above for soils and surface waters can be modified for use in sampling sediments, particularly those that are located above the water surface or in shallow water. For sampling of sediments located in deep water, specialized samplers such as Eckman or Ponar dredges are used.

As part of an investigation where sediments are sampled for chemical analysis, it is equally important to characterize physical properties of the sediment, since these properties will effect contaminant fate and transport. Physical characteristics which should be established include particle size distribution, organic carbon and total solids. Sediments might also be analyzed for chemical parameters such as pH, oxidation-reduction potential, salinity, sulfide, and reactive iron and manganese. Refer to EPA (1991 and 2001) for more specific information regarding sediment sampling.

2.3.4.11 Groundwater

Groundwater is characterized through installation and sampling of monitoring wells (Selby, 1991). In order to characterize horizontal flow direction and determine background chemicals of concern, three groundwater wells are needed, with one upgradient of the contaminant source. Installation of three wells allows for determination of the hydraulic gradient, and thus the direction of groundwater flow. Monitoring wells are usually constructed of 2-inch to 4-inch diameter PVC, and screened in the saturated zone to allow water to enter the casing. Stainless steel or Teflon casing may be used where it is feared that adsorption of organic contaminants will distort the results or where a contaminant or aquifer characteristics could damage conventional well materials. A properly constructed and developed well causes minimal disturbance to the formation. Contamination of groundwater samples by inadequate sealing or leaching of chemicals from well construction materials must be prevented.

At some sites, contaminant concentrations in groundwater can be highly variable because of changes in water flow patterns and adsorption and desorption of contaminants by different soil strata. Recommendations regarding groundwater must be based on trends and not on single sampling events.

Samples are collected from groundwater wells after they are purged to remove water which has accumulated within the well casing. Purging ensures that water sampled is representative of that in the water-bearing formation. Wells are typically sampled after three or more well volumes of water have been removed by a bailer or pump and parameters such as temperature, pH, and specific conductance have stabilized. Sampling must sometimes be done in wells which produce very little water. A low yield well is typically purged dry twice, with the well allowed to recover 80% of its volume between purgings, and then sampled.

Sampling apparatus includes bailers, suction-lift pumps, submersible pumps, air lift samplers, and gas-operated squeeze pumps. The appropriate sampling method for groundwater is dependent on the analytical parameters being determined and the regulatory protocols being implemented. For example, a bailer is the preferred method for collecting samples to be analyzed for VOCs because volatilization losses are minimized. Alternatively, sensors can be placed within the well to provide data continuously or at fixed intervals.

Alternative groundwater sampling methods are being utilized with greater frequency. A common alternative is the use of grab sampling in which a casing having a retractable tip is driven into the saturated zone. At the selected depth, the tip is retracted, exposing a porous plate which allows water to pass into narrow diameter tubing which can be purged and sampled. Grab groundwater sampling is typically used to provide screening data, allowing mapping of a contaminant plume for better well placement and site delineation. Analytical results obtained from grab samples may not be directly comparable to those obtained from permanent monitoring wells. In addition, a disadvantage of this method is that a groundwater sample from a location cannot be easily repeated for verification at a later date.

2.3.4.12 Air Monitoring

Air monitoring can be used to assess potential impacts to site workers, provide a relative indicator of contamination within soil samples, assess potential airborne contaminants associated with a release, and assess the migration of airborne contaminants from a site.

Air can be monitored with portable field instruments or by collecting samples for laboratory analysis (Waxman, 1996). The most common devices used in site characterization investigations measure organic vapor concentrations. The wide variety of organic vapor analyzers available differ in the method used to quantify vapors. Flame ionization detectors (FID) and photoionization detectors (PID) are common. Both of these provide results in the parts per million by volume range. The choice of instrument for a particular application will depend on the contaminants and site characteristics.

Other instruments have specific applications in site investigations. Combustible gas meters monitor concentrations of flammables such as methane gas in wells or near landfills. Oxygen meters are used in environments where it may be depleted. Particulate dust meters can be used to monitor migration of contaminated dust during drilling, excavation, or other activities that may raise dust. Lower explosive limit meters determine whether explosive gases are accumulating at concentrations that threaten safety.

Samplers can be used to monitor air over longer periods. In active sampling, air is moved through the collection medium. Active samplers usually consist of a pump, sample inlet, and a sampler containing an appropriate collection medium (filter paper, activated carbon, gas absorbers, sample bags, etc.). At the end of the sampling period, the collection medium is transported to a laboratory for analysis. A flow rate record must be accurately maintained to obtain data that can be compared with exposure limits or other threshold value.

Passive sampling relies on diffusion to bring compounds to the sensor. In addition to their typical use as personal monitoring devices, they have had some use as part of a site investigation. Passive samplers include dosimeters and diffusion samplers.

2.3.4.13 Data Collection for Engineering Evaluations

During the site characterization, it is desirable to collect data which will eventually support alternatives evaluation and remediation design. Much feasibility and preliminary design level data can be collected inexpensively at this stage, rather than at additional cost later on. A borehole drilled for site characterization, for example, can also yield samples for remediation design. If these are collected during the site characterization, fewer expensive borings may be needed during the design phase.

During the planning stages, the planning professional should solicit input from the remedial design team to identify data needs that are typical of the likely remedies for the site. Many of these data needs overlap with those collected for site characterization. Typical available data at the planning stage describe site topography, subsurface soil properties, groundwater levels, and groundwater geochemistry.

Site topographic data usually can be obtained inexpensively from aerial photography. The topographic map can also be used as a base map for locating the site characterization investigation points. In addition, use of stereopairs from this and other flights often provides useful soil and bedrock interpretation for use in geotechnical and hydrogeological evaluations.

Testing selected soil samples for basic geotechnical index tests, i.e. moisture content, grain-size, Atterberg Limits, and organic content can provide site specific knowledge of the soil and rock types. These data can be correlated with global or regional values in the literature to provide a framework for understanding the engineering properties of the materials at the site. These tests are excellent for evaluating the variability of materials at the site, and therefore in evaluating the level of risk uncertainty inherent in the feasibility level cost estimates. These data are also useful in evaluating hydrogeological issues, such as the potential for fissuring and fracturing in clay layers.

Most of the groundwater data collected during the site characterization are intended for contaminant transport evaluations. Occasionally, additional piezometric data from critical areas or from critical seasonal periods are needed to properly evaluate the engineering issues related to the feasibility of pump and treat, containment, or capping systems.

Some basic groundwater geochemistry analysis from selected samples during the site characterization can be very useful in evaluating major cost elements affecting the feasibility of water treatment or intrinsic remediation. Typical water chemistry analyses for evaluating water treatment issues include total suspended solids, total iron, total hardness, total organic carbon, pH, and temperature. Typical chemistry analyses for evaluating the potential for aerobic and anaerobic intrinsic remediation include nitrate, total and dissolved iron, total and dissolved manganese, sulfate, sulfide, chloride, carbon dioxide, methane, phospholipid fatty acids, pH, temperature, and redox potential. .

2.3.5 Analysis, Data Evaluation and Reporting

Samples not tested in the field should be preserved as required for the requested analysis, and logged and tracked from the point of collection to the laboratory using appropriate chain-of-custody documentation. Information regarding EPA-approved analytical methods and sampling considerations is provided by EPA (1986). This document (SW-846) can also be accessed through the internet at www.epa.gov/ epaoswer/hazwaste/test/txmain.htm. Refer to SW-846 for information regarding sampling techniques, analytical methods, preservation of samples, proper volume and appropriate containers and sample holding times. If applicable, state or local regulatory requirements or guidelines should be followed because they may be more stringent than EPA requirements.

Following collection, samples are transported to a laboratory for chemical analysis along with appropriate quality assurance/quality control (QA/QC) samples, such as duplicates and blanks. The purpose of blank samples is to assess the extent to which any constituents identified in the environmental samples might be attributable to external conditions such as impure source water or incomplete decontamination. Rinsate blanks can be collected to monitor the effectiveness of field equipment decontamination, field and ambient blanks to determine the effects of site conditions, and source blanks to document the quality of each source of water used.

The laboratory must have an established quality control program which ensures that samples are accurately analyzed and that data are defensible as representative of site conditions. If applicable, the laboratory should be certified within the state in which the investigation is being conducted. Beyond certifications, an evaluation of laboratory quality control and its capability and capacity to perform the requested analyses must be made. Laboratory staff should be interviewed to determine their methods, the availability of quality control data, current capacity, and the ability to perform the required analyses within guidelines established by EPA.

Upon receipt of analytical data from the laboratory, the report should be scrutinized to ensure that the data are accurate and reflect site conditions. The report should indicate that samples were analyzed within accepted holding times. Trends in the data should be reviewed to determine whether they are intuitively reasonable. Laboratory detection limits should be reviewed for accordance with agreed limits, project objectives, and regulatory guidance. Quality assurance and quality control data should also be reviewed to verify that data are within acceptable limits. Any anomalies should be discussed with the laboratory. Depending upon project objectives, more sophisticated data evaluation such as intercomparisons, data plots, regression analysis, and tests for fitness, can be conducted to validate the data. Data should be reported in a form that clearly describes what was done and shows trends in the data.

EPA has prepared guidance manuals for data evaluation. These manuals should be consulted for additional information regarding data evaluation.

2.4 Site Characterization Report

The site characterization report summarizes the findings obtained during the environmental site characterization process. This report also serves as closure for the remedial investigation (Table 2-4).

As for the environmental site assessment, the site characterization report should be prepared utilizing a standard format. This format should follow a systematic approach as reflected in the environmental site assessment and extend from initial planning, through implementation and concluding with achievement of the intended goals.

2.5 Development of a site conceptual model

Building on the historical site and sampling data for both soil and groundwater, a conceptual hydrogeological and contaminant model can be developed. This is a representation of the stratigraphical, geological, and environmental system of the contaminated site and a representation of the possible physical, chemical, and biological processes that determine the transport dynamics of contaminants from sources to receptors within the system (ASTM, 1995). The conceptual model aids in integrating site data and in determining whether additional data should be collected. The conceptual model should include thorough descriptions of the site geology and the hydrogeologic environment. This would include:

- Regional and local hydrogeology, including drinking water aquifers, regional confining units, and industrial, agricultural and domestic wells with their patterns of use;
- Lithology and stratigraphic relationships, including transmissive and non-transmissive units;
- Aquifer hydraulic conductivity measurements and physical, chemical properties;
- Groundwater flow velocity, gradients and piezometric or water surface maps over several seasons;
- Preferential flow paths;
- Interactions between groundwater and surface water bodies; and
- Rates of infiltration and recharge.

The conceptual model may also include soil and groundwater quality data, such as:

Table 2-4. Elements of the site characterization report

History of the site
Site characteristics
Specific purpose of the Environmental Site Characterization (ESC)
Methodology for performing the ESC
Sampling Procedures and Data Quality Objectives
Analytical procedures
Quality Assurance/Quality Control
Health and Safety Measures
Contaminants of concern, migration pathways and potential receptors
Risk assessment
Health Risk Assessment (exposure and toxicity assessment)
Ecological Risk Assessment
Discussion of Preliminary Remediation Goals (PRGs) or Regulatory Standards
Interim remedial measures and source control
Treatability studies
Systems design
Systems implementation
Results of interim remedial measures and source control
Closure of the site characterization
Adequacy of the environmental site assessment
Deficiencies to be addressed

- Three dimensional distribution of mobile and residual non-aqueous phase liquids (NAPL) and dissolved phase contaminants. The mass and distribution of mobile and residual NAPL will be used to define the dissolved phase plume source area and potential for a continuing release of contaminants;
- Groundwater and soil geochemical data, including data on background (uncontaminated) samples;
- Historic water quality data showing variations in contaminant concentrations over time;

- Chemical and physical characteristics of the contaminants and phases (NAPL, dissolved, vapor, absorbed to soil) and
- Potential for biodegradation of the contaminants.

The model should also identify the pathways and locate the receptors to be protected. Locations of discharge points down gradient of the site should be identified.

Upon completion of the conceptual model, data gaps can be identified. This step will then be followed by modifications to the original work or supplemental work plans as a prelude to collecting additional data. The plan's goals are to collect site-specific data sufficient to allow estimation, within an acceptable level of confidence, of both the rate of attenuation processes and the anticipated time required to achieve remediation objectives (EPA, 1999). Data supporting the lines of evidence demonstrating the effectiveness of natural attenuation processes discussed in Chapter 1 (ASTM, 1998) are also to be included.

The primary line of evidence for remediation by natural attenuation is provided by observed reductions in plume geometry and observed concentration reductions for the constituents of concern at the site. Is the plume shrinking, stable or expanding? In general, a shrinking or stable plume is strong evidence that the natural attenuation mechanisms are at work on a site.

Geochemical indicators of naturally occurring degradation and estimates of attenuation rates provide secondary lines of evidence. These parameters are indirect evidence to support potential aerobic and anaerobic biodegradation. Characterization data may be used to quantify the rates of contaminant sorption, dilution, or to demonstrate and quantify the rates of biological degradation processes occurring at the site (EPA, 1999).

Microbiological studies, solute transport modeling, and estimates of assimilative capacity, if required, can provide additional lines of evidence. Data from field or laboratory studies can demonstrate that a natural attenuation process is occurring at the site and its potential to degrade the contaminants of concern.

After natural attenuation is accepted as a remedy, it is necessary to continue collecting data to provide confirmation that natural attenuation is working. Such performance monitoring is of greater importance for natural attenuation than for other types of remedies because the remediation period may be longer and because of the potential for ongoing contaminant migration and other uncertainties associated with its use. Chapter 5 discusses monitoring for natural attenuation on a remedial technology.

References

American Society for Testing and Materials, (1995). Standard Guide for Developing Conceptual Site Models for Contaminated Sites, (E 1689-95), Philadelphia, PA.

American Society for Testing and Materials ASTM, (1998). Standard Guide for Remediation of Ground Water by Natural Attenuation at Petroleum Release Sites, (E 1943-98), Philadelphia, PA.

American Society for Testing and Materials, (2000). Standard Practice for Environmental Site Assessments: Phase I Environmental Site Assessment Process (ASTM, E 1527-00).

Devinny, J.S., Everett, L., Lu, J.C.S., and Stollar, R.L., (1990). Subsurface Migration of Hazardous Waste, Van Nostrand Reinhold and Company, New York, N.Y., 387 pp.

Dorrance, W.W., Wilson, L.G., Everett, L.G. and Cullen, S.J., (1995). A compendium of soil samplers for the vadose zone. In: Handbook of Vadose Zone Characterization & Monitoring, L.G. Wilson, L.G. Everett, and S.S. Cullen (eds.), Lewis Publishers, Boca Raton, Florida, pp. 401-428.

Lewis, T.E. and Wilson, L.G., (1995). Soil sampling for volatile organic compounds. In: Handbook of Vadose Zone Characterization & Monitoring, L.G. Wilson, L.G. Everett, and S.S. Cullen (eds.), Lewis Publishers, Boca Raton, Florida, pp. 429-476.

Selby, D.A., (1991). A critical review of site assessment methodologies. In Hydrocarbon Contaminated Soils and Groundwater, P.T. Kostecki and E.J. Calabrese (eds.), Lewis Publishers, Chelsea, Michigan, 1:149-160.

Warman, M.F., (1996). Monitoring. In : Hazardous Waste Site Operations, Wiley, New York, pp. 227-289.

Ullom, W.L., (1995). Soil gas sampling. In: Handbook of Vadose Zone Characterization & Monitoring, L.G. Wilson, L.G. Everett, and S.S. Cullen (eds.), Lewis Publishers, Boca Raton, Florida, pp. 555-567.

U.S. Department Of Health And Human Services, Agency for Toxic Substances and Disease Registry, (1994). Environmental Data Needed for Public Health Assessments – A Guidance Manual. U.S. Department Of Health And Human Services, Agency for Toxic Substances and Disease Registry, Washington, D.C..

U.S. EPA, (1986). SW-846 - Test Methods for Evaluating Solid Waste, 3rd Edition, U.S. Environmental Protection Agency, Washington, D.C.

U.S. EPA, (1991). Remediation of Contaminated Sediments, EPA/625/6-91/028, U.S. Environmental Protection Agency, Washington, D.C..

U.S. EPA, (1994). Final Guidance for the Data Quality Objectives Process. EPA QA/G-4. Quality Assurance Management Staff, U.S. Environmental Protection Agency, Washington, DC.

U.S. EPA, (1999). Use of Monitored Natural Attenuation at Superfund, RCRA Corrective Action, and Underground Storage Tanks, Directive Number 9200, 4-17P, U.S. Environmental Protection Agency, Washington, D.C.

U.S. EPA, (2001). Methods for Collection, Storage and Manipulation of Sediments for Chemical and Toxicological Analyses: Technical Manual. EPA 823-B-01-002. U.S. Environmental Protection Agency, Washington, D.C.

CHAPTER 3

Natural Attenuation Processes

To evaluate the implementation or effectiveness of remediation by natural attenuation, it is important to be able to predict the time of arrival and concentration of contaminants at the receptor(s) of concern (e.g., water supply well, surface water body). Thus, it is necessary to understand the transport and transformation processes that affect the contaminants. The goal of this chapter is to review the various naturally occurring physical, chemical, and biological processes in soils, sediments and groundwater that contribute to remediation by natural attenuation. Specifically, Section 3.1. focuses on the physical processes, primarily in the groundwater environment, Section 3.2. describes the relevant chemical processes, and Section 3.3. covers the biological processes of interest. Finally, Section 3.4. provides a review of the mathematical modeling of these natural attenuation processes.

3.1   Physical Processes

3.1.1   Introduction

It is important to note that the physical attenuation processes result in a reduction of the dissolved concentration and/or mobility of a chemical, but do not influence the total mass of the chemical present in the system; thus, they are "nondestructive" attenuation mechanisms (Wiedemeier et al., 1995, 1996; ASTM, 1998). The key physical attenuation processes discussed in this section are divided into two categories: (1) hydrologic processes (e.g., advection, hydrodynamic dispersion, infiltration/recharge), which act to influence the dissolved contaminant concentration through movement of water; and (2) interphase partitioning processes (e.g., sorption, volatilization), which influence the dissolved contaminant concentration by transfer to/from other phases. The relative impact of each of these physical processes on the environmental transport and fate of a chemical is a function of the compound's physical/chemical properties and the nature of the media through which the contaminant is migrating (Wiedemeier et al., 1996).

3.1.2   Conceptual Framework and Mass Balance

A logical starting point for discussing physical attenuation processes is to look at a mass balance on the system. A commonly used expression of the mass balance on dissolved reactive constituents in saturated, isotropic porous media is the advection-dispersion-reaction (ADR) equation. The one-dimensional form of the ADR equation for homogeneous media with steady, uniform flow in the x-direction can be written as follows (Freeze and Cherry, 1979) (Fig. 3-1):

$$\frac{\partial C}{\partial t} = D_L \frac{\partial^2 C}{\partial x^2} - v\frac{\partial C}{\partial x} \pm G \qquad (3.1.1)$$

where: C = solute concentration [ML^{-3}], t = time [T], D_L = longitudinal dispersion coefficient [L^{2}T^{-1}], x = distance in direction of flow [L], v = average pore water velocity [LT^{-1}], and G = source-sink term [ML^{-3}T^{-1}]. The first two processes on the right-hand side of Eq. 3.1.1 represent longitudinal hydrodynamic dispersion and advection, respectively, the two most important hydrologic processes described below (not shown directly in this equation are the hydrologic processes of infiltration and recharge, which are also covered below). The third term represents the physical, chemical and biochemical processes that result in contaminant concentration changes in the groundwater.

Although Eq. 3.1.1 has a physical basis and has been used to predict solute migration in uniformly packed laboratory columns, the current application of the advection-dispersion model to field problems is essentially empirical (Gillham and Cherry, 1982). This is largely due to limitations in the current knowledge of the physical processes occurring within the aquifer.

3.1.3   Hydrologic Processes

3.1.3.1   Advection

Advection is the movement of dissolved contaminants due to the bulk flow of groundwater. Based on Darcy's law, the specific discharge, q, [LT^{-1}] in a process model is proportional to the hydraulic gradient dh/dx, with hydraulic conductivity K [LT^{-1}] being the constant of proportionality. (Domenico and Schwartz, 1998):

$$\frac{Q}{A} = q = -K\frac{dh}{dx} \qquad (3.1.2)$$

A more realistic velocity, the average pore-water velocity, v (also known as average linear velocity, average solution velocity, and seepage velocity), may be calculated as (Domenico and Schwartz, 1998):

$$v = \frac{Q}{n_e A} = \frac{q}{n_e} = -\frac{K}{n_e}\frac{dh}{dx} \qquad (3.1.3)$$

where: $n_e A$ = the effective area of flow; and n_e = effective porosity, which is defined as the fraction of interconnected pore space.

The magnitude of the hydraulic conductivity, K, for a given geologic formation is a function of a variety of physical factors, including porosity, particle size and size

Table 3-1. Ranges of K values commonly observed for various geological materials

Material	Hydraulic conductivity (m/s)
Sedimentary	
Gravel	$3\times10^{-4} - 3\times10^{-2}$
Coarse sand	$9\times10^{-7} - 6\times10^{-3}$
Medium sand	$9\times10^{-7} - 5\times10^{-4}$
Fine sand	$2\times10^{-7} - 2\times10^{-4}$
Silt, loess	$1\times10^{-9} - 2\times10^{-5}$
Till	$1\times10^{-12} - 2\times10^{-6}$
Clay	$1\times10^{-11} - 4.7\times10^{-9}$
Unweathered marine clay	$8\times10^{-13} - 2\times10^{-9}$
Sedimentary Rocks	
Karst and reef limestone	$1\times10^{-6} - 2\times10^{-2}$
Limestone, dolomite	$1\times10^{-9} - 6\times10^{-6}$
Sandstone	$3\times10^{-10} - 6\times10^{-6}$
Siltstone	$1\times10^{-11} - 1.4\times10^{-8}$
Salt	$1\times10^{-12} - 1\times10^{-10}$
Anhydrite	$4\times10^{-13} - 2\times10^{-8}$
Shale	$1\times10^{-13} - 2\times10^{-9}$
Crystalline Rocks	
Permeable basalt	$4\times10^{-7} - 2\times10^{-2}$
Fractured igneous and metamorphic rock	$8\times10^{-9} - 3\times10^{-4}$
Weathered granite	$3.3\times10^{-6} - 5.2\times10^{-5}$
Weathered gabbro	$5.5\times10^{-7} - 3.8\times10^{-6}$
Basalt	$2\times10^{-11} - 4.2\times10^{-7}$
Unfractured igneous and metamorphic rocks	$3\times10^{-14} - 2\times10^{-10}$

Source: values from Table 3.2, p. 39, Domenico and Schwartz, 1998.

distribution, particle shape, particle arrangement (i.e., packing), and secondary features such as fracturing and dissolution (USEPA, 1987). Ranges of K values commonly observed for various geological materials are presented in Table 3-1. Based on Table 3-1, it can be concluded that K varies widely, covering over 12–13 orders of magnitude with fine-grained, clayey materials exhibiting lower values of K, and coarse-grained sandy materials typically exhibiting higher values.

Table 3-2. Ranges of total porosity values commonly observed for various geological materials

Material	Porosity (%)
Sedimentary	
Gravel, coarse	24–36
Gravel, fine	25–38
Sand, coarse	31–46
Sand, fine	26–53
Silt	34–61
Clay	34–60
Sedimentary Rocks	
Sandstone	5–30
Siltstone	21–41
Limestone, dolomite	0–40
Karst limestone	0–40
Shale	0–10
Crystalline Rocks:	
Fractured crystalline rocks	0–10
Dense crystalline rocks	0–5
Basalt	3–35
Weathered granite	34–57
Weathered gabbro	42–45

Source: values from Table 2.1, p.14, Domenico and Schwartz, 1998.

The effective porosity, n_e, is generally estimated as being somewhat less than the total porosity, n, which is calculated from the ratio of the total void volume, V_v, of a soil or rock, divided by the total unit volume, V_T (Domenico and Schwartz, 1998). Ranges of total porosity values commonly observed for various geologic materials are summarized in Table 3-2. Effective porosities may be over an order of magnitude smaller than the total porosity (Domenico and Schwartz, 1998). However, for coarse-grained materials that drain freely, the effective porosity essentially equals the total porosity (USEPA, 1987).

For most practical situations in which the density of the contaminated groundwater is not significantly different from the ambient groundwater, the flow of water and dissolved mass will move at the same rate (in the absence of other processes) and in the same direction (Domenico and Schwartz, 1998). Thus, the average pore-water velocity, v (Eq. 3.1.3), has been used to determine the advective component of groundwater flow and provide a conservative estimate of the rate of migration of dissolved constituents (USEPA, 1987).

To demonstrate movement of the front of a plume of dissolved contaminants by the process of advection alone, the relative concentration of a dissolved nonreactive contaminant emanating from a constant source (Fig. 3-1) is presented as a function of distance along the 1-dimensional flow path in Fig. 3-2a. Similarly, the plot of the relative concentration of a dissolved nonreactive contaminant plume produced by a single slug of contaminant from the source is presented in Fig. 3-2b. Note that for the one-dimensional displacement of miscible fluids, when advection is the only transporting mechanism, a sharp solute concentration front characteristic of plug flow is maintained between the initial and displacing fluids (Gillham and Cherry, 1982). In both cases, the dissolved non-reactive conservative constituent front moves with the groundwater at the average pore-water velocity, with no diminution of the concentration (USEPA, 1987); thus, advection is not a natural attenuation mechanism per se, although it may contribute to contaminant attenuation as described below. When advective transport is the main force behind contaminant migration, this may be a reasonable approximation of solute migration; however, the other key components (e.g., dispersion, sorption, biodegradation, etc.) in the ADR equation may be used to more completely describe solute transport.

3.1.3.2 Hydrodynamic Dispersion

Although nonreactive solutes are carried at an average rate equal to the average pore-water velocity, v, the solute tends to spread out from the path expected based on the advective hydraulics alone (Freeze and Cherry, 1979). This phenomenon is referred to as hydrodynamic dispersion and is an important consideration when evaluating remediation by natural attenuation (Wiedemeier et al., 1996). Specifically, hydrodynamic dispersion is important because it may result in mixing of the contaminant plume with electron acceptors and donors present in the surrounding relatively pristine portions of the aquifer and in dilution of the contaminant.

Based on early, small-scale column studies with isotropic media (e.g., Perkins and Johnston, 1963; Fried and Combarnous, 1971), it was observed that the parameter describing hydrodynamic dispersion, the hydrodynamic dispersion coefficient, D_L, (see Eq. 3.1.1), is relatively constant at low velocities, but increases linearly with velocity at higher velocities. These observations led to the definition of the hydrodynamic dispersion coefficient as the sum of the effective diffusion coefficient, D_d, and the mechanical dispersion coefficient, D_m (Palmer and Johnson, 1989):

$$D_L = D_d + D_m \qquad (3.1.4)$$

Diffusion, which is the process by which ionic or molecular constituents move under the influence of thermal-kinetic energy in the direction of their concentration gradient, is an important dispersion process only at low velocities (Freeze and Cherry, 1979). The effective diffusion coefficient, D_d, for the diffusion of the solute in a porous medium, is less than the diffusion coefficient in solution, D_o, because of the tortuosity of the diffusion pathways. Several empirical approaches have been used to

Table 3-3. Values for the diffusion coefficient in water, D_o, for several organic contaminants of concern

Organic compound	Average diffusion coefficients measured in water (cm²/s)	Temperature (°C)[1]	Reference
Alkanes			
n-Butane	0.89×10^{-5}	20	Witherspoon and Bonoli (1969)
n-Pentane	0.84×10^{-5}	20	"
Aromatics			
Benzene	1.02×10^{-5}	20	"
Toluene	0.85×10^{-5}	20	"
Ethylbenzene	0.81×10^{-5}	20	"
Cycloalkanes			
Cyclopentane	0.93×10^{-5}	20	"
Cyclohexane	0.84×10^{-5}	20	"
Chlorinated Aliphatics			
Vinyl Chloride	1.34×10^{-5}	25	Hayduk and Laudie (1974)

[1] Values at other temperatures are available in the original references.

describe D_d mathematically, so caution should be exercised when interpreting effective diffusion coefficient data from the literature (Shackelford and Daniel, 1991; Domenico and Schwartz, 1998). One commonly used approach is (Bear, 1972):

$$D_d = \tau D_o \qquad (3.1.5)$$

where: τ = tortuosity of the medium, which is a factor accounting for the increased distance a diffusing molecule must travel to go around the sand grains. Typical values for τ in granular porous media range from 0.6 to 0.7 (Perkins and Johnston, 1963; Fried and Combarnous, 1971).

The diffusion coefficients in solution, D_o, for the major ions in groundwater (e.g., Na^+, K^+, Mg^{2+}, Ca^{2+}, Cl^-, HCO_3^-, SO_4^{2-}) fall in the range of approximately 0.7×10^{-5} to 2×10^{-5} cm²/s at 25°C (Li and Gregory, 1974). Although the effect of ionic strength is small, these values are temperature dependent, with coefficients at 5°C, for example, about 50% smaller than at 25°C (Freeze and Cherry, 1979). Measured values for the diffusion coefficient for organic contaminants in water, D_o, are on the order of 10^{-5} cm²/s in magnitude. Experimentally measured D_o values for several organic contaminants are summarized in Table 3-3. A compilation of effective diffusion coefficients from the literature, primarily for inorganic ions, can be found in Shackelford (1991).

At high velocities, the dominant dispersive process is mechanical mixing, which is caused by local variations in velocity resulting from heterogeneities in the porous medium (Domenico and Schwartz, 1998). The heterogeneities responsible for mechanical dispersion occur at several scales, ranging from the microscopic (i.e., pore to pore) scale, to the macroscopic (i.e., well to well or intraformational) scale and the megascopic (i.e., formational) scale.

Based on the small-scale laboratory studies with homogeneous isotropic media, the mechanical dispersion coefficient, D_m, is proportional to the average pore-water velocity:

$$D_m = \alpha\, v \quad (3.1.6)$$

where α = the dispersivity [L], which is a characteristic property of a medium. At these small scales of measurement, there are several microscopic scale effects resulting from heterogeneities that contribute to mechanical dispersion (Gillham and Cherry, 1982): (1) variations in velocity in a pore, (2) different pore geometries, and (3) the divergence of flow lines around the grains of the porous medium. Based on about 2500 column dispersion tests, Klotz and Moser (1974) concluded that values of the longitudinal dispersivity depend on the porous medium's grain size and uniformity coefficient, with less significant effects due to grain shape, grain roughness, grain angularity, and compactness. Although some authors indicate that a more accurate form of Eq. 3.1.6 is αv^m, where m is an empirical constant ranging from 1 to 2, laboratory studies indicate that it can generally be assumed that $m = 1$ for granular geologic materials (Freeze and Cherry, 1979). Column studies also indicate that the relationship in Eq. 3.1.6 is the same in the longitudinal and transverse directions, although the longitudinal dispersivity values, α_L, in laboratory studies are greater than the transverse dispersivity, α_T, with typical ratios of α_L/α_T of 10 to 30 (Gillham and Cherry, 1982).

For the simple one-dimensional system (Fig. 3-1), the solutions of Eq. 3.1.1 for the concentration breakthrough curves resulting from advection and dispersion with a continuous contaminant source ($G = 0$) and a slug release source are shown in Figs. 3.-2a and 3-2b, respectively. With the constant source (Fig. 3-2a), dispersion results in part of the solute front traveling at a rate faster than the average pore-water velocity, and part of the solute front traveling slower, resulting in the front appearing "smeared," with a normal (or Gaussian) distribution. The larger the medium's dispersivity, the greater the mixing of the advancing solute front (Freeze and Cherry, 1979). Although the total mass of dissolved contaminant remains the same as in the case of advection only, the mass is distributed over a larger volume, effectively diluting the concentration at any point along the flow path (USEPA, 1987).

The same concepts can be applied to two- and three-dimensional scenarios, in which case there is additional spreading in the directions perpendicular to the groundwater flow, known as vertical and horizontal transverse dispersion (Palmer and Johnson, 1989). Examples of two-dimensional scenarios with a continuous-steady source and a slug source in an isotropic medium with a unidirectional flow field are presented schematically in Figs. 3-3a and 3-3b, respectively. The tracer spreads and is normally distributed in longitudinal and transverse directions in the horizontal plane

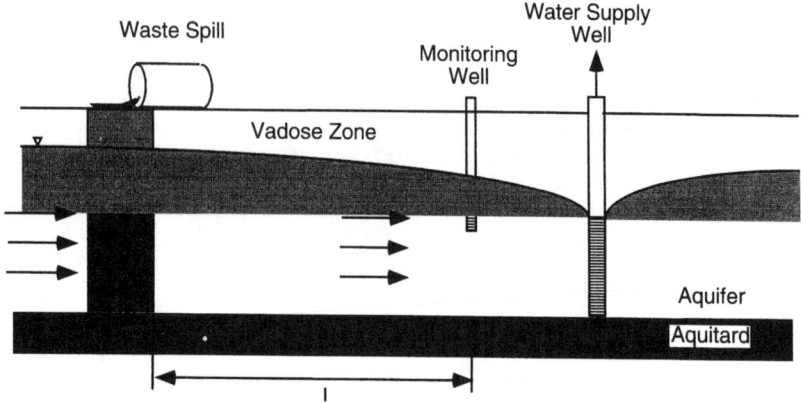

Figure 3-1. A simple one-dimensional site scenario containing a contaminant source, monitoring well, and receptor of concern

as it is transported along the flow path, resulting in the mass occupying an increasing volume of the porous medium, even though the total mass of tracer does not change (Freeze and Cherry, 1979). Although the porous medium in these examples is isotropic with respect to its textural properties and hydraulic conductivity, the anisotropic nature of dispersion (i.e., longitudinal dispersion stronger than transverse) results in the tracer zone developing an elliptical shape. For the three-dimensional case, concentration distributions form ellipsoids of revolution (i.e., football shapes) if the horizontal and vertical transverse dispersion are the same and the longitudinal dispersion is larger, and a surfboard shape if the vertical transverse dispersion is small (Domenico and Schwartz, 1998).

Eq. 3.1.1 and two- and three-dimensional versions of it have been applied to simulate transport of dissolved contaminants, but discrepancies between the theoretical predictions and laboratory experiments have also been observed and attributed to various mechanisms including immobile zones of water (e.g., dead-end pores), solution-solid interfacial processes, anion exclusion, and diffusion in and out of aggregates (Gillham and Cherry, 1982). An apparent scale dependency in the magnitude of measured values of the dispersion coefficient has also been observed. Small values of longitudinal dispersivity (0.0001 to 0.01 m) are generally observed in laboratory-scale experiments, while larger values (10 to 100 m) have been obtained in field tracer tests and from model calibration of contaminant plumes (Palmer and Johnson, 1989; Gillham and Cherry, 1982; Anderson, 1979) (Table 3-4). In a critical review of dispersivity values estimated from field experiments, Gelhar et al. (1992) evaluated data from 59 sites at which field experiments had yielded some 106 values of longitudinal dispersivity, ranging from 0.01 m to 5500 m at testing scales ranging from 0.75 to 100 km. However, although the data taken as an aggregate indicated a trend of increasing longitudinal dispersivity with increasing scale, the data that were deemed to be highly reliable did not demonstrate a clear trend with scale.

NATURAL ATTENUATION OF HAZARDOUS WASTES 37

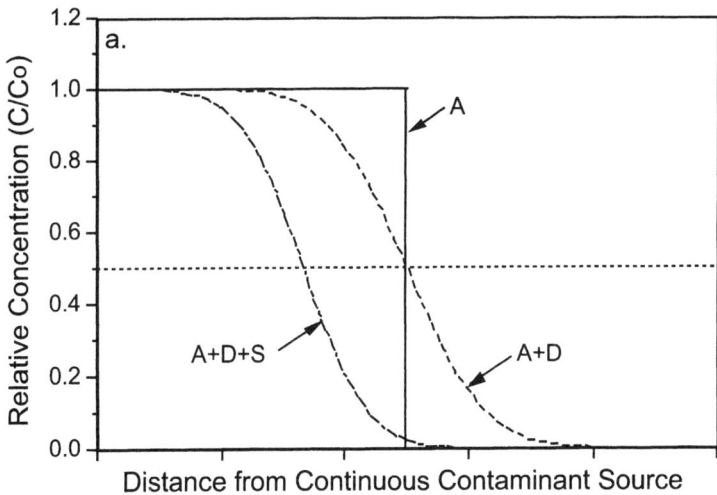

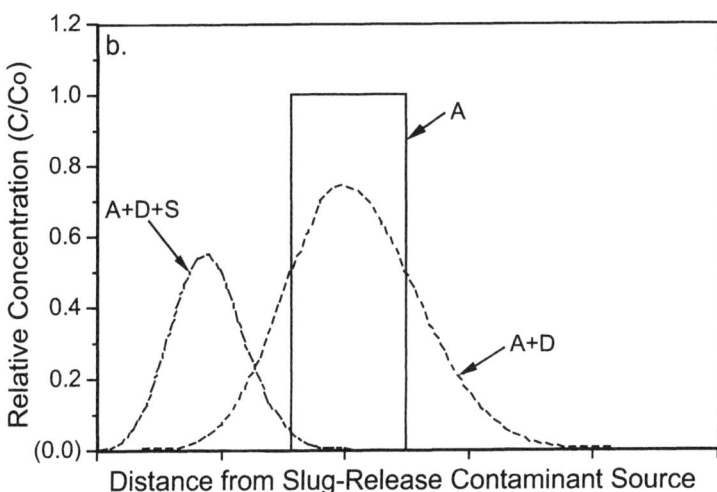

Figure 3-2. The influence of advection (A), dispersion (D), and sorption (S) on the contaminant levels downgradient from (a) continuous and (b) slug-release sources, for a one-dimensional scenario

Source: adapted from Fig. 1-14, p. 13, USEPA, 1987.

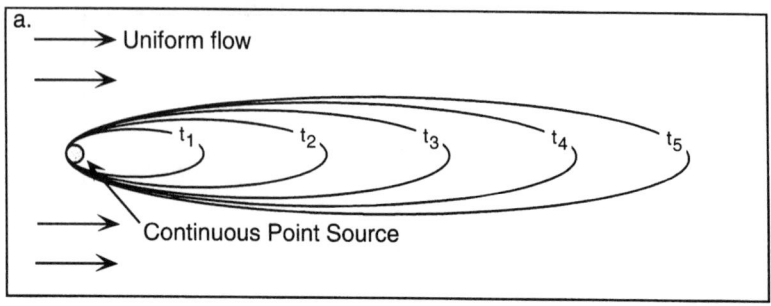

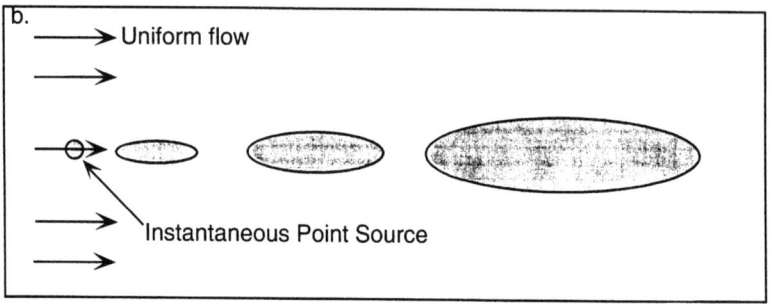

Figure 3-3. The influence of advection and dispersion on the spreading of a tracer in a two-dimensional uniform flow field in an isotropic porous medium for (a) continuous and (b) slug or instantaneous point sources (b)
Source: adapted from Fig. 9.5, p. 394, Freeze and Cherry, 1979.

Typically, longitudinal dispersion measured in the field is much stronger than transverse dispersion, because the process of mechanical dispersion is anisotropic (Freeze and Cherry, 1979), which is consistent with the findings at the laboratory scale. For example, the data compiled by Gelhar et al. (1992) indicate that field horizontal transverse dispersivities are at least an order of magnitude smaller than longitudinal values, with the field vertical transverse dispersivities typically 1 to 2 orders of magnitude smaller than the horizontal transverse values. Indeed, field observations based on detailed three-dimensional monitoring (e.g., McFarlane et al., 1983), indicate that contaminant plumes are often long and thin (Palmer and Johnson, 1989), which suggests that vertical transverse dispersivity values must be very small, or even close to zero (e.g., Sudicky, 1986; Frind and Hokkanen, 1987). Thus, the vertical transverse dispersion coefficient is diffusion controlled until the velocity is very high (see Eqs. 3.1.6 and 3.1.8) (Freeze and Cherry, 1979).

Table 3-4. Longitudinal dispersivity values measured in experiments performed at various scales in unconsolidated geologic materials

Type of test	Range of dispersivity values (m)		
	Longitudinal	Horizontal transverse	Vertical transverse
Laboratory tests	0.0001–0.01	$\approx \alpha_L/10 - \alpha_L/30$	—
Natural gradient tracer tests	0.01–2	0.01–0.3	<0.1
Single-well tests	0.03–0.3	—	—
Radial and two-well tests	0.5–15	—	—
Model calibration of contaminant plumes	3–100	1–30	0.2–0.6

Source: adapted from Table 1, p. 46, Gillham and Cherry (1982) and Table 1, p. 9, Palmer and Johnson (1989).

The observations of apparently increasing magnitudes of the longitudinal dispersivity with increasing scale indicate that the assumptions involved in applying the classic advection-dispersion theory may not be applicable in natural geologic materials (Palmer and Johnson, 1989). The view that has emerged based on the large longitudinal dispersion coefficients observed in field experiments and obtained from model calibration of contaminant plumes is that heterogeneities at the macroscopic scale, such as stratification and permeability characteristics, contribute significantly to dispersion due to the creation of local-scale variability in velocity (Domenico and Schwartz, 1998).

All geologic formations are heterogeneous to some degree. Indeed, small-scale heterogeneities are ubiquitous in granular media aquifers (Freeze and Cherry, 1979). Almost unrecognizable variations in grain-size characteristics (e.g., a change of silt or clay content of a few percent in a sandy zone) can result in hydraulic conductivity contrasts as large as an order of magnitude or more. Other macroscopic features that may contribute to contaminant dispersion include fingered beds and lenses of higher conductivity, solution channeling and fracturing (Freeze and Cherry, 1979; USEPA, 1987).

In addition to the scale-dependence in the magnitude of the measured dispersion coefficient values, field-scale apparent longitudinal dispersivity values have also been observed to increase with increasing transport distance, even in a single tracer experiment (e.g., Freyberg, 1986). One way to explain such nonconstant dispersivity values is to consider them to be a result of incomplete spatial averaging (Domenico and Schwartz, 1998). For example, if a dispersion experiment were run in a heterogeneous aquifer with relatively large lenses, the spread of tracer from a source to a nearby observation point in the same lens should yield a relatively small dispersivity, comparable to that in a laboratory column. However, as the tracer moves further, it will interact sufficiently with the heterogeneity to produce macroscale

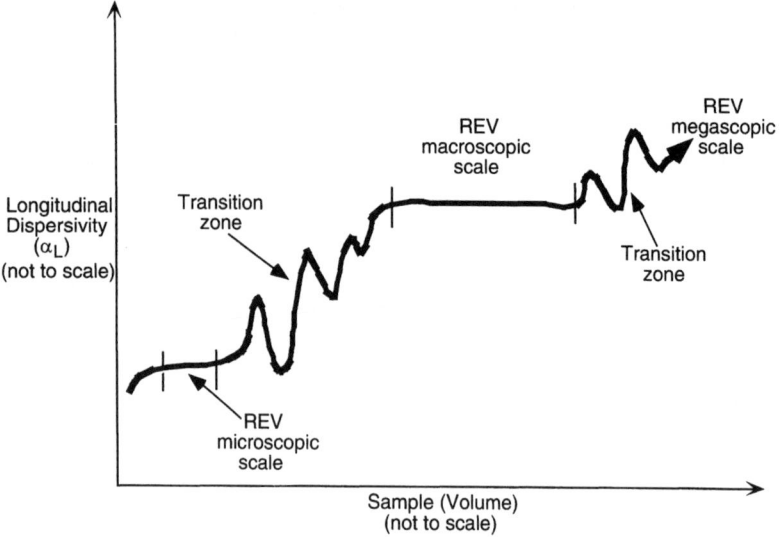

Figure 3-4. An example schematic representation of the relationship between longitudinal dispersivity and sample volume, showing REVs defined for the microscopic and macroscopic scales
Source: adapted from Fig. 10.9, p. 222, Domenico and Schwartz, 1998.

mixing. In this way dispersivity values increase with distance (e.g., 10s to 100s of meters) from a source, approaching a constant macroscale dispersivity, also known as the asymptotic macrodispersivity (Gelhar et al., 1979).

Alternatively, the spatially varying dispersivity values can be understood in terms of some representative elemental volume (REV) for dispersivity (see Fig. 3-4) (Domenico and Schwartz, 1998). As the tracer spreads out, there will be a transition from the microscale to the macroscale and eventually megascopic scale, with some REV for each scale for which a continuum value for dispersivity can be defined. In the transition zones between these scales, tracer spreading may not have encompassed a volume equivalent to an REV, in which case small changes in volume could result in significant variability in the spreading and, hence, the dispersivity value.

Unfortunately, a detailed evaluation of aquifer properties cannot be done routinely at contaminated sites. Nevertheless, the heterogeneity in aquifer properties can be represented by geostatistical models. Specifically, the asymptotic dispersivity can be quantitatively related to geostatistical models of hydraulic conductivity (Domenico and Schwartz, 1998). A geostatistical description treats the hydraulic conductivity of the aquifer as a random process and represents it by a small number of statistical parameters, such as the mean, variance, and the autocorrelation function. Such statistical analyses were performed by Sudicky (1986) on the Borden aquifer (Ontario, Canada) where permeameter measurements were analyzed on 1,279 samples. Although the aquifer material was described as being relatively homogeneous (Freyberg, 1986), the range of hydraulic conductivity values measured spanned more than two orders of magnitude (Sudicky, 1986). Applying the stochastic theory of Gelhar and Axness (1983), Sudicky (1986) calculated an asymptotic longitudinal macrodispersivity of 0.61 m, which is close to the value of 0.43 m estimated by Freyberg (1986) using large scale field tracer test data at the site. The estimated vertical transverse dispersion coefficient was on the order of the effective diffusion coefficient, consistent with the discussion above.

In addition to the effects of aquifer heterogeneity reviewed above, a number of other processes also contribute to the spread of contaminants (Palmer and Johnson, 1989). For example, diverging advective flow lines can spread contaminants over a larger cross-section of the aquifer. In addition, the direction of groundwater flow and the lateral spread of contaminants can be changed by temporal variations in the water table. Apparent longitudinal dispersion can also be caused by variations in the contaminant concentration at the source (Frind and Hokkanen, 1987).

Many field dispersion experiments have also suffered from errors committed during the data collection and interpretation, which typically cause inferred dispersivity values to be larger than the actual values (Domenico and Schwartz, 1998; Palmer and Johnson, 1989). Some of the errors are due to: (1) poor definition of the plume as a result of few monitoring wells and nonpoint sampling; (2) failure to account for temporal variations in the advective flow regime, which can contribute to lateral spreading of mass; (3) incomplete knowledge of the source loading; (4) inherent limitations in some test procedures; (5) inappropriate groundwater sampling methods, which can also result in apparent spreading of contaminant plumes (e.g., insufficient well purging, comparing monitoring wells with different screen lengths); and (6) use of oversimplified techniques for data interpretation.

Despite the limitations described above, macroscale dispersivities remain a useful conceptual model for a complicated process (Domenico and Schwartz, 1998). Nevertheless, it is important to be aware of the limitations associated with extending the classical dispersion concepts from microscale to macroscale and larger. The observations indicating a tendency for dispersivity values to be scale-dependent impact how macroscopic dispersivities are estimated and used. For example, longitudinal dispersivity values obtained from tracer studies run over short distances may not represent the macroscopic asymptotic values. However, routine running of dispersion experiments over the scale required to yield an asymptotic value may not be practical.

There are several approaches for estimating dispersivity values at the field scale (i.e., macroscopic to megascopic scales), including: tracer tests, calibration of contaminant plumes and environmental tracers, and regression formulas. The tracers and testing strategies used depend on the scale of the study and the presence of a contaminant plume (Domenico and Schwartz, 1998). For small systems (e.g., some fraction of a kilometer), field tracer experiments may be run, which fall into four standard categories: (1) natural gradient tracer tests, (2) single-well pulse tests, (3) two-well tracer tests, and (4) single-well injection or withdrawal tests with multiple observation wells (e.g., Domenico and Schwartz, 1998; Freeze and Cherry, 1979). Example value ranges obtained are summarized in Table 3-4. Performing a tracer test is potentially the most reliable method for estimating dispersivities, but the time and cost required may be prohibitive (Wiedemeier et al., 1996).

For scales greater than a kilometer, tracer experiments are not feasible because of time constraints. For a localized or site-specific system (e.g., several kilometers), a contaminant plume that has spread over a long time period is the tracer of choice (Domenico and Schwartz, 1998). For a unit of regional extent (e.g., 10s to 100s of kilometers), studying flow and dispersion involves using either naturally occurring ions (e.g., Br^- or Cl^-), or environmental isotopes (e.g., ^2H, ^3H, or ^{18}O). For these large scale, uncontrolled "tracers", the dispersivity values are typically determined by fitting a solute transport model to the data (Gelhar et al., 1992). Advection and dispersion estimates based on contaminant plumes and environmental tracers are less reliable than controlled field tracer tests, due to the poorly defined tracer input (both quantity and temporal distribution), and the often limited measurement of the tracer concentration in space.

Another alternative for estimating dispersivity values is to use regression formulas from the literature, which relate dispersivity and field scale (e.g., Neuman, 1990; Xu and Eckstein, 1995). However, Gelhar et al. (1992) have cautioned against routinely adopting dispersivities from their compilation and criticized single regressions to the data as having inadequately considered the reliability of the data, and having ignored the fact that different aquifers will have different degrees of heterogeneity at a given scale. Based on their reliability assessment, Gelhar et al. (1992, 1993) suggest that if data from their compilation are to be applied for selecting dispersivities for use in site-specific deterministic modeling of existing or potential contamination releases when no site specific information is available, dispersivities should be selected from the lower half of the observed range in their review (i.e., α_L on the order of 10^{-2} to 10^1 m). Furthermore, they indicate that for very heterogeneous aquifers and large sources, somewhat larger dispersivities are appropriate, and that any site specific information should be considered (Gelhar et al., 1993).

With respect to remediation by natural attenuation, hydrodynamic dispersion is important because it is the primary mixing process for solutes in the subsurface. Thus, mass-transport via advection and dispersion can significantly impact contaminant distribution and the availability of substrates, nutrients, and electron acceptors to microorganisms (Sturman et al., 1995), which may, in turn, drive the distribution and activity of microbes in the subsurface (Brockman and Murray, 1997; McMahon and Chapelle, 1991). Therefore, because dispersion influences the rates of substrate and nutrient flux and mixing, redox conditions, and microbial ecology, it impacts *in situ* biodegradation.

For example, in the simple one-dimensional scenario presented in Fig. 3-2b, dispersive spreading of concentration peaks for two compounds (e.g., the electron-donor and electron-acceptor substrates) introduced at different points or times may lead to overlapping and mixing of the substrates (Cirpka et al., 1999). Another case of this phenomenon is the hydraulic mixing at the interface of the dissolved contaminant plume and the "clean" groundwater, which is induced by subsurface heterogeneities in the hydraulic conductivity, K. In particular, vertical dispersion is an important process for developing zones of mixing (Domenico and Schwartz, 1998; Sudicky et al., 1985). For example, laboratory and modeling studies with two-dimensional (vertical) stratified systems (Szecsody et al., 1994; Odencrantz, 1992; Wood et al., 1994; Yang et al., 1994) and small-scale K heterogeneity (Murphy et al., 1997; MacQuarrie and Sudicky, 1990; Murphy et al., 1997; Cirpka et al., 1999), under dual-substrate limitation, demonstrate that K heterogeneity can create regions of increased hydraulic mixing between waters carrying different substrates due to dispersion. This, in turn, was shown to result in enhanced microbial activity and growth. However, other modeling studies indicate that aquifer heterogeneities may result in a decrease in contaminant biodegradation, e.g., by preventing the electron acceptor and donor from coming into contact (Schäfer and Kinzelbach, 1992).

Evaluating whether or not dispersion results in dilution of the contaminant is also an important component of remediation by natural attenuation. Thus, as demonstrated by the modeling studies of Kitanidis (1994), it is key to distinguish spreading—"the stretching and deformation of a contaminant plume"—from dilution—"the increase in volume of the fluid occupied by the solute." This distinction is not necessary under the idealized conditions described by the advection-dispersion equation (Eq. 3.1.9), because the dilution and spreading of the Gaussian plume in a homogeneous porous medium with constant velocity (e.g., Fig. 3-2) are related in a simple way, and both are characterized by the same parameter, the dispersion coefficients. In fact, in the case of a plume in a large formation, the most complete state of dilution is given by the normal or Gaussian distribution of concentration (e.g., Fig. 3-2).

However, in field applications, heterogeneous geological formations result in irregularly shaped, highly-non Gaussian plumes, indicating a less complete state of dilution (Thierrin and Kitanidis, 1994). As Kitanidis (1994) demonstrates, the scale-dependent macroscopic dispersion coefficients discussed above are not reliable measures of dilution in heterogeneous formations because the rates of dilution and spreading can be quite different. Rather, the rate of increase of dilution depends only on the local dispersion and the shape of the plume. For example, in a stratified formation of low- and high-conductivity layers, the nonuniformity of velocity distorts the plume, resulting in spreading and apparent macroscopic dispersion. However, although a plume in such a formation may be spreading relatively quickly, the dilution of the plume does not increase at the same rate because the plume's volume will increase more slowly at early times. Nevertheless, in the long term, such geologic heterogeneity should increase dilution because the resulting irregularity in the shape of the plume increases the surface area over which the mass transfer due to local dispersion can occur and, thus, increases the dilution rate.

3.1.3.3 Infiltration and Groundwater Recharge

Infiltration and groundwater recharge are interrelated physical processes in the unsaturated and saturated zones, respectively, that may contribute to natural attenuation. Freeze and Cherry (1979) define infiltration as "the entry into the soil of water made available at the ground surface, together with the associated flow away from the ground surface within the unsaturated zone" and groundwater recharge as "the entry into the saturated zone of water made available at the water-table surface, together with the associated flow away from the water table within the saturated zone." Although groundwater typically discharges to surface water bodies, under some conditions, the hydraulic gradient is such that the surface water has a higher potential than the groundwater (e.g., during flood stages), resulting in flow in the reverse direction (USEPA, 1987). The combined processes of infiltration and recharge can potentially influence remediation by natural attenuation in several ways that can be broken down into two categories: (1) dilution, and (2) transfer of materials from the unsaturated zone to the saturated zone.

One way that groundwater recharge resulting from precipitation or leakage from surface water bodies contributes to natural attenuation is by introducing additional water mixing in the groundwater flow. This results in dilution of dissolved contaminants, reducing the concentration in the plume (ASTM, 1998). However, because vertical transverse dispersion is a weak process in groundwater systems, there may be little mixing between the overlaying zone of recharge water and the contaminant plume (e.g., MacFarlane et al., 1983).

Infiltration can also result in the transfer of materials from the unsaturated zone to the saturated zone, because as the water moves downward through the unsaturated zone under the influence of gravity, it dissolves materials with which it comes into contact (USEPA, 1987). Thus, infiltration may dissolve soil organic matter, a potential source of electron donors. In addition, infiltration may dissolve electron acceptors (e.g., dissolved oxygen, nitrate, sulfate), which then may enter the groundwater system and increase the overall electron-accepting capacity within the contaminated plume and change the predominant terminal electron accepting processes (TEAPs) (Wiedemeier et al., 1996). For example, Vroblesky and Chapelle (1994) presented data for a petroleum hydrocarbon-contaminated aquifer, which indicated that during times of low aquifer recharge, methanogenesis, the least efficient TEAP for metabolism of petroleum contaminants, occurred in the most contaminated parts of the aquifer. However, addition of the more microbially efficient electron acceptors oxygen and sulfate by rainfall caused a shift in the aquifer to a more efficient TEAP, potentially increasing biodegradation.

Similarly, seasonal water table fluctuations resulting from aquifer recharge and discharge can trap soil gas and can also supply additional dissolved oxygen to the saturated zone (ASTM, 1998). The effect of a trapped gas phase on the distribution and transport of dissolved oxygen was demonstrated experimentally in a large-scale physical aquifer model by Fry et al. (1995). This may be an important phenomenon in natural systems. For example, anonymously large rises in water levels in observation wells in shallow unconfined aquifers have often been observed during heavy rainstorms as a result of air entrapment in the unsaturated zone (Freeze and Cherry, 1979).

In addition to supplying dissolved inorganic electron acceptors, the chemical properties of the infiltrating water may also result in geochemical changes in the aquifer (Wiedemeier et al., 1996). For example, Vroblesky and Chapelle (1994) describe a site at which oxidation of Fe(II) by aerobic recharge water resulted in the reprecipitation of Fe(III) onto the mineral grains, which caused a shift of the TEAP from sulfate reduction to Fe(III) reduction at one shallow location, and from methanogenesis to Fe(III) reduction at another location. In a situation where the contaminants are used by microorganisms as electron donors, e.g., fuel hydrocarbons or vinyl chloride, such a shift in the dominant terminal electron acceptor available may be beneficial for removal by biodegradation; however, in the case of reductive dehalogenation such a shift could result in less favorable conditions (Wiedemeier et al., 1996). It is also possible that dissolution of materials in the unsaturated zone by infiltration may exacerbate the contamination situation, if the unsaturated zone is contaminated. Indeed, contamination of groundwater by infiltration is probably the most common groundwater contamination mechanism (USEPA, 1987). Similarly, recharge from contaminated surface water may also enter and contaminate the groundwater system.

Wiedemeier et al. (1996) note that it may be difficult to evaluate the effects of recharge on natural attenuation. For example, it may be possible to estimate the effects of dilution if a detailed water budget has been constructed for the system, but with plumes of significant vertical extent, it may be difficult to know what proportion of the plume mass is being diluted by recharge. Separating out the effects of dilution is made even more difficult by the fact that dispersivity, sorption, and biodegradation are often not well quantified. The effects of the addition of electron acceptors in infiltrating precipitation may be discernable qualitatively in the area of recharge based on elevated electron acceptor concentrations, or different patterns in microbial electron-acceptor consumption or byproduct formation. Nevertheless, because of the typically intermittent nature of precipitation events, it may be difficult to understand the effects of short-term variation in such systems. In the case of recharge from surface water, for which the influx of mass and electron acceptors is more steady, it may be easier to quantify the effects of dilution and to identify (if not quantify) the effects of electron acceptor replenishment.

3.1.4 Interphase Partitioning Processes

3.1.4.1 Sorption/Desorption

Sorption refers to the interphase partitioning of a dissolved contaminant (i.e., the sorbate) between the water phase (i.e., the solvent) and a solid phase material (i.e., the sorbent). This process is included here with the physical processes because the mechanism itself does not result in irreversible change of the compound; however, it can also be considered a chemical process because it can be controlled by chemical forces and interactions (ASTM, 1998). Although sorption is a nondestructive attenuation mechanism, it is an important consideration in evaluating natural attenuation because it affects the contaminant's mobility and dissolved concentration, which in turn affects the rates of other transport and transformation reactions (ASTM, 1998).

Sorption processes involve a complex array of phenomena that can alter the contaminant distribution between and among the various phases and interfaces in environmental systems (Weber et al., 1991). Two broad categories of sorption phenomena can be defined: adsorption, in which solute accumulation is generally restricted to a surface or the interface between solution and adsorbent, and absorption, in which the solute interpenetrates the sorbent phase.

A variety of different types of attractive forces between solute, solvent, and sorbent molecules result in sorption (Weber et al., 1991). These forces typically act together, although their contributions vary in magnitude. Distribution between the phases is a result of relative affinity of the solute for each phase, which is in turn related to the nature of the intermolecular forces between the phases. The primary driving force for adsorption may be a result of the solvent-disliking character of the solute (i.e., solvent-motivated sorption), or of a specific affinity of the solute for the solid (i.e., sorbent-motivated sorption) (Weber, 1972; Weber et al., 1991). Usually sorption results from a combination of these two forces. The most significant factor in determining solvent-motivated sorption is the degree of solubility of a dissolved substance (Weber, 1972). For example, in aqueous systems, a more hydrophilic substance is less likely to move toward an interface and be adsorbed, and a more hydrophobic substance is more likely to be adsorbed.

Three categories of sorbent-motivated adsorption are loosely defined, based on the class of attractive forces that predominates (Weber, 1972; Weber et al., 1991), as illustrated schematically in Fig. 3-5 for 4-chloroaniline: (1) chemical sorption or chemisorption, which involves solute-sorbent interactions with the characteristics of true chemical bonds (e.g., covalent bonds); (2) electrostatic or exchange adsorption, which involves electrical attraction between oppositely charged species (e.g., ion exchange); and (3) physical adsorption, which results from the action of Van der Waals forces. Typically, adsorption phenomena are a combination of these three types of adsorption. In the case of hydrophobic molecules, the relatively weak bonding forces associated with physical sorption are often amplified by tthe substantial thermodynamic drive for repulsion from aqueous solution (Weber et al., 1991). This combined effect is often referred to as "hydrophobic bonding" (Hamaker and Thompson, 1972).

The contaminant mass distribution at equilibrium can be quantified in batch experiments, which are performed by mixing aqueous solutions of various contaminant concentrations with soil, allowing time for equilibrium to be established, and then measuring the contaminant concentration remaining in solution (Devinny, 1990). At equilibrium, the constant-temperature relationship between the mass of adsorbate accumulated per unit mass adsorbent, S_e [MM^{-1}], and the equilibrium concentration of adsorbate in solution, C_e [ML^{-3}], is called the adsorption isotherm (Snoeyink, 1988). Three of the more commonly applied phenomenological models for characterizing adsorption are: the linear equation, the Langmuir equation, and the Freundlich equation.

The linear model assumes the accumulation of solute by the sorbent is directly proportional to the solution phase concentration (Weber et al., 1991):

$$S_e = K_D C_e \qquad (3.1.7)$$

where: K_D = the constant of proportionality or distribution coefficient (also known as a partition coefficient) = (volume solution/mass solid) [$L^3 M^{-1}$] (See Fig. 3-6). Weber et al. (1991) indicate that the linear isotherm equation "is appropriate for sorption relationships in which the energetics of sorption are uniform with increasing concentration and the loading of the sorbent is low." Specifically, the linear equation

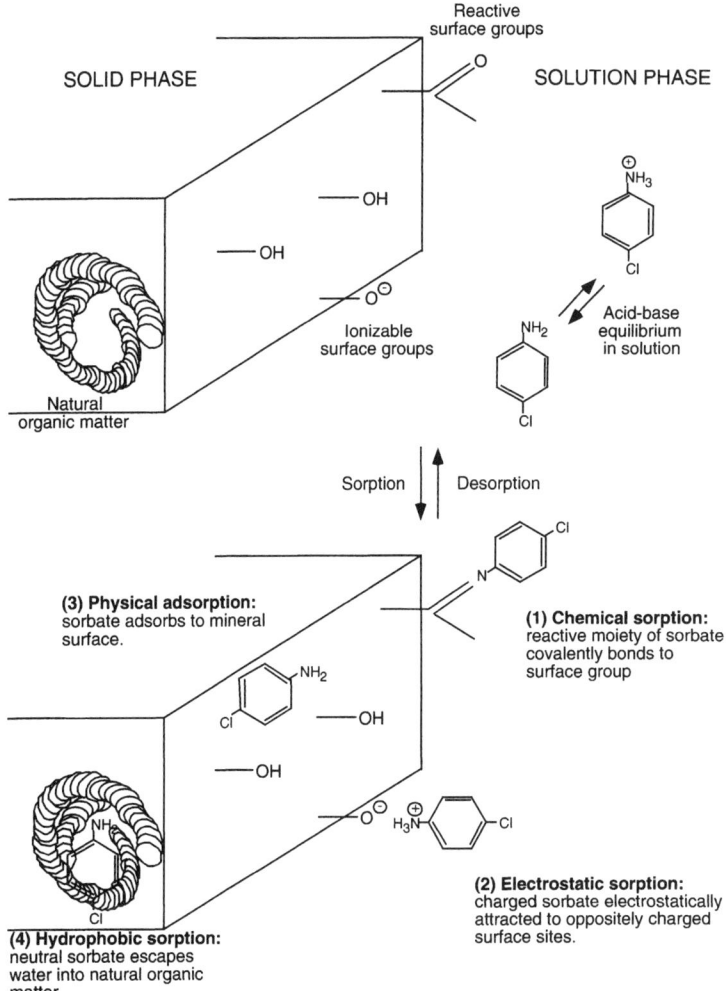

Figure 3-5. Schematic representation of key categories of sorbent-sorbate interactions

Source: adapted from Fig. 11.2, p. 257, Schwarzenbach et al., 1993.

accurately describes absorption, and may adequately describe certain cases of adsorption, in particular at very low solute concentrations (e.g., either $\leq 10^{-5}$ M, or less than half the solubility, whichever is lower) and for solids with low potential for sorption (Johnson et al., 1989; Weber et al., 1991).

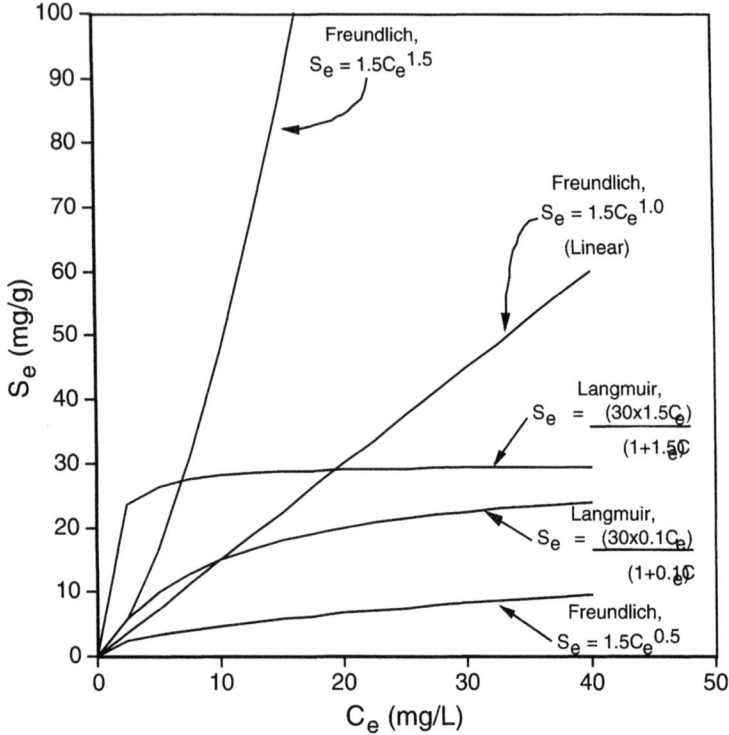

Figure 3-6. Example linear, Langmuir, and Freundlich isotherms
Source: adapted from Fig. 12.7, p. 267, Domenico and Schwartz, 1998.

The Langmuir model is a non-linear equation, which originally was developed for systems in which sorption results in deposition of a single solute molecule layer on the sorbent surface (Weber et al., 1991):

$$S_e = \frac{Q^o b C_e}{1 + b C_e} \qquad (3.1.8)$$

where: Q^o = the sorbed solute concentration on the sorbent at monolayer coverage (i.e., the maximum value S_e that can be achieved as C_e is increased), and b = a sorption coefficient, related to the energy of adsorption, which increases as the strength of the adsorption bond increases (Snoeyink, 1988) (see Fig. 3-6). Eq. 3.1.8 reduces to a linear relationship at low surface coverage (Weber et al., 1991).

Probably the most widely used nonlinear sorption equilibrium model is the Freundlich isotherm, which is frequently applied to describe and predict adsorption of organic chemicals from groundwater (Devinny, 1990). The development and applications of the Freundlich isotherm are for the most part empirical; however, the model can be demonstrated to be thermodynamically rigorous for special cases (Weber et al., 1991). The general form of the Freundlich equation is (Snoeyink, 1988):

$$S_e = K_F C_e^{1/n} \tag{3.1.9}$$

where: K_F is related primarily to the capacity of the adsorbent for the adsorbate, and $1/n$ is a function of the strength of adsorption (See Fig. 3-6). When $n = 1$, the Freundlich isotherm reduces to the linear case, but exponents $(1/n) < 1$ are often observed for low-molecular weight organic compounds and natural aquifer solids (e.g., Ball and Roberts, 1991; Weber et al., 1992; Young and Ball, 1994).

One constraint on application of the preceding isotherm equations is that they are developed for single solutes, whereas contaminated groundwater typically has multiple solutes present. In adsorption of such a mixture, there are three possible outcomes: the compounds mutually enhance their respective adsorptions, they act independently, or they mutually inhibit their respective adsorptions (Weber and DiGiano, 1996). It can be predicted that mutual inhibition of adsorption or competition will occur if: (1) adsorption is confined to a single or a few molecular layers, (2) the solute adsorption affinities do not differ by several orders of magnitude, and (3) there are no specific solute interactions to enhance adsorption. Research reviewed by Lyman et al. (1992) indicates that such competitive sorption may occur for organic cations, chlorinated phenols, and inorganic ions, and between immobile adsorbates and mobile colloidal adsorbates, but that it may not be important for neutral organics. Several models for describing competitive multi-solute sorption have been used. Models based on adaptations of the Langmuir and Freundlich isotherms have been developed, but application is restricted by the limiting assumptions of the basic models (Weber et al., 1991). Alternatively, the ideal adsorbed solution theory (IAST) often provides a more accurate characterization of multi-component adsorption equilibria (see discussion in Weber and DiGiano, 1996).

Mechanistic models of ion exchange, surface complexation, and hydrophobic sorption reactions have also been developed (see Weber et al., 1991). The focus here is on hydrophobic sorption, which comprises a large category of sorption reactions in soil and sediments. A linear equilibrium sorption model (Eq. 3.1.10) has proven to often be useful for predicting the sorption equilibria of relatively nonpolar, neutral organic solutes with soils (Weber et al., 1991).

Empirically, K_D for hydrophobic compounds often are proportional to a soil-specific factor—f_{oc}, the organic content of the solid (mass organic carbon/mass solid) [MM^{-1}]—and a compound-specific factor related to hydrophobicity of the compound—K_{oc}, the organic carbon partition coefficient {(mass sorbate/mass organic carbon)/(mass solute/volume solution)} [L^3M^{-1}] (Devinny, 1990; Weber et al., 1991):

$$K_D = f_{oc} K_{oc} \tag{3.1.10}$$

Thus, when values of f_{oc} and K_{oc} are available, Eq. 3.1.10 can be used to determine the K_D value and predict the amount of partitioning. Because K_D is often a function of the organic carbon content of the soil, a common approach is to conceptualize the sorption of nonpolar hydrophobic compounds as the partitioning of the compound into an organic phase consisting of polymeric natural organic materials associated with a mineral phase (Fig. 3-5) (Chiou et al., 1983).

Johnson et al. (1989) note that the approach described by Eq. 3.1.13 works reasonably well for a wide range of soils as long as the soil organic carbon content is sufficiently high (i.e., $f_{oc} > 0.001$); however, the correlation between K_D and f_{oc} is generally weak for $f_{oc} < 0.001$ (e.g., MacIntyre et al., 1991). Although f_{oc} values for porous medium samples can be measured in the laboratory, it is not a well characterized parameter and the range of values reported for geologic materials is extremely variable (Domenico and Schwartz, 1998). Furthermore, the f_{oc} value measured may vary depending on the technique used (e.g., Weber et al., 1992; Powell et al., 1989).

Sorption of hydrophobic solutes onto inorganic surfaces can also occur (e.g., Rogers et al., 1980). As reviewed by Ball et al. (1997), surface adsorption to mineral surfaces may be as important as partitioning into a natural organic phase in some situations, including: when very little organic matter is present (e.g., Ball and Roberts, 1991); when an especially strong adsorbing phase, such as kerogen or shale is present (e.g., Weber et al., 1992); or when the organic chemical of concern has polar or ionic structural groups that may engage in surface complexation or ion exchange reactions with the mineral surface (e.g., Means et al., 1982; Zachara et al., 1990). Thus, with lower organic carbon content soils, the sorption of neutral organics onto the mineral phase can result in errors in the estimate of K_D (Chiou et al., 1985).

McCarty et al. (1981) proposed that below a critical level of organic matter, $(f_{oc})^*$, sorption is dominated by mineral-solute interactions, and above $(f_{oc})^*$ sorption is dominated by organic carbon-solute interactions. The following empirical equation was developed by McCarty et al. (1981) to estimate this value for solute i as a function of octanol/water partition coefficient, K_{ow} [L^{3}L^{-3}], and the surface area of the mineralogical component of the aquifer matrix, A_s, [m^2/g]:

$$\left(f_{oc}\right)_i^* = \frac{A_s}{200} \frac{1}{K_{ow}^{0.84}} \tag{3.1.11}$$

Based on this relationship, the critical level is smaller for organic compounds with a high affinity for organic carbon (high K_{ow}) than for those with a low K_{ow}. Furthermore, the critical level is greater for a medium with a large surface area (e.g., clay).

K_{oc} values for the organic contaminants covered in this book are presented in Sections 5.1, 5.2, and 5.3. There have been numerous empirical observations that K_{oc} values for particular solutes on various natural sorbents can be related to more commonly available chemical properties, such as the solute's water solubility or K_{ow} value, which are also presented in Sections 5.1, 5.2, and 5.3. Most recently, Seth et al. (1999) suggested using a correlation between K_{oc} and K_{ow} of:

$$K_{oc} = 0.33 K_{ow} \qquad (3.1.12)$$

With an upper and lower limit of:

$$0.14 K_{ow} < K_{oc} < 0.89 K_{ow} \qquad (3.1.13)$$

Variability in the composition of organic matter in soils and experimental difficulties and constraints in K_{oc} measurements result in uncertainties in K_{oc} values. Because of these uncertainties, Seth et al. (1999) suggested that a K_{oc} estimate obtained using a correlation should be viewed as having a range of values.

To estimate K_D for a particular aquifer/contaminant system, one option is to use tabulations of K_D values, or alternatively relationships such as Eqs. 3.1.12 and 3.1.13 for K_{oc}, in conjunction with f_{oc} values for the porous medium. However, such values and relationships are only approximate and are strictly valid only for the concentration range of the original data and the particular soil for which they were derived (e.g., Hamaker and Thompson, 1972; Garbarini and Lion, 1986). Domenico and Schwartz (1998) suggest that relationships such as Eqs. 3.1.12 and 3.1.13 can be used to provide an estimate of K_{oc} in the absence of other data, but that in field studies, such estimates must be refined with field or laboratory experiments.

Field experiments include determination of the degree of contaminant partitioning during passage of contaminant solutions through a small segment of the groundwater system followed by monitoring, or evaluations of existing sites at which contamination has already occurred (Freeze and Cherry, 1979). Field data operate on more realistic scales, time frames and conditions than typically reproduced in the laboratory, but the data are often expensive, difficult, and time-consuming to gather (Johnson et al., 1989).

Laboratory experiments performed to estimate the distribution coefficient include two types: batch isotherm experiments, such as described above; and column breakthrough-curve experiments, using uncontaminated aquifer material and reactive (e.g., the contaminants) and nonreactive tracer solutions. Recommended procedures for batch-type sorption isotherms are described by Roy et al. (1991). Wiedemeier et al. (1996) suggest that properly conducted batch tests will provide sufficiently accurate results for modeling purposes, if sensitivity analyses for retardation are also conducted. Data obtained in column experiments can provide an indication of what will occur in the field, if the experiments are carefully performed (e.g., flow rate and input water chemistry approximate field conditions, and disturbance of the sample prior to emplacement in the column has not significantly changed the material's properties) (Freeze and Cherry, 1979).

Gillham and Cherry (1982) note that there are often large differences between batch-determined and column-determined sorption coefficients. There are several possible explanations for cases where significant differences have been observed, including: insufficient time allowed for batch equilibration to occur, failure to attain local equilibrium in columns, loss of sorbent particles through column ends, wall- and end-effects on column flow, channeling of column flow, and nonlinearity of sorption isotherms (MacIntyre et al., 1991). Nevertheless, in some cases, agreement has been found between sorption coefficients determined by batch and column experiments.

For example, several studies have demonstrated good agreement between sorption coefficients measured by batch, column, and sand-tank methods for organic compounds on low-organic carbon aquifer material (e.g., MacIntyre et al., 1991; Larsen et al., 1992).

The linear equilibrium approach can easily be implemented in the ADR equation. For example, for the case of homogeneous saturated porous media with steady-state flow, the one-dimensional form of the ADR equation (Eq. 3.1.1) can be rewritten to incorporate the influence of sorption (Freeze and Cherry, 1979) as follows:

$$\frac{\partial C}{\partial t} = D_L \frac{\partial^2 C}{\partial x^2} - v \frac{\partial C}{\partial x} - \frac{\rho_b}{n} \frac{\partial S}{\partial t} \qquad (3.1.14)$$

where: ρ_b is the bulk mass density of the porous medium, $\partial S/\partial t$ is the rate at which the chemical is adsorbed [MM^{-1}T^{-1}], and $(\rho_b/n)\partial S/\partial t$ gives the change in dissolved concentration caused by adsorption or desorption. Assuming that the amount of sorbed contaminant is a function of the solute concentration (Freeze and Cherry, 1979):

$$\frac{\partial S}{\partial t} = \frac{\partial S}{\partial C} \frac{\partial C}{\partial t} \qquad (3.1.15)$$

where: $\partial S/\partial C$ represents the partitioning of the contaminant between solution and solids. If the time for the sorption process to reach thermodynamic equilibrium is much faster than the time scales for competing processes (e.g., advection and dispersion), then an equilibrium approach can be applied, i.e., the time rate of change of the sorbed concentration, S, at any point is assumed to be instantaneously reflected in the time rate of change of the solution phase concentration, C, at that point (Weber et al., 1991). This local equilibrium model can be further simplified by assuming a linear relationship between S and C (e.g., Eq. 3.1.7), thus (Freeze and Cherry, 1979):

$$\frac{\partial S}{\partial C} = K_D \qquad (3.1.16)$$

Rearranging Eq. 3.1.14 and substituting in the relationships in Eq. 3.1.15 and 3.1.16, gives:

$$\left(1 + \frac{\rho_b}{n} K_D\right)\left(\frac{\partial C}{\partial t}\right) = R\left(\frac{\partial C}{\partial t}\right) = D_L \frac{\partial^2 C}{\partial x^2} - v \frac{\partial C}{\partial x} \qquad (3.1.17)$$

where the term $R = (1 + (\rho_b/n)K_D)$ is referred to as the retardation factor. By dividing Eq. 3.1.17 through by R, the retarded transport velocity of a sorbing chemical, v_c, can be calculated using the so-called retardation equation (Domenico and Schwartz, 1998):

$$v_c = \frac{v}{R} = \frac{v}{1+\frac{\rho_b}{n}K_D} \qquad (3.1.18)$$

Fig. 3-2a illustrates the movement of a concentration front by advection with dispersion and local linear equilibrium sorption. Note how the concentration profile for the sorbed contaminant also spreads out, but travels behind the front of the nonsorbed species, i.e., it is retarded. Fig. 3-2b illustrates the combined effects of advection, dispersion and sorption on a slug of contaminant, in which case the sorbed contaminant is retarded and the peak concentration is reduced.

Despite its utility for describing sorption equilibria of nonpolar hydrophobic compounds, the simple linear equilibrium approach described above is not appropriate for some real world conditions. For example, K_D may not be a constant. Data for TCE sorption onto glacial till analyzed by Johnson et al. (1989) demonstrate a change in K_D of greater than 50-fold over a range of final aqueous concentrations of 1-4 ppb. In addition, there are other complications for hydrophobic compounds. For example, the sorption potential is expected to decrease, yielding less sorption than predicted, if dissolved organic matter or colloidal particles are present in the aqueous phase. On the other hand, the desorption potential will increase if organic co-solvents such as methanol, methyl *tert*-butyl ether (MTBE), or acetone are also present, which influence a chemical's solubility (Lyman et al., 1992; Schwarzenbach et al., 1993).

Another limitation of the model of hydrophobic sorption relates to desorption. Adsorption in most cases is not permanent, i.e., it is a reversible reaction, and at any given solute concentration, a portion of the dissolved contaminant is partitioning to the solid phase, while some portion of the sorbed contaminant is also desorbing from the solid phase back into solution. Nevertheless, various studies indicate that the desorption isotherm differs from that for adsorption. For example, Hamaker and Thompson, (1972) reviewed several studies of the sorption of pesticides to soils, which demonstrated that desorption differed from adsorption, being slower, with a portion of the sorbed material very difficult to remove. They noted that this portion tends to increase with time (i.e., aging) and certain soil treatments (e.g., drying and rewetting).

Furthermore, in practical systems, the time scales associated with attainment of equilibrium may approach or exceed the time scales associated with solute concentration changes due to macroscopic transport processes such as advection and dispersion. In such cases, the rates at which equilibrium is approached may significantly affect the process and the contaminant distribution among the system phases (Weber et al., 1991). For example, adsorption and desorption may occur over time scales of days to months for many very hydrophobic organics (e.g., Karickhoff and Morris, 1985; Coates and Elzerman, 1986; Wu and Gschwend, 1986). In fact, nonequilibrium sorption conditions have been observed in laboratory column experiments (e.g., Schwarzenbach and Westall, 1981) and under field conditions (e.g., Glotz and Roberts, 1986; Brusseau and Rao, 1989). In such a case, if equilibrium is incorrectly assumed, the amount of sorption will be overestimated.

For these reasons, predicting the effect of sorption processes on contaminant fate and transport may require characterization and quantification of not only the contaminant equilibrium mass distributions, but also the rates at which such distributions are approached (Weber et al., 1991). Under nonequilibrium conditions, the rates of adsorption at the microscopic level may be controlled by the sorption reaction rate, or local mass-transfer (e.g., diffusion) (Weber et al., 1991; Schwarzenbach et al., 1993). Several investigators have applied a simple approach to describe the mass-transfer kinetics between soil aggregates and the mobile water phase, with a linear driving force and a lumped first order mass-transfer coefficient (Lapidus and Amundson, 1952):

$$\frac{\partial S}{\partial t} = K_s \left(K_D C - S \right) \qquad (3.1.19)$$

where: K_s = the lumped sorption mass-transfer-rate coefficient [T^{-1}]. At equilibrium (i.e., $\partial S/\partial t = 0$), Eq. 3.1.19 reduces to Eq. 3.1.7. This approach is justified when mass transport is fast, i.e. when the diffusional lengths are extremely short. Although beyond the scope of this report, more complicated models, which incorporate internal (e.g., Szecsody and Bales, 1989; Wu and Gschwend, 1986) or external (e.g., Weber and Smith, 1987) mass transfer resistances, may be required to describe some systems.

In terms of remediation by natural attenuation, sorption is important because it results in retardation (i.e., slowing) of the migration of chemicals in an aquifer relative to the advective groundwater velocity, v, and reduces the contaminant concentration dissolved in the groundwater, although it usually results in high concentration in the solid phase. Differences in the retardation of chemicals may also lead to chromatographic effects, which can contribute to mixing (Cirpka et al., 1999). For example, for the one-dimensional scenario in Fig. 3-1, if a nonsorbing electron acceptor (e.g., O_2) is injected into the domain, which contains an initial background concentration of a sorbing electron donor, the displaced donor front travels at a retarded velocity, v_c, while the injected acceptor front travels at v. As a result of the different velocities, a large spatial zone develops where electron donor and acceptor are simultaneously present, potentially resulting in enhanced microbial activity (Oya and Valocchi, 1997).

In addition to the retardation effects, sorption can also impact the relative rates of other natural attenuation processes, such as the rates of volatilization, chemical reactions, and biodegradation, by reducing the dissolved concentration (ASTM, 1998; Lyman et al., 1992). The effect of sorption on biodegradation is particularly complex, and not well understood (e.g., Rittmann et al., 1994; Alexander, 1994). Indeed, a variety of negative and positive influences of sorption on biodegradation have been observed, including: decreased biodegradation because of decreased solute concentration (e.g., Ogram et al., 1985), limitation of biodegradation rates by desorption rates (e.g., Rijnaarts et al., 1990), inhibition of biodegradation by irreversible sorption of substrate (e.g., Weber and Coble, 1968), increased biodegradation of toxic substrates because of decreased solute concentration (e.g., Ehrhardt and Rehm, 1985; Morsen and Rehm, 1987; Apajalahti and Salkinoja-Salonen, 1984), and increased biological growth when sorbed substrate is later made available for biodegradation by desorption (e.g., Ehrhardt and Rehm, 1985; Morsen and Rehm, 1987). The type, and degree, of effect sorption has on biodegradation is a function of the characteristics of the contaminant, the solid surface(s) and sorption mechanism(s), and the microorganisms (e.g., their capacity to use the sorbed compound) (Rittmann et al., 1994; Alexander, 1994).

3.1.4.2 Volatilization and Gas Solution/Exsolution

Another interphase partitioning process that may be an important consideration in certain field situations is the volatilization from the subsurface of contaminants dissolved in water or nonaqueous phase liquids (NAPLs) (ASTM, 1998). Volatilization is a nondestructive attenuation mechanism, but it does remove contaminants from the groundwater phase (ASTM, 1998; Wiedemeier et al., 1996). ASTM (1998) notes that volatilization is likely to be an important mechanism for: fresh spills of petroleum products, older but highly volatile petroleum constituents (e.g., jet fuel), free phase product, and high dissolved contaminant concentrations. However, the importance of this mechanism is likely to decrease with increasing time since discharge.

Organic contaminants that are sufficiently volatile can partition from the liquid phase into the soil gas/vapor phase, after which they can be transported through the unsaturated zone and eventually may diffuse into the atmosphere (Palmer and Johnson, 1989). As a result, the volatilization rate and the amount of mass loss depend on several factors, including chemical-specific factors (e.g., contaminant concentration, Henry's law constant, and diffusion coefficient), and media-specific properties (e.g., climate, depth to water, mass-transport coefficient for the contaminant in water and soil gas, sorption, temperature, effective porosity, and soil type) (ASTM, 1998; Wiedemeier et al., 1996). Two steps are involved in estimating contaminant volatilization from soil (Mercer and Cohen, 1990): (1) estimating the partitioning of the contaminant between water and air, and NAPL and air; and (2) estimating the contaminant vapor transport from the soil.

The equilibrium partitioning of a volatile chemical from dilute solutions in water into the soil gas can be described using Henry's law (Kavanaugh and Trussell, 1980).

Henry's constant, K_{H1}, values for the organic contaminants covered in this book are presented in Sections 5.1, 5.2, and 5.3. Using this definition, the larger the Henry's constant, the greater the tendency for the chemical to volatilize. These values range over several orders of magnitude, indicating a wide range in potential for this attenuation mechanism. Importantly, the values are not valid for all conditions, because Henry's constants are also a function of environmental variables, such as temperature and the concentration of other dissolved gases and solids. For example, in the case of hydrocarbons, Henry's constant is expected to: increase with increasing temperature, increase slightly with increasing water salinity, and decrease with increasing concentration of dissolved organic carbon in water. The temperature increase in Henry's constant for most volatile hydrocarbons is about twofold for every 10°C temperature rise (Kavanaugh and Trussell, 1980).

Other factors also can limit the mass-transfer rates from the groundwater into the soil gas. For example, using laboratory experiments and mathematical modeling, McCarthy and Johnson (1993) observed that aqueous phase diffusion controls mass transport from the saturated zone to the unsaturated zone and, thus, mass transport across the capillary fringe will be relatively small. In addition, field experiments and modeling performed by Rivett (1995) indicate that for dissolved-phase plumes emanating from pools or residual saturation of DNAPL (i.e., dense NAPL) located about 1 m or more below the water table, there will be little, if any, contaminant concentration detectable in the soil gas. This is due to the weakness of vertical transverse dispersion in the groundwater and the downward groundwater velocity in the vicinity of the water table during recharge. Based on these observations, little if any mass will be lost via volatilization for the portions of plumes > 1 m below the water table (Wiedemeier et al., 1996).

In addition to volatilization of dissolved contaminants from the groundwater, volatilization can also affect NAPLs in the subsurface (ASTM, 1998), especially when the NAPL has a very high vapor pressure, or where NAPL exists close to the ground surface or in dry pervious sandy soils (Mercer and Cohen, 1990). When sufficient NAPL is spilled or leaked into the unsaturated zone, the NAPL migrates downward through the pore spaces under the influence of gravity. In the process, significant volumes of NAPL become entrapped in the porous medium in the form of ganglia or pools due to capillary forces (e.g., Schwille, 1967; Schwille, 1984; Mercer and Cohen, 1990). NAPLs that are highly volatile, such as gasoline and certain chlorinated hydrocarbons, volatilize very rapidly in the unsaturated zone, resulting in the formation of a zone of vapors in the surrounding pore space, with decreasing concentrations moving outwards from the NAPL (Schwille, 1984).

Laboratory studies of soil vapor extraction suggest that interphase mass transfer between the NAPL contaminant and the vapor phase can be described by a local equilibrium assumption (e.g., Baehr et al., 1989). Raoult's law given below has been used to quantify the ideal reference state for equilibrium between the NAPL and air phase (e.g., Corapcioglu and Baehr, 1987):

$$P_i = X_i P_i^\circ \qquad (3.1.20)$$

where the equilibrium partial pressure, P_i, of a volatile organic i in the air NAPL solution is proportional to the P_i°, and X_i is the mole fraction of component i in the NAPL (Mercer and Cohen, 1990).

However, laboratory and modeling studies have suggested that the NAPL/vapor phase mass exchange is mass-transfer limited under some conditions, e.g., at increasing pore gas phase velocities, such as found in soil vapor extraction (e.g., Kearl et al., 1991; Wilkins et al., 1995; Sleep and Sykes, 1989). One approach that has been applied for modeling rate-limited NAPL-vapor phase mass exchange is to represent the process by a first-order expression in which the mass-transfer driving force is proportional to the difference between equilibrium and actual concentrations (e.g., Sleep and Sykes, 1989; Wilkins et al., 1995):

$$\frac{\partial C_g}{\partial t} = K_v \left(C_{ge} - C_g \right) \qquad (3.1.21)$$

where: C_{ge} = the vapor phase concentration in thermodynamic equilibrium with the NAPL [ML^{-3}], and K_v = the lumped overall mass-transfer coefficient for volatilization of the NAPL [T^{-1}]. A similar expression could be used to represent mass-transfer limited exchange between the dissolved and gaseous phases.

Once the dissolved or NAPL contaminant has volatilized into the vadose zone, the key physical processes affecting the transport of gases are diffusion and advection, of which diffusion plays the largest role due to the large diffusion coefficients for gases ($\approx 10^{-1}$ cm^2/s) compared to solutes ($\approx 10^{-5}$ cm^2/s) (Palmer and Johnson, 1989). Several other processes will also affect the transport of volatile organics through the soil-gas phase, including partitioning of the gas phase into soil water, sorption, and biodegradation (Ong and Lion, 1991a and 1991b). Furthermore, other factors may limit diffusion of the gas phase into the atmosphere. For example, many volatile organic chemicals have equilibrium concentrations that are sufficiently high to increase the density of the vapor phase to 1.5 g/cm^3, which, in theory, should cause these vapors to sink to the capillary fringe. In addition, constructed features such as parking lots, streets, and foundations, can also limit the exchange of gases with the atmosphere.

Given the constraints described above, coupled with the low Henry's law constants for the BTEX compounds and chlorinated solvents (Sections 5.1, 5.2, and 5.3) and the small surface area of groundwater exposed to the soil gas, the volatilization of these contaminants from groundwater into the soil gas is a relatively slow process that can generally be neglected in conservative modeling estimates of biodegradation (Wiedemeier et al., 1996). For example, in a study of natural attenuation at a petroleum-product impacted sandy aquifer, Chiang et al. (1989) estimated that volatilization accounted for only 5% of the total mass loss of dissolved benzene in the groundwater. Nevertheless, at sites with a shallow or highly fluctuating water table, volatilization may make a more significant contribution to the total mass loss from the groundwater (ASTM, 1998).

Other examples of gas-transfer processes that may influence remediation by natural attenuation are gas dissolution and release from solution. These processes can transfer significant quantities of mass between soil gases and groundwater and may play a major role in controlling the chemistry of groundwater (Domenico and Schwartz, 1998). Typically, these processes are modeled using the equilibrium relationship of Henry's Law. In the strict sense, Henry's law cannot be applied to gases that react in solution, such as CO_2 or NH_3, but in the case of CO_2, so little $CO_{2(aq)}$ reacts that Henry's law can be used to approximate the gas distribution between the two phases (Domenico and Schwartz, 1998). An example of gas solution that may influence remediation by natural attenuation is transfer of oxygen across the capillary fringe/water table interface into the groundwater. This may be an especially critical process for LNAPLs (light NAPLs), which are commonly found at or near the water table. Theoretical and field studies (Borden and Bedient, 1986; Borden et al., 1986; Molz and Widdowson, 1988) indicate that the vertical transverse dispersion coefficient can have a significant impact on oxygen exchange and the contaminant aerobic decay rate. However, the weak vertical dispersion/diffusion process combined with the microbial oxygen demand results in steep vertical oxygen gradients and often

Table 3-5. Some organic groups or functional groups resistant or potentially susceptible to hydrolysis

Resistant to hydrolysis	Susceptible to hydrolysis
Alkane	Alkyl halide
Alkene	Amide
Alkyne	Amine
Benzene/biphenyl	Carbamate
Polycyclic aromatic hydrocarbon	Carboxylic acid esters
Heterocyclic polycyclic aromatic hydrocarbons	Epoxides
	Nitriles
Halogenated aromatics/PCBs	Phosphonic acid esters
Aromatic nitro compounds	Sulfonic acid esters
Aromatic amines	
Phenols	

Source: Lyman et al., 1990.

limits this impact to relatively shallow plumes (< 2–3 meters below the water table) (MacQuarrie and Sudicky, 1990; Smith et al., 1991a; Smith et al., 1991b).

3.2 Chemical Processes

3.2.1 Hydrolysis

Hydrolysis of organic compounds may be important transformation reactions in aqueous environments and may affect the environmental fate of the organic compounds. Hydrolysis is a chemical reaction in which a hydroxyl group (–OH) is added to an organic compound:

$$R-X + H_2O \rightarrow ROH + X^- + H^+$$

In a typical hydrolysis reaction, a nucleophile (water or hydroxide ion) attacks an electrophile (example, carbon atom) and displaces a leaving group (example, chloride). Many compounds are relatively inert to hydrolysis but some compounds may hydrolyze under certain environmental conditions. Table 3-5 lists some of the compounds that are resistant to hydrolysis and compounds that are susceptible to hydrolysis. The hydrolysis of the compounds is commonly modeled assuming first-order kinetics:

Table 3-6. Hydrolysis half-life for various organic compounds in water at pH 7 and 25°C

Group of compounds	Compounds	Average half lives
Organohalides	$CH_3CH_2CH_2Br$	26 days
	CH_3Cl	0.9 years
	$CHCl_3$	3,500 years
Esters	$HCOOCH_3$	2 days
	$CH_3COOCH_2CH_3$	2 years
Amides	CH_3CONH_2	3,950 years
Carbamates	$C_6H_5OCONHC_6H_5$	170 days
	$CH_3CH_2OCONHC_6H_5$	6,700 years
Organophorus compounds	$CH_3P(O)(OCH_3)_2$	88 years
	$C_6H_5P(O)(OCH_2CH_3)_2$	440 years

Source: Lyman et al., 1990.

$$\frac{d[RX]}{dt} = -k_T[RX]$$

where k_T is the hydrolysis rate constant (1/T).

However, assumption of first-order kinetics may be an oversimplification if several hydrolysis reactions contribute towards the transformation of the compound and the reaction is pH dependent. These reactions may include acid-catalyzed or base-catalyzed hydrolysis as well as nucleophile attack by water. Hydrolysis half-lives may range from as low as several hours to as long as several hundred years (Table 3-6). In recent years, various methods have been developed to estimate the hydrolysis rate constants of various organic compounds. Some of the prediction methods used include the Hammett Correlation and the Taft Correlation (Lyman et al., 1990). However, these correlations are useful for a certain class or group of compounds.

Halogenated aliphatic compounds are susceptible to hydrolysis in the absence of inorganic and biochemical catalysts (Table 3-6). Hydrolysis of these compounds is potentially faster at higher pH because hydroxide ion acts as a nucleophile. However, hydrolysis of more halogenated compounds is slower. For example, hydrolysis of trichloromethane is much slower than that of monochlorinated compounds and hydrolysis may be less important than dechlorination of the compound under reducing conditions or by anaerobic degradation. If the half-lives of the hydrolysis of the compounds are in the order of several tens of years, natural attenuation by hydrolysis may be negligible.

3.2.2 Dehydrohalogenation

In dehydrogenation, a hydrogen and a halide atom are removed from the compound:

$$-\underset{\underset{H}{|}}{\overset{|}{C}}-\underset{\underset{X}{|}}{\overset{|}{C}}- \rightarrow \overset{|}{\underset{|}{C}}=\overset{|}{\underset{|}{C}} + HX$$

These reactions are common amongst halogenated aliphatic compounds, especially at high pH (Mabey et al., 1980). In general, dehydrohalogenation tends to follow a bimolecular kinetic rate with the pH of the solution playing an important role in the reaction rates. Compounds with more halogens tend to exhibit increased dehydrohalogenation, in contrast to the trend for hydrolysis. A few compounds may undergo simultaneous hydrolysis and dehydrohalogenation. For example, 1,1,1-trichloroethane undergoes hydrolysis to acetate and dehydrohalogenation to 1,1-dichloroethylene (Vogel and McCarthy, 1987).

3.2.3 Precipitation and Dissolution

Precipitation and dissolution of solids are two important processes that affect the transport of contaminants in groundwater. This is especially true for inorganic species. Factors that influenced precipitation and dissolution are the pH and redox potential of the solution, presence of other chemical species and complexing agents, and the presence of other solids. Both equilibrium considerations and rates of reactions are important and must be considered. The extent to which a chemical species is favored in the dissolution process can be assessed based on the solubility of the solid phase. The solubility product describes the extent of dissolution of the solids and in turn the concentrations and fate of the chemical species in solution. Table 3-7 provides a listing of the solubility products of several compounds of concern at contaminated sites. As indicated earlier, the solubilities of the solids are affected by the presence of complexing agents such as chelating agents, humic acid and inorganic anions (ammonia and cyanide). The solubility products of metal sulfides are very much lower than the solubility products of carbonates or hydroxides. Under sulfate-reducing conditions in the aquifer, metal sulfides may precipitate and the mobility and transport of these metals will be greatly reduced.

3.2.4 Models for Geochemical Reactions

Geochemical models may be used to predict chemical species concentrations in groundwater and in the presence of various solids. In addition, mineral/solution interfacial reactions such as surface complexation, cation exchange, co-precipitation and redox reactions, kinetic reactions and, sometimes, colloidal reactions can be modeled, providing information on the mobility of the various chemical species in the environment. Most of the currently known geochemical models assume local thermodynamic equilibrium (Westall et al., 1986; Bethke, 1994; van der Lee, 1998)

Table 3-7 Solubility products of various metal solids of environmental interest

Metal	Reaction	log K_{so}
Pb	$Pb(OH)_2\ (s) = Pb^{2+} + 2OH^-$	−14.3
	$PbCO_3\ (s) = Pb^{2+} + CO_3^{2-}$	−12.8
	$Pb_3(PO_4)_2\ (s) = 3\ Pb^{2+} + 2PO_4^{3-}$	−32.0
	$PbS\ (s) = Pb^{2+} + S^{2-}$	−28.2
Cu	$Cu(OH)_2\ (s) = Cu^{2+} + 2OH^-$	−18.8
	$CuCO_3\ (s) = Cu^{2+} + CO_3^{2-}$	−9.6
	$CuS\ (s) = Cu^{2+} + S^{2-}$	−48.9
Cd	$Cd(OH)_2\ (s) = Cd^{2+} + 2OH-$	−13.6
	$CdCO_3\ (s) = Cd^{2+} + CO_3^{2-}$	−11.3
	$CdS\ (s) + Cd^{2+} + S^{2-}$	−28.0
Cr	$Cr(OH)_3\ (s) = Cr^{3+} + 3OH^-$	−30.0
Zn	$Zn(OH)_2\ (s) = Zn^{2+} + 2OH^-$	−16.3
	$ZnCO_3\ (s) = Zn^{2+} + CO_3^{2-}$	−10.5
	$ZnS\ (s) = Zn^{2+} + S^{2-}$	−22.7

Source: Benefield et al., 1982.

because the aqueous reactions in most natural waters are fast. However, precipitation and dissolution including interfacial reactions may be kinetically controlled. Some recent models have been developed to include reactions with first- or second-order kinetics (Soetaert et al., 1996). A mix of thermodynamic equilibrium reactions and kinetically controlled reactions are also available for aquatic system. Geochemical models have traditionally been developed independently of the hydrodynamic processes. The lack of "marriage" between geochemical and hydrodynamic processes is generally due to their distinct scientific disciplines. However, over the last decade, geochemical and hydrodynamic models have been combined to reflect the impact and effect of geochemical processes on hydrodynamic processes (Theis and Iyer, 1994, Mercurio et al., 1999). For example, porosity and permeability of a porous media may change when minerals precipitate or dissolve—clogging the interparticle pores. However, there are many difficulties synthesizing the mass-transport, thermodynamic, and kinetic models into hydrodynamic models. Krainov (1999) found that for models to be optimal in their predictions they should incorporate mass-transport, thermodynamic, and kinetic processes as separate submodels. An abbreviated chronology of the development of geochemical models over the last three decades is

Table 3-8. Selected geochemical and transport models

Geochemistry model	Transport model	Dimensions[1]	References
CHESS	SUBIEF	1D	Ollar et al., 1997 van der Lee, 1998 Moulin, 1996
PHREEQC	—	1D	Parkhurst et al., 1980
NEWKIN	SARIP	1D	Gallo et al., 1998 Bildstein, 1998 Thiez and Lemonnier, 1990
CHESS	RT1D, METIS	1,2D	van der Lee, 1998 van der Lee, 1997 Goblet, 1989
MINTEQA2	PLUME2D	1,2D	Walter et al., 1994 Allison et al., 1990
REACT	TOUGH2	1,2,3D	Stoessell, 1988 Pruess, 1991
PHREEQC	MODFLOW	1,2D	Mercurio et al., 1999

Source: adapted from van der Lee and De Windt, 2001.
[1]Number of dimensions in the transport model.

presented in Basset and Melchior (1990). A list of geochemical models and their associated transport models is presented in Table 3-8.

Some of the more common geochemical models are MINEQL+, MINTEQA2 and PHREEQC. MINEQL+ is a chemical equilibrium model capable of calculating aqueous speciation, solid phase saturation states, precipitation-dissolution, and adsorption. An extensive thermodynamic database is included in the model (Westall, 1986). MINTEQA2, a geochemical equilibrium speciation model for dilute aqueous systems, is an update of MINTEQ that was developed by combining the fundamental mathematical structure of MINEQL with the thermodynamic database of WATEQ3 (Allison et al., 1991). MINTEQA2 is accompanied by an interactive program PRODEFA2, that is used to create MINTEQA2 input files. With PRODEFA2, the user can use data available in MINTEQA2 thermodynamic database and define other aqueous, solid, and/or adsorption species and constants not present in the database. MINTEQA2 can be used to calculate the mass distribution between dissolved, adsorbed, and multiple solid phases under a variety of conditions. Another geochemical model, AQUACHEM, is a fully-integrated software package developed specifically for graphical and numerical analysis of geochemical data sets. AQUACHEM features a graphical interface to the popular geochemical modeling program PHREEQC for calculating equilibrium concentrations (or activities) of chemical species in solution and saturation indices of solid phases in equilibrium with a solution (Appelo and Parkhurst, 1998). These models are useful in the analysis of geochemical reactions such as precipitation/dissolution, ion exchange, adsorption, surface metal and ligand complexation. They also estimate chemical species concentrations that may have an impact on the attenuation and fate of inorganic species in the subsurface.

3.3 Biological Processes

3.3.1 Biodegradation

Destructive attenuation mechanisms are capable of reducing the mass of the chemicals of concern in a contaminated medium, such as soil, sediments, or groundwater. Of the available destructive mechanisms, biodegradation is the most important for most organic contaminants. Biodegradation is a term that is used very loosely among environmental engineers and scientists, usually implying conversion of a harmful organic compound to harmless products, especially inorganic products, such as carbon dioxide and water. In its strictest sense, however, biodegradation describes the biological conversion of organic compounds to products that are lower in free energy, and it implies nothing regarding the extent of the transformation or its desirability (Alexander, 1973; Grady, 1985; Egli, 1992). In fact, the biological conversion of organic compounds to products that are more harmful than the parent substrate is well known (e.g., transformation of tetrachloroethene to vinyl chloride). A more precise terminology should distinguish biodegradation reactions based on the extent to which the target molecule is chemically altered by microbial metabolism. At the extremes of this scale are the processes of biotransformation and mineralization.

Biotransformation is the conversion of organic molecules by a single biochemical reaction or by multiple reactions that do not result in significant changes in the structure of the parent compound. Examples of biotransformation reactions include many reductive dehalogenation reactions (e.g., conversion of tetrachloroethene to trichloroethene), oxygenase-catalyzed hydroxylations of aromatic or aliphatic hydrocarbons (e.g., formation of cresols from toluene), and simple hydrolyses (e.g., cleavage of nitrate esters of nitroglycerin). The extent to which biotransformation reactions reduce the hazard associated with contaminated media depends on the relative toxicity of the substrates and products. Some biotransformation reactions reduce risk, some increase it, and others are risk neutral. In addition, the products of biotransformation reactions may have significantly altered solubility and mobility which should be considered in fate and transport models supporting natural attenuation. Since the products of biotransformation reactions sometimes are not detected by the same analytical procedures that are used for measuring the concentrations of the parent compounds, disappearance of contaminants of concern should not be equated with reduction in hazard unless the products are known and appropriate analytical procedures are used to detect their formation.

Mineralization is the conversion of organic substrates to inorganic products, such as carbon dioxide, water, and chloride. Since these products are generally harmless at the concentrations that are likely to be formed by microbial metabolism, mineralization will usually eliminate the hazard associated with the chemicals of concern.

Metabolic pathways that produce extensive structural changes in the chemicals of concern, but which do not lead to complete conversion to inorganic products, are considered partial biodegradation processes. Like biotransformation, partial biodegradation may or may not reduce the hazard associated with the contaminated media.

The definitions of biotransformation and mineralization that are given above are based on the fate of the organic substrate, specifically on the extent of structural alteration that occurs during biodegradation. It is also useful to classify biodegradation reactions based on their ability to support microbial growth, however, because microorganisms are the catalysts for these reactions. As a result, the biodegradation reactions for substrates that support growth can be autocatalytic, because the quantity of catalyst (i.e., microbial biomass) that is available increases as the reaction proceeds.

Based on their ability to support microbial growth, substrates can be classified as either *primary* or *secondary*. Primary substrates support microbial growth, whereas secondary substrates do not. Primary substrates may support growth either by serving as sources of energy or as precursors of biomass synthesis. Substrates that are used in energy-generating pathways are either electron donors or electron acceptors, depending on whether they are oxidized or reduced during metabolism. Biomass synthesis requires carbon sources; macronutrients such as nitrogen, phosphorus, and sulfur; and micronutrients such as vitamins and biologically important transition metals (e.g., iron, copper, and cobalt). Heterotrophic bacteria usually use the same compound as the primary electron-donor substrate for energy generation and the source of carbon for biosynthesis.

Secondary substrates may be incapable of supporting microbial growth for one of two reasons: (1) microorganisms may be incapable of converting the compound to intermediates that can enter energy-generating or biosynthetic pathways, or (2) the compound may be present at a concentration that is too low to supply enough energy to meet the microorganism's maintenance energy requirements and drive biosynthesis. Secondary utilization for the latter reason differs from primary utilization only with respect to the quantity of energy or biomass-precursor compounds that can be harvested by the biodegrading organism per unit time. This is purely a function of the bioavailable substrate concentration. Secondary utilization for the former reason is fundamentally different, however. Substrates that are biodegradable but do not support microbial growth are said to be degraded by *cometabolism* (Horvath, 1972; Fewson, 1988; Criddle, 1993). This process usually results from fortuitous activity of an enzyme that is involved in the normal catabolic pathways of the cometabolizing organism. The target contaminant binds to the active site and is transformed by a reaction that is analogous to that normally catalyzed by the enzyme. Although cometabolism has been defined to be the biological transformation of a non-utilizable substrate in the obligatory presence of a growth-supporting substrate (Dalton and Stirling, 1982), it is more commonly used to describe the metabolism of substrates that do not support growth by (a) organisms that are actively growing on a growth-supporting substrate, (b) resting (i.e., nongrowing) cells of microorganisms that were grown on a growth-supporting substrate which has since been completely removed (Alvarez-Cohen and McCarty, 1991a,b,c; Hamamura et al., 1997), or (c) cells that were grown on a growth-supporting substrate and which are carrying out the transformation in the presence of an exogenous electron-donor substrate whether or not that substrate is capable of supporting growth (Horvath, 1972; Fewson, 1988; Alvarez-Cohen and McCarty, 1991a,b,c; Henry and Grbic-Galic, 1991; Criddle, 1993). The broader sense in which cometabolism is commonly used includes what is sometimes called gratuitous or fortuitous metabolism, but the distinction is unimportant in a mechanistic sense and for most practical purposes.

At least two conditions must be met for biodegradation by cometabolism to occur. First, an enzyme that is capable of binding and transforming the target contaminant must exist in the microbial population. Second, that enzyme must be present at a sufficiently high concentration for the reaction to proceed at an observable rate. In addition, appropriate cosubstrates (e.g., molecular oxygen or electron-donor substrates) are often required. Depending on the specific reaction mechanism, these cosubstrates may participate directly in the reaction, be required to activate the enzyme, or to supply reducing power to drive the reaction. The requirement for the existence of an enzyme with appropriate substrate-binding ability and reactivity usually implies that the target substrate must be a structural analog of the normal substrate. For example, cometabolic transformation of halogenated aromatic compounds, such as chlorophenols and chlorobenzoates, is often catalyzed by microorgansisms capable of growth on phenol or benzoate (Horvath, 1972). In these cases, the enzyme responsible for catalyzing the initial hydroxylation reactions are able to accommodate the bulky, electron-withdrawing halogen substituent, but subsequent enzymes in the metabolic pathway are not. Similarly, many PAHs, including some important high molecular weight compounds (e.g., benzo[a]pyrene), can be cometabolized by bacteria growing on other substrates (Barnsley, 1975; Dean-Raymond and Bartha, 1975; Boldrin et al., 1993; Selifonov et al., 1998). One or more of the enzymes catalyzing early steps in the metabolic pathway for another PAH have relaxed substrate specificity, allowing them to accommodate PAHs with a different number of aromatic rings or differences in the relative orientation of the rings, but all of the enzymes in the pathway do not have sufficiently broad specificity. So the alternative substrate cannot be degraded to central metabolic intermediates, and partially oxidized products accumulate.

Some enzymes, such as the soluble methane monooxygenase, have extremely broad substrate specificities, and therefore, they are able to catalyze reactions with substrates that bear little superficial resemblance to their normal substrate. For example, the soluble methane monooxygenase is able to oxidize halogenated methanes, halogenated ethenes, halogenated alkanes as large as chlorobutane, and normal alkanes up to octane (Colby et al., 1977; Patel et al., 1982; Fox et al., 1990). Similarly varied activity is observed with some oxygenase enzymes involved in metabolism of monoaromatic substrates, such as toluenc and phenol. So, cometabolism is not necessarily restricted to substrates that bear a close resemblance to natural substrates. Oxygenases, in particular, appear to be particularly common catalysts of cometabolic reactions (Reineke, 1984; Ensley, 1991; Janssen and Witholt, 1992), but broad specificity reductases are also known. For example, a variety of common metal-containing tetrapyrrole cofactors, such as heme, cobalamin, and F_{430}, and enzymes that contain these cofactors are known to catalyze reductive dehalogenation reactions (Wade and Castro, 1973; Wood et al., 1968; Castro et al., 1985; Krone et al., 1989a,b; Wackett and Schanke, 1992; Janssen and Witholt, 1992; Picardal et al., 1993).

Many cometabolic biodegradation processes are also biotransformations, but these two descriptions differ in that biotransformation focuses on the fate of the chemical and cometabolism focuses on the fate of the catalyst (i.e., the microorganism). Whereas biotransformation reactions, by definition, result in only minor changes in the structure of the organic substrate, cometabolic reactions can produce dramatic alterations in molecular structure, sometimes leading to mineralization (Little et al., 1988; Malachowsk et al., 1994; Kim et al., 1997). Alternatively, some biotransformation reactions—for example, the reductive dechlorination of tetrachloroethene to *cis*-1,2-dichloroethene—support microbial growth (Holliger et al., 1993; Scholz-Murumatsu et al., 1995; Gerritse et al., 1996; Wild et al., 1996). So, a similar overlapping, but noncoincident relationship exists between primary metabolism and mineralization.

3.3.2 Categorization of Potential for Natural Attenuation

By combining two criteria—the extent of conversion and the ability of the substrate to support growth—we can establish a classification system for biodegradation reactions that can be used to determine the likelihood that specific contaminants will be amenable to treatment by natural attenuation. Figure 3-7 shows this natural attenuation matrix along with examples of biodegradation reactions that fit into each category. Each of the four categories has differing potential for natural attenuation. Contaminants with the least potential for natural attenuation fall into the "biotransformation and cometabolism" category, and contaminants with the greatest potential fall into the "mineralization and primary substrate category".

The natural attenuation matrix can be interpreted by considering two criteria: the toxicity of the products and the microbial ecology of biodegradation. First, successful natural attenuation requires the contaminants of concern to be converted to harmless products. Inorganic products are the most obviously acceptable end products of biodegradation, but certain organic products may also be acceptable. Second, the rate of biodegradation must be fast enough to convert the contaminants of concern to acceptable products within a reasonable period of time or within an acceptable distance from the source in a groundwater plume. Biodegradation rate is commonly directly proportional to the size of the competent microbial population, because the microorganisms are the catalysts for these reactions. Therefore, the ability of a contaminant to support microbial growth is important, because the size of the microbial population will increase in proportion to the amount of contaminant that has been consumed if the contaminant is a primary substrate, but the size of the population will be determined by other factors if the contaminant is metabolized as a secondary substrate. So, assuming that a finite population of *competent* microorgan-

	Mineralization	Biotransformation
Primary Substrate	contaminants are mineralized and support microbial growth: aerobic biodegradation of petroleum hydrocarbons (e.g., BTEX) biodegradation of some chlorinated and oxygenated solvents (e.g., methylene chloride, acetone)	contaminants are biotransformed but support microbial growth: reductive dechlorination of PCE to *cis*-1,2-DCE acetogenesis ethanol and lactate fermentations
Secondary Substrate (cometabolism)	contaminants are mineralized but do not support growth: oxygenase-catalyzed oxidation of TCE mineralization of cyclohexane by *Mycobacterium convolutum* R-22 during growth on n-alkanes	contaminants are biotransformed and do not support microbial growth: reductive dechlorination of most chlorinated solvents and PCBs reduction of nitroaromatics oxidation of many high molecular weight PAHs

Figure 3-7. Natural attenuation matrix: a system for classifying biodegradation reactions according to their potential to contribute to successful natural attenuation

Note: The examples provided for each category are intended to be illustrative, not exhaustive.

isms is initially present at a contaminated site, primary substrates exert a selective pressure that favors the development of a large competent microbial population, whereas secondary substrates do not.

The left-hand column of the natural attenuation matrix shown in Figure 3-7 contains processes that will produce predominantly harmless products, whereas the right-hand column contains processes for which the harmfulness of the product must be evaluated. Similarly, the first row contains processes for which ecological principals favor the establishment of a large population of bacteria with the ability to degrade the contaminants of concern, whereas the second row contains processes for which selective pressures may operate for or against development of a competent population, but favorable selective pressures are likely to be independent of the presence of the contaminant. Therefore, those contaminants that can be classified in the upper left corner of the natural attenuation matrix will have a high potential for remediation by natural attenuation. Aerobic biodegradation of many petroleum hydrocarbons (e.g., normal alkanes, BETX) and oxygenated solvents (e.g., acetone, 1- and 2-propanol) falls into this category. Conversely, because of the unfavorable microbial ecology and the uncertain acceptability of the metabolic products, contaminants that are classified in the lower right hand corner of the natural attenuation matrix will have a low potential for remediation by natural attenuation. This category includes many important pollutants, such as high molecular weight PAHs and PCBs (Bedard et al., 1986; Boyle et al., 1992; Boldrin et al., 1993; Commandeur et al., 1995; Seto et al., 1995; Schneider et al., 1996; Selifonov et al., 1998). Although PAHs and PCBs are relatively immobile in soils, the "biotransformation/secondary substrate" category may also contain mobile compounds like MTBE and other ethers that are used as solvents or gasoline oxygenates (Mo et al., 1997; Hardison et al., 1997; Steffan et al., 1997; Hur et al., 1997).

The natural attenuation potential of contaminants that are categorized in either of the remaining groups is highly site-specific and will depend on the composition of the microbial population, the geochemistry of the contaminated site, and the presence of cosubstrates (which may be cocontaminants). For example, TCE and less-chlorinated ethenes can be mineralized by a wide variety of aerobic bacteria during—or subsequent to—growth on electron-donor substrates such as methane, propane, butane, toluene, phenol, and ammonia (Folsom et al., 1990; Ensley, 1991; Janssen and Witholt, 1992; Dabrock et al., 1992; Malachowsky et al., 1994; Fries et al., 1997b; Hamamura et al., 1997). This process is catalyzed by broad-specificity monooxygenases and has been demonstrated in the field as a pilot-scale example of engineered bioremediation using methane and phenol as the cosubstrates (Semprini et al., 1990; Semprini et al., 1991; Hopkins et al., 1993a,b; Hopkins and McCarty, 1995; Sutfin and Ramey, 1997; McCarty et al., 1998).

Because some of these substrates are common contaminants (e.g., toluene) or common products of contaminant biodegradation (e.g., methane), it is easy to imagine conditions under which natural attenuation of TCE by this cometabolic mechanism could occur. Unfortunately, such conditions are rare because of the requirement for the simultaneous presence of oxygen (a reactant in the cometabolic oxidation of chlorinated ethenes by monooxygenases) and a highly reduced electron-donor substrate. The generally low reaeration rate of subsurface and sedimentary environments means that oxygen and electron donors will probably not be present together except in a small band at the periphery of the contaminant plume. Another limitation to the potential for natural attenuation by this mechanism is the formation of highly reactive intermediates. These compounds can react with and inactivate the monooxygenase that catalyzes the reaction, as well as with other important macromolecules. The net result is inactivation of the oxygenase and reduction in the growth yield of the degrading organisms (Alvarez-Cohen and McCarty, 1991a,b; Henry and Grbic-Galic, 1991; Dolan and McCarty, 1995; Chang and Alvarez-Cohen, 1996; Anderson and McCarty, 1997; Hamamura et al., 1997; van Hylckama Vlieg et al., 1997; Chu and Alvarez-Cohen, 1999), both of which provide a negative selective pressure for TCE-degrading organisms. Thus, the toxicity associated with cometabolic oxidation of TCE by organisms utilizing broad-specificity monooxygenases provides a selective pressure that favors growth of bacteria possessing more selective enzymes; enzymes that will oxidize the energy substrate, but not the contaminant of concern (Fries et al., 1997a; Mars et al., 1998). Therefore, both solute transport and ecological considerations work against successful natural attenuation of chlorinated ethenes by this oxidative mechanism.

Reductive dechlorination of PCE and TCE is an example of the final category in the natural attenuation matrix: substrates that support growth but which are not mineralized. Because these compounds can serve as primary electron-acceptor substrates, selective pressure favors the development of a competent microbial population. Because the growth-supporting reactions do not result in complete dechlorination, however, the hazard associated with the contamination will not necessarily be reduced by the population that develops. In fact, vinyl chloride is a much more hazardous chemical than are either PCE or TCE, because it is a known human carcinogen. Fortunately, cis-1,2-DCE and vinyl chloride can be mineralized under aerobic conditions and, to a lesser extent, under Fe(III)- and sulfate-reducing conditions (Bradley and Chapelle, 1998a,b). Therefore, a spatially separated sequential biodegradation process can be envisioned in which one group of organisms reductively dechlorinates PCE and TCE and a second group oxidizes cis-1,2-DCE and vinyl chloride to carbon dioxide. Thus far, however, very few organisms have been described that can use vinyl chloride as a growth substrate (Hartmans and Debont, 1992; Hauschild et al., 1994; Verce et al., 2000). Therefore, the second step would likely fall into the "cometabolic mineralization" category and would be subject to the same caveats as were described above for cometabolic biodegradation of TCE. Nevertheless, given the right set of hydrological, geochemical, and microbiological conditions, natural attenuation of PCE and TCE could occur. Because of the additional complexity of this scenario over one in which the contaminant of concern can support growth and be mineralized by the same organism or consortium of organisms, however, it is less likely to occur.

In summary, natural attenuation of compounds that are mineralized in the process of supporting microbial growth is favored, because the selective pressure will promote development of a competent microbial community and the products of biodegradation are environmentally acceptable. For compounds in this category, site monitoring data demonstrating contaminant disappearance may be sufficient evidence that natural attenuation is occurring. For contaminants in the other three categories, however, more site-specific evidence should be obtained to support the hypothesis that natural attenuation will be an acceptable response alternative. This evidence should be directed either at demonstrating that a sustainable destructive attenuation mechanism is operating at the site and/or that the destructive process reduces the risk associated with exposure to the contaminated media (i.e., that the biodegradation products are environmentally acceptable).

3.3.3 Alternative Biodegradation Pathways

When Figure 3-7 is used to categorize the natural attenuation potential for contaminants at a specific site, it is important to remember that classification is based on the characteristics of the biodegradation process that is thought to be occurring, not on the identity of the chemicals of concern. As a result, the same contaminant can be categorized differently at different sites depending on the geochemical conditions that prevail. An important implication of this is that a biodegradation pathway that may be common, complete, and growth supporting under one set of conditions (e.g., aerobic) may be none of the above under another (e.g., when oxygen is absent). Even when alternative biodegradation pathways exist, the rate and extent of transformation may be significantly different among alternatives. Biodegradation of petroleum hydrocarbons provides a good example of this phenomenon.

Whereas many petroleum hydrocarbons are mineralized and support microbial growth under aerobic conditions (Smith, 1990; Watkinson and Morgan, 1990; Cerniglia, 1992), and the organisms that catalyze aerobic biodegradation are ubiquitous (Mulkins-Phillips and Stewart, 1974; Roubal and Atlas, 1978; Focht et al., 1989; Berardesco et al., 1998; Langworthy et al., 1998), biodegradation of these compounds under anaerobic conditions is more limited and much slower. Anaerobic biodegradation of hydrocarbons requires use of an alternative pathway because, under aerobic conditions, oxygen functions both as a high-energy electron acceptor and as a reactant that converts the otherwise nonreactive hydrocarbons into a form that is more readily oxidized by microbial enzymes (e.g., in the monooxygenase-catalyzed hydroxylation of benzene to phenol). Since this type of initial priming reaction is not available when oxygen is absent, completely different biochemical pathways must be used for hydrocarbon degradation under anaerobic conditions.

The hydrocarbon that is mineralized most commonly, and which supports growth most reliably, under anaerobic conditions is toluene (Wilson et al., 1986; Evans et al., 1991; Hutchins et al., 1991; Bellar et al., 1992; Edwards et al., 1992; Edwards and Grbic-Galic, 1994; Edwards et al., 1994; Lovley et al., 1994; Lovley et al., 1995; Lagenhoff et al., 1996; Haner et al., 1997; Anderson et al., 1998). Other aromatic hydrocarbons, such as benzene, ethylbenzene, and the xylene isomers, are degraded in the absence of oxygen in some studies but not in others (Wilson et al., 1986; Evans et al., 1991; Hutchins et al., 1991; Edwards et al., 1992; Bellar et al., 1992; Edwards and Grbic-Galic, 1994; Rabus and Widdel, 1995; Haner et al., 1997). Benzene, in particular, has often been observed to persist in the absence of oxygen even when other aromatic hydrocarbons are degraded rapidly. Anaerobic benzene oxidation has been reported much more frequently in recent years (Lovley et al., 1994; Lovley et al., 1995; Lovley et al., 1996; Kazumi et al., 1997; Anderson et al., 1998; Weiner and Lovley, 1998a,b; Caldwell and Suflita, 2000), but long lag periods frequently precede benzene consumption in these studies or the reaction rate was stimulated by addition of exogenous chemicals (e.g., synthetic chelators to increase the bioavailability of Fe(III) as a primary electron acceptor). So, it is not clear how common benzene mineralization is in anaerobic petroleum-contaminated environments. Anaerobic biodegradation of low molecular weight PAHs, such as naphthalene and phenanthrene, which was also once thought to be impossible, has been reported with increasing frequency in recent years (e.g., Mihelcic and Luthy, 1988 a,b; Al-Bashir et al., 1990; Thierrin et al., 1995; Coates et al., 1996; Lagenhoff et al., 1996; Meckenstock et al., 2000), but anaerobic biodegradation of higher molecular weight PAHs, especially four- and five-ring compounds, has not yet been reported. Finally, although anaerobic mineralization of aliphatic hydrocarbons has been recently discovered, and in some cases these substrates have been shown to support microbial growth (Aeckersberg et al., 1991; Rueter et al., 1994; Bregnard et al., 1997; Coates et al., 1997; Hunkeler et al., 1998; Rabus et al., 1999; So and Young, 1999), in general, alkanes are recalcitrant in the absence of oxygen. Branched and cyclic alkanes are particularly resistant to anaerobic biodegradation (Rueter et al., 1994; Hunkeler et al., 1998; Rabus et al., 1999), and aliphatic hydrocarbon biodegradation has never been observed in the absence of an exogenous electron acceptor, such as nitrate or sulfate.

In summary, although aerobic biodegradation of many petroleum hydrocarbons is a process that clearly falls into the "mineralization/primary substrate" category in Figure 3-7, anaerobic hydrocarbon biodegradation is more variable. Anaerobic biodegradation is probably a viable natural attenuation mechanism for simple aromatic hydrocarbons, especially toluene, ethylbenzene, xylenes, and naphthalene, but it appears to be highly site-specific for benzene. High molecular weight PAHs and aliphatic hydrocarbons should probably be considered to be recalcitrant in the absence of oxygen.

3.3.4 Requirement for Cosubstrates

Many biodegradation processes require substrates in addition to the target contaminant. These cosubstrates may be required to complete a redox reaction (e.g., accept electrons from substrates that are degraded by oxidation or provide electrons to substrates that are degraded by reduction), induce or activate the enzyme(s) required to catalyze the reaction of interest, or as a reactant in bimolecular reactions (e.g., oxygen is a coreactant in oxygenase-catalyzed reactions). On a quantitative basis, the requirement for cosubstrates that serve as a source or sink of electrons is the most important, because the amount required is based on the stoichiometry of the net reaction to terminal products. The requirement for coreactants, such as oxygen, is also stoichiometric, but the amount required is usually based on the involvement of the cosubstrate in one or two reactions in a larger metabolic pathway. In the case of oxygen, however, the requirement for oxygen as an electron acceptor can limit the amount that is available for direct reaction, and therefore, the total amount required to sustain biodegradation may be much greater than that required to meet the stoichiometric demand as a coreactant. At least in principle, the requirement for inducers is the smallest, but if the inducer is also a substrate for one of the enzymes in the induced pathway, it will probably be consumed by microbial activity. Continued induction of such a pathway will depend on the availability of a continuous source of the inducer; so, this type of cosubstrate may also have a relatively large demand for sustained operation of destructive natural attenuation processes. Because they are typically required in the largest amounts, the requirements for electron donors and acceptors to complete biodegradation redox reactions are often important elements in the evaluation of natural attenuation feasibility (Wiedemeier et al., 1999).

Since many important environmental contaminants are degraded by microbial oxidation, including petroleum hydrocarbons and oxygen-containing solvents, electron acceptors are often given special attention. Electron acceptors that can support microbial oxidation of organic pollutants include molecular oxygen, nitrate, manganese and iron, sulfate, and carbon dioxide (Zehnder and Stumm, 1988; Lovley and Goodwin, 1988). Although these are listed in order of decreasing free energy of the reduction half reaction, and therefore the approximate order in which they are expected to be consumed, this is not necessarily the order of importance for natural attenuation. This is particularly true in contaminated subsurface soils, groundwater, and sediments, where transport of electron acceptors into the contaminated medium is usually slow relative to the rate of microbial consumption. Therefore, high-energy electron acceptors, such as oxygen and nitrate, which must be transported into the contaminated environment from the surroundings are usually less important than are endogenous (i.e., derived from the contaminated geological medium) electron acceptors, such as Mn(IV), Fe(III), and carbon dioxide, despite the fact that reduction of these species yields less energy per electron equivalent that is transferred. In many cases, carbon dioxide, which has the lowest free energy of reduction of all, is the most important electron acceptor because its supply is virtually unlimited.

Contaminants that are degraded via reductive mechanisms require electron-donor cosubstrates. Examples of such reactions include reductive dechlorination of chlorinated aliphatic hydrocarbons (e.g., PCE, carbon tetrachloride) and chlorinated aromatic compounds (e.g., PCP, PCBs) and many biotransformations of nitroaromatic compounds (e.g., 2,4,6-TNT, 2,4-DNT). Some of these compounds, such as PCE and TCE, can serve as primary electron acceptors, whereas others (e.g., TNT and PCBs) appear to be degraded primarily by cometabolism. Regardless of whether the biological transformation reaction supports microbial growth, however, the electron donor requirement is determined by the relative oxidation states of the substrates and products. For example, complete reduction of PCE to ethene requires 8 electron equivalents per mole (48 meq/g PCE), and complete reduction of TNT to 2,4,6-triaminotoluene (TAT) requires 18 electron equivalents per mole (79 meq/g TNT). Both of these compounds can also be incompletely reduced (e.g., conversion of PCE to TCE or reduction of TNT to aminodinitrotoluene) which will reduce the stoichiometric demand for reducing equivalents.

Several simple organic and inorganic compounds have been shown to be good electron donors for reductive dehalogenation reactions. Effective electron donors include H_2, formate, acetate, lactate, butyrate, methanol, ethanol, and glucose (Freedman and Gossett, 1989; Gibson et al., 1994; Bagley and Gossett, 1995; Fennell et al., 1995; Smatlak et al., 1996; Wrenn and Rittmann, 1996; Ballapragada et al., 1997; Fennell et al., 1997; Gao et al., 1997; Gerritse et al., 1999). A similar suite of simple substrates has been reported to support reduction of nitroaromatic compounds (Schackmann and Muller, 1991; Boopathy et al., 1993a,b; Preuss et al., 1993; Drzyzga et al., 1999; Tharakan and Gordon, 1999; Vanderloop et al., 1999). In contaminated aquifers and sediments, H_2 is a likely common electron donor, because it can be produced during metabolism of many other substrates. In these environments, competition between reductive dechlorinators, or nitroaromatic reducers, and H_2-oxidizing bacteria that use other terminal electron acceptors, such as carbon dioxide, sulfate, or Fe(III), will be important in determining the extent of contaminant biotransformation. At least in the case of reductive dechlorination of chloroethenes, the half-saturation constants for hydrogen utilization are lower for dechlorinators than for methanogens or sulfate reducers (Robinson and Tiedje, 1983; Smatlak et al., 1996; Ballapragada et al., 1997). This suggests that reductive dechlorination can be competitively advantageous in the presence of low steady-state concentrations of hydrogen, such as are expected in many anaerobic habitats (Lovley and Goodwin, 1988).

3.4 Models of Natural Attenuation

3.4.1 Introduction

Modeling natural attenuation is not fundamentally different than conventional modeling of groundwater flow and transport. The core of any attenuation model is the mass balance equation. For water flow, the mass balance is hydrological, with a linear Darcy type representation of flow. Boundary conditions include either a specified head, a specified gradient or both to represent flow through lateral boundaries, flow through streams, recharge, and injection and extraction wells. The transport of dissolved contaminants in the flowing water is also represented by a mass balance equation, with advective and dispersive fluxes, a specified concentration or specified mass flux to represent the constituent source and a loss term representing, in most cases, constituent loss through either chemical or biologically mediated chemical reactions.

Modeling begins with development of a conceptual model of the simulated domain and selection of a numerical model or analytical solution. If a numerical technique is warranted, the domain is discretized, field data are entered and the model is calibrated by varying parameters until the simulated system satisfactorily represents the physical system according to some predetermined criteria. Where there is no density-driven flow, the flow model is developed and calibrated first, followed by development and calibration of the transport model. Calibration of the transport model may require modifications to the flow model from which the advective velocities are derived.

Many good references are available, such as Spitz and Moreno (1996), Bear and Verruijt (1987) and Anderson and Woessner (1992), describing numerical methods, mathematical formulation of the flow and transport equations, and application of numerical methods to groundwater flow and transport. The focus of this section is the mathematical representation and simulation of degradation of organic compounds. The chapter begins with a discussion of modeling objectives, and description of the limitations of modeling of complex natural systems. Next, the mathematical representation of transport is described with particular emphasis on methods of representation and the applicable analytical forms for losses by natural attenuation. Methods used in practice for estimation of loss rate model parameters are described and a case study is presented that illustrates the issues.

3.4.2 Modeling Objectives

It is important at the outset to note the limitations of models in simulating transport and natural attenuation and to choose appropriate objectives. As noted by Schwarzenbach et al. (1993), "A model is an imitation of reality which stresses those aspects that are assumed to be important and omits all properties considered to be nonessential." In the case under discussion, the set of omitted properties should be expanded to include *unknown* as well as nonessential properties. An aquifer is typically an extremely heterogeneous medium, in which properties such as permeability may vary by orders of magnitude over relatively short distances. It is not atypical for properties to vary over distances less than that of the measurement techniques at our disposal and practical levels of model discretization. In addition, the magnitude and timing of contaminant release is generally not known and must be inferred by the spatial distribution and chemical composition of the constituent plume. Nonconservative compounds introduce another set of challenges, including identification of the dominant chemical processes and quantifying the impact of toxicity, nonlinear sorption, diffusive transport, and site-specific geochemical conditions that may be expected to vary spatially and be impacted by the presence of the constituent itself.

Given these challenges it is important to define an achievable set of modeling objectives, which warrants the level of effort anticipated. The following objectives are introduced in this regard:

1. Guide and inform site investigations in the development of a conceptual model of constituent transport and natural attenuation. Ideally, modeling in some form is initiated prior to the beginning of field investigations and is used to integrate data as it is gathered in the field. This encourages the thoughtful collection of data, by placing information into context with the larger objectives of the investigation and making evident gaps in the understanding of how the system operates. The location or depth of additional monitoring wells, the analysis of additional compounds representing either potential degradation products or electron acceptors and other investigatory activities should be guided and informed by the ongoing modeling activities.

2. Estimate field scale properties. Models provide a mechanism for estimation of transport and decay rate parameters, through both trial and error and formal optimization techniques. Estimates of natural attenuation rate parameters are useful only if the process has been correctly characterized. Zero order decay rates estimated under conditions of first order decay may provide a good fit to the measured data, but will result in poor forecasts of future concentration. Also, other unknown quantities such as dispersion and the timing and magnitude of sources may undermine attempts to accurately estimate decay parameters.

3. Qualified predictions of anticipated concentration at downgradient receptors. Models cannot by themselves tell with any degree of certainty whether natural attenuation will reduce contaminant concentrations at receptors to an acceptable level. They can, however, support the effort of developing a comprehensive description of the system, which is consistent with field data, lab scale findings and a reasonable physical description of the attenuation and transport processes. With this system description in place the model may be used to make qualified predictions, with the responsibility imposed on the modeler of ensuring that the modeled

processes and properties are physically reasonable and the results adequately qualified to convey to the end user some sense of the forecast uncertainty. Quantitative methods exist to evaluate the uncertainty (Schafer and Kinzelbach, 1992 and Minsker and Shoemaker, 1998), although qualitative descriptions of the likely magnitude of model error may be of equal value.

In conclusion, it is interesting to consider the following observations on modeling by Schwarzenbach et al. (1993), "Real observations are always better than outputs produced by some model. But model results can help us examine various scenarios ... and likely future outcomes ... regarding the behavior of environmental systems for which real data would never be collected as frequently and ubiquitously as we may need them." Models should not be used as predictive tools without the awareness of the likelihood of forecast errors and an understanding that forecast error does not necessarily constitute a failure on the part of the modeler.

3.4.3 Mathematical Representation of Transport and Constituent Losses

The differential equation describing the transport of dissolved constituents in groundwater flow is simply a mass balance, equating the rate of increase in stored mass in a control volume to the net mass flux into the control volume minus internal sources and sinks. A general form of Eq. 3.1.1:

$$\frac{\partial C}{\partial t} = -v_i \frac{\partial C}{\partial x_i} + \frac{\partial}{\partial x_i}\left(D_{i,j} \frac{\partial C}{\partial x_j}\right) - f \qquad (3.4.1)$$

where
- C = fluid concentration (M/V)
- v_i = average pore scale velocity (L/T)
- D_{ij} = dispersion coefficient (L^2/T)
- f = losses (M/V/T)

The terms on the right-hand side represent from left to right the advective transport, the dispersive transport, and losses which may be some unspecified function of concentration, location, or concentration of some other constituent which controls the rate of decay of the constituent of concern (Section 3.1). In cases where it is important to quantify the concentration of daughter products, as in the degradation of chlorinated hydrocarbon solvents, the problem becomes one of solving multiple coupled mass balance equations. Under these conditions terms representing parent product degradation are source terms to the daughter product mass balance equation.

Equations of the same form may be used to represent the transport of nutrients or electron acceptors, which may act as limiting factors in microbial degradation. In that case, solution of the system requires the simultaneous solution of multiple transport equations. For example, the mass balance of dissolved oxygen concentration, C_O, is

$$\frac{\partial C_O}{\partial t} = -v_i \frac{\partial C_O}{\partial x_i} + \frac{\partial}{\partial x_i}\left(D_{i,j} \frac{\partial C_O}{\partial x_j} \right) - fF \qquad (3.4.2)$$

The rate of contaminant compound mass loss, f, is a function of oxygen concentration, while fF is the consequent loss of oxygen mass accounting for reaction stoichiometry. Appropriate values for BTEX compounds are discussed in some detail in Section 5.1. Anaerobic decay occurs with a sequence of electron receptors generally utilized in order of decreasing energy value. In principal, solution of this problem is the same as that described for oxygen-limited problems, where in this case the distribution of each postulated electron acceptor is represented by a unique mass balance equation and a unique loss term in the constituent mass balance equation. More detailed descriptions of formulations of anaerobic decay as implemented in the RT3D and BIOMOC codes are presented in Clement et al. (1998) and Essaid et al. (1995).

3.4.4 Degradation Kinetics

In general, the uptake and transformation of organic compounds is represented as a function of contaminant concentrations, microbial abundance, and the concentration of electron acceptors and other limiting nutrients. Both classical techniques related to the availability of substrates and limiting nutrients and transport-limited formulations are described. In some cases, functions describing bacterial growth in the presence of the compound of concern are represented explicitly. Here, an instantaneous reaction formulation is considered first, followed by the four most commonly employed models of microbial degradation—zero order decay, first order decay, and Monod kinetics. Lastly, a transport limited formulation is described, in which the diffusive transport of nutrients to microbes is assumed to be the rate limiting step.

When groundwater velocities are high relative to the rate of decay, system kinetics are important because they describe the decay of organic compounds over finite and possibly significant distances. In cases of low to moderate velocities, an understanding of kinetics may be of lesser importance, as the system may be adequately represented assuming that reactions occur instantaneously. In the latter condition degradation may be thought of as occurring instantaneously relative to the attenuated time scales over which groundwater passes some point. In practice, constituent losses are simply computed by subtracting the availability of some rate limiting compound such as oxygen, divided by a factor representing the oxygen consumed in metabolizing a unit mass of the degrading constituent.

Borden and Bedient (1986) performed simulations of a hydrocarbon plume under both Monod kinetics and instantaneous decay conditions to test the practical application of the latter as an approximation of the more complicated kinetic formulation. For a constant source, with a groundwater velocity of approximately 0.015 m/d, the concentration profile 11,000 days after the source began was nearly identical in both cases. Careful consideration is necessary before applying an instantaneous kinetic model to a given site. The model cannot simulate the initial period of adaptation when a growth inducing substrate is introduced into the aquifer.

Also, plumes with relatively short release duration or with velocities high relative to the rate of decay are likely to show more sensitivity to the form of the kinetic model. This might include, for example, more recalcitrant compounds with lower decay rates.

First-order decay processes are those for which losses occur at a rate proportional to the availability of the compound.

$$\frac{\partial C}{\partial t} = -\lambda C \qquad (3.4.3)$$

where $\lambda(1/T)$ is the degradation rate parameter. The solution in a batch experiment, for an initial concentration, C_o, is the negative exponential $C = C_o e^{-\lambda t}$, with a characteristic half-life, $t_{1/2}$, describing the time for the concentration to reduce to half its initial value. The half-life may be written as a function of the degradation rate as $(\ln 2)/\lambda$.

In some cases, more complex models of degradation may be approximated as first order under a given set of environmental conditions. For example, the rate of degradation may be controlled by the random collision of the solute molecules and microbial cells. It is not inconceivable under these conditions that the degradation process would be well represented by a degradation model of the form:

$$\frac{\partial C}{\partial t} = -\lambda CB \qquad (3.4.4)$$

where B [M/L^3] represents the concentration of microbes able to degrade the modeled constituent. If the concentration of microbes is nearly constant, as would be the case for low solute concentration, then the microbial count might be incorporated into a first-order rate constant $\lambda = \lambda B$ and the system represented as first-order. Real physical systems that may be adequately represented as first order are generally in fact higher order systems that are approximated as first order. Psuedo-first order rate constants are understandably site-specific as they incorporate information about site-specific environmental conditions.

Consider now the case of a constituent of concern whose presence induces the growth of bacteria capable of metabolizing the compound of concern. This requires high concentrations and an energy yield high enough so that it is worth the effort of the cell to metabolize the compound. If the initial microbial abundance is low, then the initial degradation rate will be low. It is typical to see exponential bacterial growth during this early period, during which the rate of degradation increases. In order to represent the rate of mass loss under these conditions, therefore, it is important to describe the dynamics of bacterial growth.

Monod (Monod, 1949 cited in Shwarzenbach, et al., 1993) related cell growth to the cell count, B, and the limiting substrate concentration, C, as:

$$\frac{\partial B}{\partial t} = \frac{\mu_{max} CB}{K_m + C}$$

For C large relative to K_m, as might be the case initially after a sudden large release, bacterial growth will be exponential with a growth rate constant of μ_{max}. Conversely, as C diminishes to small values relative to K_m the rate of cell growth will be proportional to C.

Now defining the yield Y as the cells grown per unit mass of degraded constituent:

$$\frac{\partial B}{\partial t} = -Y\frac{\partial C}{\partial t}$$

the rate of contaminant loss may be expressed as

$$\frac{\partial C}{\partial t} = -\frac{\mu_{max}CB}{Y(K_m + C)}$$

For a system in which the bacterial population is not significantly impacted by the introduction of the contaminant of interest, the Monod kinetic model is an empirical representation of the Michelis-Menten model used to describe lab scale microbial enzyme kinetics. Imposing the condition of stable microbial populations (B constant), the Monod model reduces to the familiar zero order and first order systems at large and small values of C:

$$\frac{\partial C}{\partial t} = \begin{cases} -\lambda_0 = -\dfrac{\mu_{max}B}{Y}; & C \gg K_m \\ -\lambda_1 C = -\dfrac{\mu_{max}B}{YK_m}C; & C \ll K_m \end{cases}$$

This model is typically used to describe a condition whereby the bacterial population and the rate of contaminant loss are both initially small. Bacterial population growth is stimulated by the release of the contaminant and the rate of contaminant loss accelerates. Thereafter, the bacterial population is stable and the contaminant concentration falls off exponentially as a first order system with decay rate $(\mu_{max}B)/(YK_m)$. In so far as the initial bacterial population is a limiting factor on the rate of contaminant loss, the Monod kinetic model may be used to describe the initial period commonly observed on introduction of organic compounds to the environment. If a lag period exists, then it may be represented explicitly as in Wood et al. (1994) by ramping up the rate of decay and population growth from zero initially to their full values over some specified time interval.

The microbial population dynamics expressed in the Monod formulation are those of a microbe population with a single growth-limiting substance. Many systems operate in a condition where more than one substance may be growth limiting under different conditions. One example is that of oxygen-limited degradation of hydrocarbons. In these systems the bacterial growth rate is a function of both oxygen and the availability of hydrocarbons. Two approaches have been used to represent situations with more than one growth-limiting substance. In the "noninteractive" model (Bader, 1982) only one substrate is assumed to be limiting under any given set of environmental conditions. For example, in a "noninteractive" model of oxygen limited degradation, where oxygen concentrations are low, the rate of microbial growth is assumed to be only a function of the oxygen concentration. A noninteractive approach was implemented by Essaid et al. (1995) and referred to as the minimum Monod formulation, whereby the growth rate of bacteria as a function of two substrate concentrations, C_1 and C_2 is given by:

$$\frac{\partial C_1}{\partial t} = \frac{\mu_{max} B}{Y} \min\left\{\left(\frac{C_1}{K_m + C_1}\right), \frac{1}{\beta_2}\left(\frac{C_2}{K_m + C_2}\right)\right\}$$

where β is the ratio of substrate 2 mass uptake to substrate 1 mass uptake based on the reaction stoichiometry.

Alternatively, in the "interactive" model as used by Chen et al. (1992), Borden and Bedient (1986), Essaid et al. (1995) and others more than one constituent is simultaneously growth limiting. In Essaid et al. (1995) for example substrate uptake is taken to be:

$$\frac{\partial C_1}{\partial t} = \frac{\mu_{max} B}{Y}\left\{\left(\frac{C_1}{K_{m,1} + C_1}\right)\left(\frac{C_2}{K_{m,2} + C_2}\right)\right\} \quad (3.4.3)$$

As noted by Borden and Bedient (1986) the ability of this modified Monod formulation to accurately represent the adaptation period has not been demonstrated. A warning advanced by Baveye and Valocchi (1989) is also noteworthy regarding the use of lab-scale analytical techniques in representation of pore-scale processes with large local fluctuations in substrate concentration and microbial response. This, they note,

> amounts to assuming that in the volume averaging process required to obtain a macroscopic picture of the system, the local microscopic fluctuations are virtually smoothed out and a functional relationship characteristic of homogeneous well-stirred laboratory reactors is obtained. This conjecture might be correct but has yet to be confirmed experimentally in a definitive way.

The same note of caution should be introduced in thinking about degradation in the context of unresolved heterogeneity at a field scale, which introduces unresolved fluctuations in environmental conditions and rates of decay.

We have assumed that the rate of microbial growth is a function of the solute concentration in the bulk water phase. But if the microbe population is on the grain surfaces then it is likely exposed to the immobile water phase concentration at the grain surface, rather than the mobile phase concentration. In some cases the cell population growth rate is assumed to be dependent on the intracellular concentration.

If the transport of compounds to the microscopic environment within the pore where degradation actually occurs is slow, then the system is said to be transport limited and accurate simulations must represent this step.

In some models, the microbes are assumed to occupy a "biofilm" distributed uniformly over the grain surface. Alternatively, microbes have been represented as microcolonies (Molz et al., 1986). In either case the interphase transfer of rate limiting substrates between the mobile and immobile phases, $J_{im,m}$, is represented as a diffusive transport through a diffusive layer separating the more homogeneous mobile and immobile phases:

$$J_{im,m} = \alpha(C_{im} - C_m)$$

where $\alpha(V/T)$ is related to the thickness of the diffusive layer, the diffusivity of the substrate of concern, and the surface area of the microbial population and C_{im} is the solute concentration in the immobile phase (M/L^3). Baveye and Valocchi (1989) describe this as a capacitance-like equation, where the flux is proportional to the potential difference across the interface thickness. Estimation of the value of the flux equation parameters is at this point based largely on conjecture. Two additional equations are added to provide model closure for this transport limited condition.

$$J_{im,m} = J_{im,b}$$
$$J_{im,b} = R$$

where $J_{im,b}$ is the interphase transfer rate between the microbe interior and the immobile phase and R is the microbial uptake. These latter expressions assume that storage of substrate in either the biomass or immobile water phases is constant. Bayeve and Valocchi (1989) stated that "the body of experimental evidence is far too scanty at this point to determine whether (differences between models explicitly representing a transport limiting step and more traditional representations) are significant in practice." A consensus on whether systems are in general transport limited has not been reached to date, however, for practicing modelers, the complexity of the transport limited representation and the lack of guidance on parameter values has favored selection of simpler models without explicit representation of interphase transport.

Wood et al. (1994) addressed this issue by computation of dimensionless numbers relating the characteristic time for advective transport, interphase transport and immobile phase degradation reactions. The ratios of these characteristic times, referred to as Damkohler numbers, are diagnostic in determining whether explicit representation of an immobile phase is necessary. Wood et al. (1994) found that even in cases with relatively long interphase transport time, if the immobile phase reaction time is significantly longer than the characteristic time for interphase transport, then the single phase representation produces an acceptably accurate result. In the example presented by Wood et al. (1994) this was the case. The interphase transport time was long relative to the characteristic time for advective transport, but the ratio of reaction time to interphase transport exceeded 100 so that the system was accurately represented using a single phase transport model.

3.4.5 Parameter Estimation

Parameter estimation is the process of fitting measured data to a mathematical representation of some physical model. Model parameters are adjusted by trial and error or through formal optimization strategies until the model predictions replicate the measured values. In most cases, the measured data are multiple point values of solute concentrations from a contaminant plume. The spill and the resultant plume are in essence an unplanned experiment whose analysis and interpretation result in estimates of attenuation rate parameters. Alternatively, and often in conjunction with analysis of field measurements, lab scale experiments may be performed in aquifer microcosms under controlled conditions.

Parameter estimation depends on selection of an appropriate, physically reasonable model of attenuation. Good fits of experimental data and theoretical results are sometimes possible even with the wrong model. If this is done, extrapolations made to predict concentrations at receptors, where the concentrations or other conditions are outside those used in estimating the parameters, will be wrong.

Consider, for example, the case described by Bekins et al. (1997), where the impact of kinetic model misidentification is considered. Michaelis-Menten kinetics, as described above approximate a first-order degradation process at low concentrations and zero order degradation at higher concentrations. If the first-order degradation model is fit to "low" concentrations within the first-order zone of the physical system, then the analytical model is probably an acceptable approximation of the more complex system. If, on the other hand, degradation occurring at high concentrations is used to fit the first-order model, then predictions using the first-order model with these parameters are likely to over-estimate the concentrations at distant receptors with large travel times.

Chapelle (1999) describes three approaches to estimation of degradation rate parameters:
1. Fit physical model of degradation to measured concentrations along a flow path
2. Implementation and analysis of a controlled tracer test
3. Analysis of lab-scale microcosm investigations

To these one might add a fourth technique in which degradation rate parameters are estimated as part of a trial and error analysis performed by matching the simulated contaminant distribution in multiple dimensions to measurements of contaminant concentration in the field. The first and second approaches differ only in the control exerted over the timing, location and magnitude of the source in controlled tracer tests. The bulk of the discussion on parameter estimation will concern the first technique involving analysis of measurements along a flow line. The controlled tracer test, less common in practice, will be discussed in passing and the analysis of microcosms will be centered on understanding the limitations of this technique and sources of uncertainty in application to field scale rates.

Microcosm investigations are beyond the scope of this section, although obviously related to the subject of parameter estimation. It may be briefly noted that microcosm investigations need to replicate in situ conditions including the availability of oxygen and other electron acceptors. A detailed discussion of their use in this context is contained in Wiedemeier et al. (1995).

In principal, better knowledge of the source timing and loading rate as in a controlled tracer test will increase the reliability of the decay rate estimates. Thierrin et al. (1995) describe a tracer test conducted with dueterated gasoline hydrocarbons (BTEX) and napthalene. A series of multiport samplers on a line perpendicular to the direction of flow were used to monitor the hydrocarbon decay. Only 68 percent of the conservative bromide tracer was accounted for, casting doubt on the accuracy of results. The technique was useful in revealing spatial variability in degradation rates due to zones of localized anaerobic conditions in the most heavily contaminated portions of the plume. The benefits of a controlled tracer test of this type must be weighed against the additional monitoring expenses and the additional effort involved in overcoming regulatory obstacles to introducing contaminants into an aquifer.

Estimation of degradation parameters by analysis of flow measurements rely on travel times as an indicator of the time a compound has been in the process of degradation. This is not an exact analog of the elapsed time in a controlled lab-scale experiment. The aquifer is heterogeneous in the distribution of contaminants and environmental geochemistry. By comparing measurement of concentrations along the flow path, one can estimate the loss rate for a given physical model, based on the concentration decline in the downgradient direction.

Most of the literature on parameter estimation focuses on first order models rather than the Monod and Michaelis-Menton kinetic models. This is because more data over a larger concentration range are needed for the more complex models. By contrast, estimates of the first-order decay rate parameter may be made based on any two measurements.

Buscheck and Alacantar (1995) estimated the first-order decay rate parameter, λ, by employing the steady-state analytical solution of one-dimensional transport model with a specified, constant source concentration, C_o

$$C(x) = C_o \exp\left[\left(\frac{x}{2\alpha_x}\right)\left(1-\left(1+\frac{4\lambda\alpha_x}{v}\right)^{1/2}\right)\right]$$

where α_x (L) is the longitudinal dispersivity and v is the average pore scale velocity in the direction of flow. The decay rate parameter is therefore

$$\lambda = \left(\frac{v}{4\alpha_x}\right)\left([1+2\alpha_x m]^2 - 1\right)$$

where m is the slope of $\ln(C/C_o)$ versus downstream distance, found by linear regression of the measured concentrations along the flow path..

NATURAL ATTENUATION OF HAZARDOUS WASTES 83

The technique should be restricted in application to plumes with one-dimensional characteristics, i.e., wide source and limited transverse dispersion except at the lateral plume boundary. Plumes resulting from the introduction of contaminant compounds at a compact source area may not be appropriately represented as one-dimensional. For these sources, the center line concentration will be influenced by the transverse and the longitudinal dispersion as in the analytical solution described in Wilson and Miller (1978).

Wilson et al. (1994) introduced a technique where measurements of conservative constituents in the contaminant plume are used to distinguish between attenuation due to hydrodynamic dispersion and that occurring as a result of decay. The equation used to "correct" the hydrocarbon compounds (Wiedemeier et al., 1995) for dilution due to dispersion is:

$$C_{B,corr} = C_B \frac{T_A}{T_B}$$

where
$C_{B,corr}$ = corrected concentration at point B (downgradient of point A)
C_B = measured concentration at point B
T_A = tracer concentration at upgradient point A
T_B = tracer concentration at downgradient point B

and given a travel time of t between points A and B, the first-order degradation rate constant is computed as

$$\lambda = \frac{1}{t} \ln \left(\frac{C_{B,corr}}{C_A} \right)$$

Wiedemeier (1995) describes a similar relationship for sorbing compounds. Wilson et al. (1994) and Wiedemeier et al. (1996) used trimethylbenzene as tracers in application of this technique and fitted a first order degradation model using the estimated "residence" time between monitoring points. The reader should be cautioned that the decay rates estimated in this instance represent those prevalent within the body of an established plume. Frequently, anaerobic conditions prevail at the core of a plume, with degradation rates characteristic of this type of process, while more rapid aerobic degradation is simultaneously occurring at a plume front, where oxygen concentration has not yet been depleted.

3.4.6 Numerical Codes

Several modeling codes for modeling natural attenuation are in common use. They are BIOSCREEN, BIOCHLOR, BIOPLUME, MT3D and RT3D. There are many other codes available with similar features, but it has been our experience that only several of these will over time be accepted as an engineering standard. BIOSCREEN is a spreadsheet-based implementation of an analytical solution for transport in a uniform flow field, while the others are numerical codes that simulate transport and decay based on numerical flow field simulations. All are available in the public domain, although proprietary versions of some codes with enhanced capabilities are in some cases also available. As of July 2001, the US Environmental Protection Agency Office of Research and Development maintained a web site, www.epa.gov/ada/csmos/models.html, from which the executables and supporting documentation of the codes mentioned above—except for RT3D—may be downloaded. RT3D is available from a web site maintained by Batelle at www.bioprocess.pnl.gov/rt3d.htm. A number of groundwater modeling interfaces are also available which simplify data-entry and interpretation tasks. The model descriptions herein will be out-of-date as new versions are developed, so the reader is cautioned to seek contemporary sources of information as to the features of particular models.

BIOSCREEN (Newell et al., 1996) and BIOCHLOR (Aziz et al., 2000) implement a three-dimensional, analytical solution of solute transport in a uniform flow field. The analytical solution (Domenico, 1987) is that of solute transport downgradient of a vertical planar source, subject to advection, dispersion, linear-equilibrium adsorption, and decay. BIOSCREEN is designed principally for analysis of attenuation of hydrocarbons, so it allows for either first-order or instantaneous decay of a single constituent. BIOCHLOR incorporates sequential first order decay, tracking the transport and eventual decay of daughter products, as occurs in the degradation of chlorinated solvents. BIOCHLOR also incorporates a scoring system for evaluation of the likelihood of natural attenuation based on environmental conditions favorable for attenuation and constituents considered to be evidence of degradation of chlorinated hydrocarbons. Both BIOSCREEN and BIOCHLOR use a spreadsheet interface providing forms for data-entry and generation of both center-line concentrations and concentration versus downgradient and transverse distance.

BIOPLUME (Rifai et al., 1998) is a numerical code to solve for transport and attenuation of hydrocarbon plumes. BIOPLUME also has an interface for data-entry and interpretation and an integrated two-dimensional, finite difference code for solution of the flow field. The transport simulation consists of advective-dispersive transport of the bulk concentration of organic compounds in the presence of a sequence of electron acceptors. Solute transport is accomplished using the method of characteristics. While earlier versions represented only instantaneous aerobic reactions, the current version, BIOPLUME III, can represent instantaneous reactions, first-order decay and monod kinetics.

The MT3D, Modular Three-Dimensional Transport Model (Zheng, 1990) and RT3D, Reactive Transport in 3-Dimensions (Clement, 1997) are both numerical codes for simulation of advective-dispersive transport that are also compatible with MODFLOW generated flow field simulations. While MT3D is often thought of as a simpler precursor to RT3D, newer versions of MT3D, such as MT3D99 and MT3DMS (Zheng and Wang, 1999), emulate many of the more complex multi-species reactions that are the hallmark of RT3D. RT3D adopted the transport portion of the MT3D code, while allowing for simulation of multiple species and multiple electron acceptors with a general formulation of the chemical and biological kinetics. The user may either code and compile a user-defined reaction module, or adopt one of the seven pre-programmed reaction modules including: instantaneous aerobic decay of BTEX, instantaneous degradation of BTEX using multiple electron acceptors, kinetic-limited degradation of BTEX using multiple electron acceptors, rate limited sorption reactions, double-monod model (noninteractive monod model), sequential decay reactions, and aerobic/anaerobic model of PCE/TCE degradation.

3.4.7 Case Study

A modeling study was conducted to address groundwater contamination at a gasoline fuel terminal to support risk assessment and remedial design. The study included groundwater flow modeling, floating fuel product (LNAPL) flow modeling and solute transport modeling. The solute transport modeling included a number of gasoline constituents. This case study focuses on simulation of the natural attenuation of dissolved benzene. Benzene exhibited the highest concentrations of the dissolved gasoline constituents at the site. The initial modeling was conducted in 1992. Since then, comparisons have been made of simulated and actual benzene transport.

The DYNSYSTEM groundwater modeling codes (Camp Dresser & McKee, 1994) were used, including DYNTRACK, a fully 3-dimensional, particle based random walk transport modeling code which links directly to DYNFLOW, a 3-dimensional finite element groundwater flow code. DYNTRACK as applied in this case study represents the processes of advection, Fickian dispersion (with optional scale dependent dispersivity), linear equilibrium adsorption and first-order exponential decay. Recent enhancements to DYNTRACK also allow simulation of non-linear and non-equilibrium adsorption. DYNBIO, an extension of DYNTRACK, which simulated oxygen-limited decay, is described below. DYNFLOW and DYNTRACK have been used since 1982 in over 150 groundwater modeling studies. Further information on these codes is available at www.dynsystem.com.

Two reported gasoline product releases were of particular concern at this site: a tank overtopping incident which resulted in the release of approximately 5,000 gallons on the east side of the terminal (east plume), and a pipe leak, subsequently repaired, which resulted in a release of more than one million gallons of product at the west side of the terminal (west plume).

3.4.7.1 East Plume: Modeling with an Exponential Decay Rate

Review of site data prior to modeling indicated that natural biodegradation was inhibiting the migration of benzene in the east plume. For example, measured benzene concentrations showed more rapid decrease with distance from the source than other constituents. This can be seen in the plots shown in Figure 3-8. The fingerprints show that benzene was predominant in monitoring wells near the source (MW-09, MW-10, MW-119), but less so at downgradient wells (MW-120, MW-125).

For modeling, benzene degradation was represented by a first order decay algorithm. A literature search indicated laboratory estimates of benzene half-lives ranging from 5 days to approximately 1 year. The rate of benzene degradation at this site was estimated by comparing model simulated distribution of concentrations with field measured data. That is, trial simulations were made using a number of different degradation rates to find the rate that produced the best fit between the model and the observations. This method of estimating the natural biodegradation rate incorporates

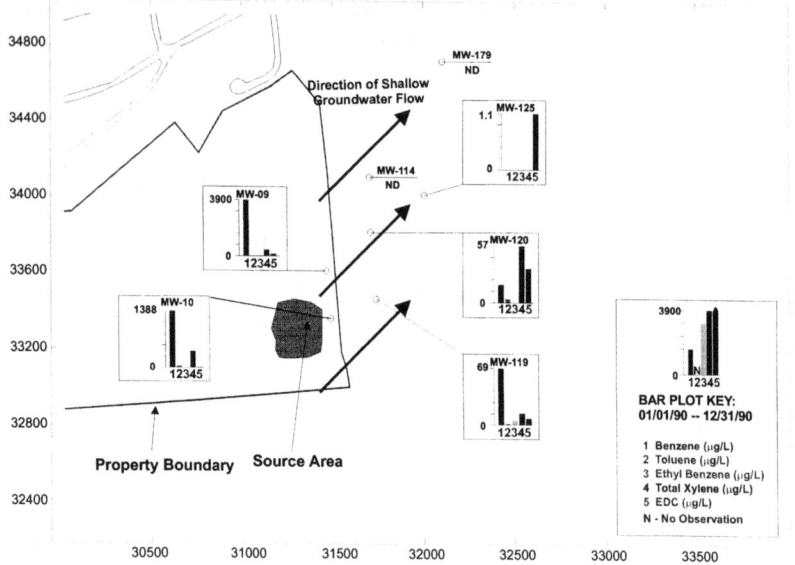

Figure 3-8. Fingerprint plot at gasoline terminal site showing 1990 plume concentration

available data related to the source, groundwater flow field, adsorptive retardation as well as groundwater quality data.

Figure 3-9 shows that a reasonable model representation of the east plume benzene concentration distribution was achieved using a degradation rate of 0.005/day (4.5 month half-life). These projections indicated generally decreasing concentrations due to natural attenuation and depletion of the source of dissolved benzene in the soil. Figure 3-10 shows the projected east plume benzene concentrations for 1995 as simulated in 1992. Also shown in Figure 3-10 are measured 1995 concentrations, the most recent data available for these wells. Both the measured data and simulation results for 1995 reflect a significant decrease in concentrations near the source, and no significant downgradient migration of the benzene plume. These data indicate that the 1992 modeling (incorporating natural biodegradation with a first-order degradation rate of 0.005/day) provided a reasonable projection of future benzene transport and attenuation. In this case, benzene transport from the source has been very significantly limited by natural biodegradation.

3.4.7.2 West Plume: Oxygen-Limited Decay Modeling

The modeling and analysis of the west plume was considerably more complicated than the east plume. Three-dimensional flow modeling was essential here because the plume migrated through multiple aquifer layers with limited potential pathways from

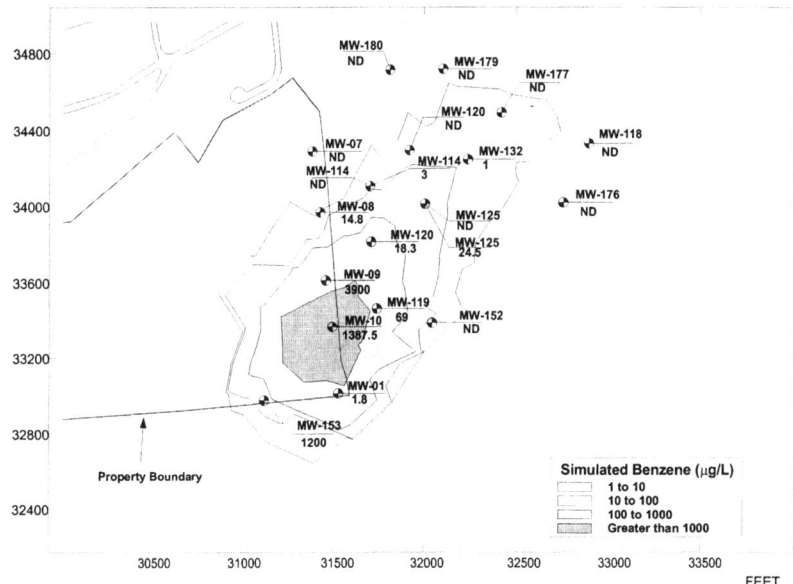

Figure 3-9. Simulated and measured 1990 benzene concentrations using first order decay rate of 0.005/day

one layer to the other. Additionally, a moving source, in the form of an LNAPL plume, had to be considered due to the large size of the release.

With respect to natural attenuation, it appeared that biodegradation was relatively inhibited near the source compared with apparent degradation rates in the east plume and areas further downgradient in the west plume. This was attributed to the limited availability of oxygen compared with the mass and concentration of benzene in the vicinity of the west plume source area. While other compounds besides oxygen may act as electron acceptors, the rate of benzene degradation is much faster when oxygen is available and utilized.

Measured west-plume benzene concentrations and simulated concentrations using a first-order decay model are indicated in Figure 3-11. The measured plume shows a distinct sharp front moving away from the source area, not adequately represented using the first-order decay model. In Figure 3-11, benzene concentrations vary over an order of magnitude between monitoring wells located at 600 feet and 1000 feet from the source area.

This sharp-front behavior is characteristic of oxygen-limited conditions, where dissolved oxygen is relatively abundant in the plume fringes and scarce in the core of the plume. In addition dissolved oxygen availability may be limited in this aquifer due to an upper confining layer serving as a barrier for oxygen transport from the ground surface.

An acceptable representation of the west benzene plume was achieved by application of the oxygen-limited model, DYNBIO. DYNBIO is an extension of

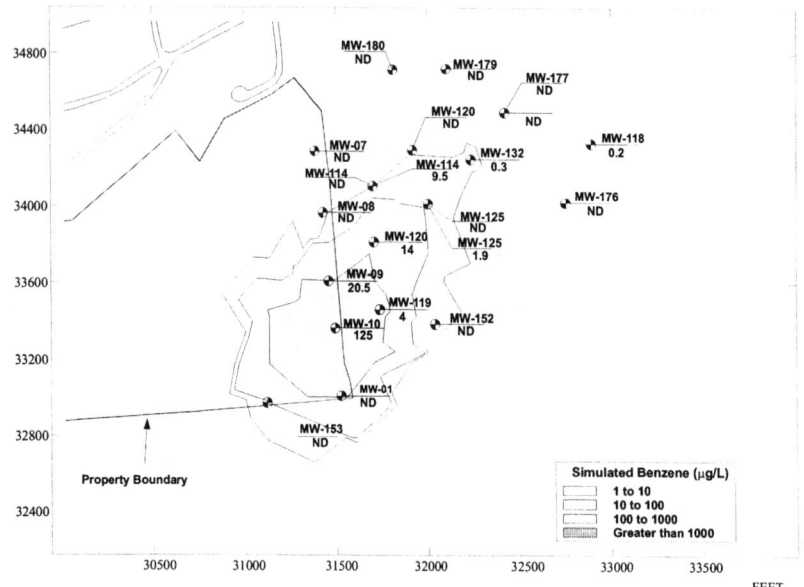

Figure 3-10. Simulated and measured 1995 benzene concentrations using first order decay rate of 0.005/day

DYNTRACK, which explicitly accounts for the effects of limited oxygen on biodegradation of hydrocarbons. DYNBIO simulates the coupled oxygen and hydrocarbon mass balance equations, with a modified form of Michaelis-Menten kinetics of the type introduced by Borden and Bedient (1986) as described above.

DYNBIO assumes a two-constituent system consisting of dissolved oxygen and the contaminant. However, the methodology can be extended to multiple constituents, considering limitations of computational storage and speed. The governing equations are solved using an operator splitting technique (Wheeler and Dawson, 1988) which estimates transport and biodegradation independently within a numerical time step. In this way, a first stage computation is performed in which the contaminant and oxidizer mass are first advected, dispersed and retarded. Following this, a second stage computation consists of the intermediate concentrations being subjected to a reduction produced by the biodegradation Monod kinetics function, over the same time step. The final result is an updated distribution of concentrations that serves as the input to the next time step, where the same two-stage numerical computations are carried out. Consistency of an operator splitting approach implies that for a small enough time discretization, the concentration field can be computed by using the transport and reaction operators separately in this staged fashion. The important modeling implication here is that the two operators may be treated with different numerical techniques, i.e., random walk for transport, Adomian polynomials for reaction.

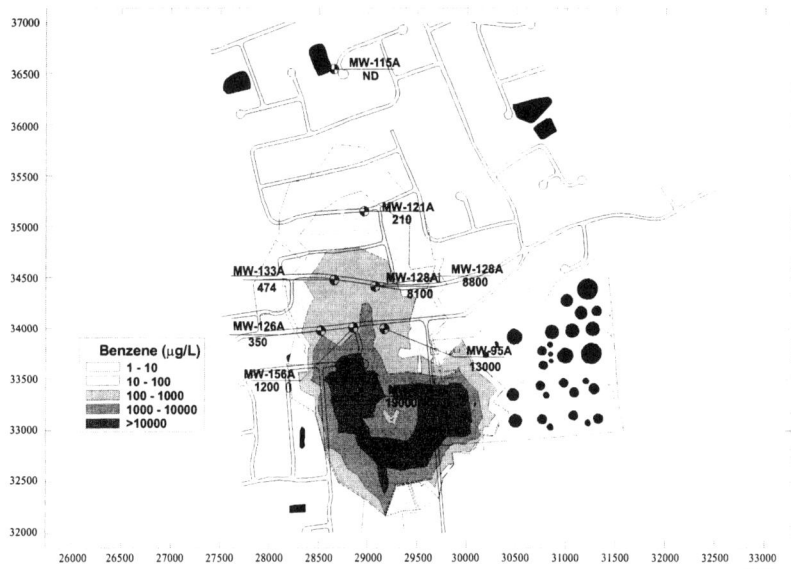

Figure 3-11. Simulated and measured 1994 benzene concentrations in the western plume using first order decay rate of 0.005/day

Transport is simulated out using a random walk technique (CDM, 1994), employing a particle discretization of the transported mass for each constituent. In the case of oxygen limited biodegradation, since dissolved oxygen can be considered a background constituent found in groundwater, it is less computationally intensive to compute concentrations of oxygen deficit about the background dissolved oxygen level. In this case, the model transports two types of particles: contaminant and oxygen deficit (OD). This oxygen deficit is defined as:

$$OD = DOBG - DO$$

where $DOBG$ is the background oxygen level [ML^{-3}], which is generally known under field conditions and DO is the dissolved oxygen. The value of $DOBG$ should be representative of background dissolved oxygen levels in the contaminated aquifer and may be determined from measurements in contaminant-free groundwater at the site.

Biodegradation is estimated using an approximate series solution to solve for the concentration decrease due to a modified Michaelis-Menten decay function (Adomian and Rach, 1986). Details of this technique are presented in Miralles-Wilhelm et al., (1994).

Biodegradation parameters for this site were obtained from literature sources, and refined during the transient calibration process. These parameters are:
- maximum biodegradation rate $\kappa = 0.08$ day^{-1}

- biomass concentration $M = 200$ ug/l
- contaminant half-saturation constant $K_S = 100$ ug/l
- dissolved oxygen half-saturation constant $K_C = 100$ ug/l
- stoichiometric ratio $F = 3$
- background dissolved oxygen concentration $DOBG = 6000$ ug/l

The values for the transport parameters were taken from previous simulations documented in (CDM, 1992). The calibration process was done with the objectives of reproducing the observed features of the Benzene western plume, explained above. Three biodegradation parameters were key in obtaining a successful calibration:

- the contaminant half-saturation constant K_S, as the parameter controlling the strength of the biodegradation process at varying benzene concentrations, and therefore controlling the sharpness of the benzene front as it migrates off the source area.
- the maximum biodegradation rate κ (or the product κM) as the parameter controlling the magnitude of the biodegradation process. For the calibration process presented here, the biomass concentration M was given a fixed value (200 ug/l), and κ was used as the calibration parameter.
- the background dissolved oxygen concentration $DOBG$, which controls the overall extent of benzene contamination by providing the level of dissolved oxygen availability in the plume fringes.

The calibration process that generated these biodegradation parameters reproduced the specific characteristic behavior of the benzene western plume. In particular, the plume shows a distinct sharp front moving northward, with relatively low concentrations (< 100 µg/l) being completely biodegraded at a much faster rate at higher concentrations. Figures 3-12, 3-13, and 3-14 show the simulated plumes for the years 1990, 1992 and 1994, which improve the agreement with the observed benzene concentrations relative to the first-order decay model results presented in Figure 3-11.

NATURAL ATTENUATION OF HAZARDOUS WASTES 91

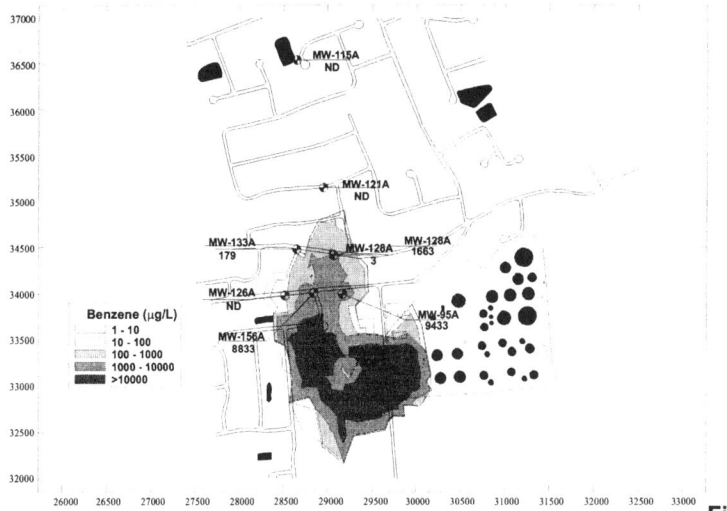

Figure 3-12. Simulated and measured benzene concentration in the western plume (1990) with simulations of oxygen limited system and monod-type kinetics

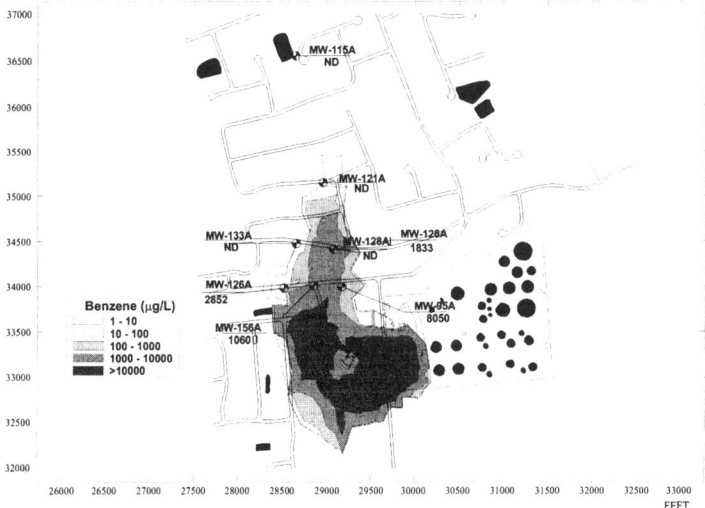

Figure 3-13. Simulated and measured benzene concentration in the western plume (1992) with simulations of oxygen limited system and monod-type kinetics

92 NATURAL ATTENUATION OF HAZARDOUS WASTES

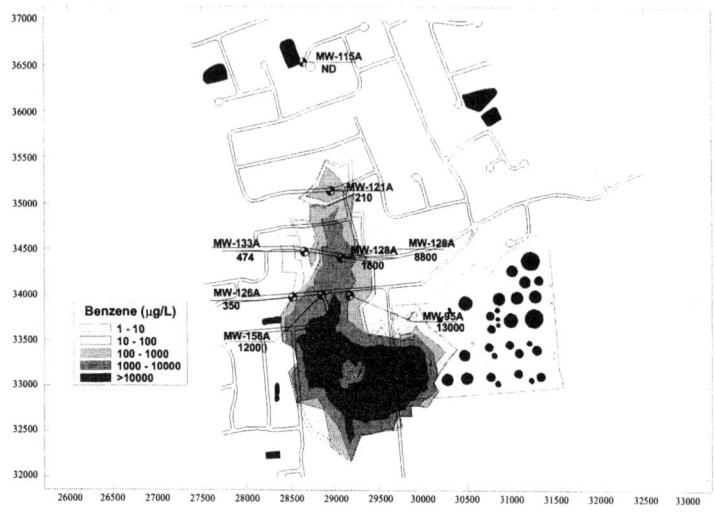

Figure 3-14. Simulated and measured benzene concentration in the western plume (1994) with simulations of oxygen limited system and monod-type kinetics

References

Adomian, G. and Rach, R. (1986). "A coupled nonlinear system" J. Math. Anal. Appl., 113, 510-513.

Aeckersberg, F., Bak, F., and Widdel, F. (1991). "Anaerobic oxidation of saturated hydrocarbons to CO_2 by a new type of sulfate-reducing bacterium." Arch. Microbiol., 156, 5-14.

Al-Bashir, B., Cseh, T., Leduc, R., and Samson, R. (1990). "Effect of soil/contaminant interactions on the biodegradation of naphthalene in flooded soil under denitrifying conditions." Appl. Microbiol. Biotechnol., 34, 414-419.

Alexander, M. (1973). "Nonbiodegradable and other recalcitrant molecules." Biotechnol. Bioengin., 15, 611-647.

Alexander, M. (1994). Biodegradation and bioremediation, Academic Press, San Diego, CA.

Allison, J.D., Brown, D.S., and Nova-Gradac, K.J. (1990). "MINTEQA2/ PRODEFA2, a geochemical assessment model for environmental systems: Version 3.0 User's Manual" US Environ. Protection Agency, Athens, GA.

Allison, J.D., Brown, D.S., and Novo-Gradac, K.J. (1991). "MINTEQA2/ PRODEFA2, A geochemical assessment model for environmental systems: Version 3.0 User's Manual" Environmental Research Laboratory, Office of Research and Development, U.S. Environ. Protection Agency, Athens, Georgia. EPA/600/3-91/021, p.106.

Alvarez-Cohen, L. and McCarty, P.L. (1991a). "A cometabolic biotransformation model for halogenated aliphatic compounds exhibiting product toxicity." Environ. Sci. Technol., 25, 1381-1387.

Alvarez-Cohen, L. and McCarty, P.L. (1991b). "Two-stage dispersed-growth treatment of halogenated aliphatic compounds by cometabolism." Environ. Sci. Technol., 25, 1387-1393.

Alvarez-Cohen, L. and McCarty, P.L. (1991c). "Effects of toxicity, aeration, and reductant supply on trichloroethylene transformation by a mixed methanotrophic culture." Appl. Environ. Microbiol., 57, 228-235. Anderson, M.P., and Woessner, W.W. (1992). Applied Groundwater Modeling – Simulation of Flow and Advective Transport. Academic Press, New York.

Anderson, M.P. (1979). "Using models to simulate the movement of contaminants through groundwater flow systems." CRC Crit. Rev. Environ. Control, 9(2), 97-156.

Anderson, M.P., and Woessner, W.W. (1992). Applied Groundwater Modeling – Simulation of Flow and Advective Transport. Academic Press, New York.

Anderson, J.E., and McCarty, P.L. (1997). "Effect of chlorinated ethenes on S_{min} for a methanotrophic mixed culture." Environ. Sci. Technol., 31, 2204-2210

Anderson, R.T., Rooney-Varga, J.N., Gaw, C.V., and Lovley, D.R. (1998). "Anaerobic benzene oxidation in the Fe(II) reduction zone of petroleum-contaminated aquifers." Environ. Sci. Technol., 32, 1222-1229.

Apajalahti, J.H.A., and Salkinoja-Salonen, M.S. (1984). "Absorption of pentachlorophenol (PCP) by bark chips and its role in microbial PCP degradation." Microb. Ecol., 10(4), 359-367.
Appelo, C., and Parkhurst, D. (1998). "Enhancements to the Geochemical Model PHREEQC-1D Transport and Reaction Kinetics" In: Arehart, G.B., Hulston, J.R. (Eds.), Water–Rock Interaction. Balkema, Rotterdam,NL, pp. 873–876.
ASTM. (1998). "Standard guide for remediation of ground water by natural attenuation at petroleum release sites." E 1943-98, West Conshohocken, PA.
Aziz, C.E., Newell, C.J., Gonzales, J.R., Haas, P., Clement, T.P. and Sun, Y. (2000). BIOCHLOR Natural Attenuation Decision Support System User's Manual, Version 1.0. U.S. Environmental Protection Agency, EPA/600/R-00/008.
Bader, F.B. (1992). Kinetics of Double-Substrate Limited Growth in Microbial Population Dynamics, edited by M.J. Bazin, CRC Press, Inc., Boca Raton, Florida.
Baehr, A.L., Hoag, G.E., and Marley, M.C. (1989). "Removing volatile contaminants from the unsaturated zone by inducing advective air-phase transport." J. Contam. Hydrol., 4(1), 1-26.
Bagley, D.M. and Gossett, J.M. (1995). "Chloroform degradation in methanogenic methanol enrichment cultures and by *Methanosarcina barkeri* 227." Appl. Environ. Microbiol., 61, 3195-3201.
Ball, W.P., and Roberts, P.V. (1991). "Long-term sorption of halogenated organic chemicals by aquifer material. 1. Equilibrium." Environ. Sci. Technol., 25(7), 1223-1237.
Ball, W.P., Curtis, G.P., and Roberts, P.V. (1997). "Physical/chemical processes affecting the subsurface fate and transport of synthetic organic materials." Subsurface restoration, C.H. Ward, J.A. Cherry, and M.R. Scalf, eds., Ann Arbor Press, Inc., Chelsea, MI, 27-58.
Ballapragada, B.G., Stensel, H.D., Puhakka, J.A., and Ferguson, J.F. (1997). "Effect of hydrogen on reductive dechlorination of chlorinated ethenes." Environ. Sci. Technol., 31, 1728-1734.
Barnsley, E.A. (1975). "The bacterial degradation of fluoranthene and benzo[a]pyrene." Can. J. Microbiol., 21, 1004-1008.
Bassett, R. L., and Melchior, D. C. (1990). "Chemical modeling of aqueous systems" In: Melchior, D. C. and Bassett, R. L. (Eds), Chemical Modeling of Aqueous Systems II, ACS Symposium Series 416, American Chemical Society, Washington, D.C., pp. 1-12.
Baveye, P., and Valocchi, A. (1989). "An evaluation of mathematical models of the transport of biologically reacting solutes in saturated soils and aquifers." Water Resources Research, 25(6), 1413-1421.
Bear, J. (1972). Dynamics of fluids in porous media, American Elsevier, New York, NY.
Bear, J., and Verruijt, A. (1987). Modeling Groundwater Flow and Pollution. D. Reidel Publishing Co., Boston.
Bedard, D.L., Unterman, R.L., Bopp, L.H., Brennan, M.J., Haberl, M.L., and Johnson, C. (1986). "Rapid assay for screening and characterizing micro-

organisms for the ability to degrade polychlorinated biophenyls." Appl. Environ. Microbiol., 51, 761-768.
Bekins, B.A., Warren, A., and Godsy, E.M. (1997). "Comparing zero- and first-order approximations to the monod model." In Situ and On-Site Bioremediation: Volume 5, Batelle Press, 547-552.
Bellar, H.R., Grbic-Galic, D., and Reinhard, M. (1992). "Microbial degradation of toluene under sulfate-reducing conditions and the influence of iron on the process." Appl. Environ. Microbiol., 58, 786-793.
Berardesco, G., Dyhrman, S., Gallagher, E., and Shiaris, M.P. (1998). "Spatial and temporal variation of phenanthrene-degrading bacteria in intertidal sediments." Appl. Environ. Microbiol., 64, 2560-2565.
Bethke, C. (1994). "The geochemist's workbench:. A User's Guide to Rxn, Act 2, Tact, React and Gtplot." University of Illinois, Urbana-Champaign.
Benefield, L.D., Judkins, J. F., and Weand, B.L., (1982). Process Chemistry for Water and Wastewater Treatment, Prentice-Hall, Inc., Englewood Cliffs, New Jersey.
Bildstein, O. (1998). "Modelisation geochimique des interactions eau–gaz–roche. Application a la diagenese minerale dans les reservoirs geologiques" Master's thesis, Universite Louis Pasteur, Strasbourg, France.
Boldrin, B., Tiehm, A. and Fritzsche, C. (1993). "Degradation of phenanthrene, fluorene, fluoranthene, and pyrene by a *Mycobacterium* sp." Appl. Environ. Microbiol., 59, 1927-1930.
Boopathy, R., Wilson, M., and Kulpa, C.F. (1993a). "Anaerobic removal of 2,4,6-trinitrotoluene (TNT) under different electron accepting conditions: laboratory study." Water Environ. Res., 65, 271-275.
Boopathy, R., Kulpa, C.F., and Wilson, M. (1993b). "Metabolism of 2,4,6-trinitrotoluene (TNT) by a *Desulfovibrio* sp. (B strain)." Appl. Microbiol. Biotechnol., 39, 270-275.
Borden, R.C., and Bedient, P.B. (1986). "Transport of dissolved hydrocarbons influenced by oxygen-limited biodegradation. 1. Theoretical development." Water Resour. Res., 22(13), 1973-1982.
Borden, R.C., Bedient, P.B., Lee, M.D., Ward, C.H., and Wilson, J.T. (1986). "Transport of dissolved hydrocarbons influenced by oxygen-limited biodegradation. 2. Field application." Water Resour. Res., 22(13), 1983-1990.
Boyle, A.W., Silvin, C.J., Hassett, J.P., Nakas, J.P., and Tanenbaum, S.W. (1992). "Bacterial PCB biodegradation." Biodegradation, 3, 285-298.
Bradley, P.M. and Chapelle, F.H. (1998a). "Effect of contaminant concentration on aerobic microbial mineralization of DCE and VC in stream-bed sediments." Environ. Sci. Technol., 32, 553-557.
Bradley, P.M. and Chapelle, F.H. (1998b). "Microbial mineralization of VC and DCE under different terminal electron accepting conditions." Anaerobe, 4, 81-87.
Bregnard, T.P.A., Haner, A., Hohener, P., Zeyer, J. (1997). "Anaerobic degradation of pristane in nitrate-reducing microcosms and enrichment cultures." Appl. Environ. Microbiol., 63, 2077-2081.
Brockman, F.J., and Murray, C.J. (1997). "Microbiological heterogeneity in the terrestrial subsurface and approaches for its description." The Microbiology of

the Terrestrial Deep Subsurface, P.S. Amy and D.L. Haldeman, eds., Lewis Publishers, Boca Raton, FL, 75-102.

Brusseau, M.L., and Rao, P.S.C. (1989). "Sorption nonideality during organic contaminant transport in porous media." CRC Crit. Rev. Environ. Control, 19, 22-99.

Buscheck, T.E., and Alcantar, C.M. (1995). "Regression techniques and analytical solutions to demonstrate intrinsic bioremediation." Intrinsic Bioremediation, R.E. Hinchee, J.T. Wilson, and D.C. Downey, eds., Batelle Press, Columbus, Ohio, 109-116.

Caldwell, M.E., and Suflita, J.M. (2000). "Detection of phenol and benzoate as intermediates of anaerobic benzene biodegradation under different terminal electron-accepting conditions." Environ. Sci. Technol., 34, 1216-1220.

Camp Dresser & McKee. (1994) DYNTRACK: A three-dimensional contaminant transport model for groundwater studies, Cambridge, MA.

Camp Dresser & McKee. (1992) Risk evaluation for groundwater exposure pathway, Consolidated Petroleum Terminal, Northville Industries Corporation, New York.

Castro, C.E., Wade, R.S., and Belser, N.O. (1985). "Biodehalogenation: Reactions of cytochrome P-450 with polyhalomethanes." Biochemistry, 24, 204-210.

Cerniglia, C.E. (1992). "Biodegradation of polycyclic aromatic hydrocarbons." Biodegradation, 3, 351-368.

Chang, H-L., and Alvarez-Cohen, L. (1996). "Biodegradation of individual and multiple chlorinated aliphatic hydrocarbons by methane-oxidizing cultures." Appl. Environ. Microbiol., 62, 3371-3377.

Chapelle, F.H. (1999). "Bioremediation of petroleum hydrocarbon-contaminated ground water: The perspectives of history and hydrology." Ground Water, 34(4), 691-698.

Chen, Y.M., Abriola, L.M., Alvarez, P.J., Anid , P.J. and Vogel, T.M. 1992. "Modeling transport and biodegradation of benzene and toluene in sandy aquifer material: Comparisons with experimental measurements." Water Resources Research, 18. 1833-1847.

Chiang, C.Y., Salanitro, J.P., Chai, E.Y., Colthart, J.D., and Klein, C.L. (1989). "Aerobic biodegradation of benzene, toluene, and xylene in a sandy aquifer--data analysis and computer modeling." Ground Water, 27(6), 823-834.

Chiou, C.T., Porter, P.E., and Schmedding, D.W. (1983). "Partition equilibria of nonionic organic compounds between soil organic matter and water." Environ. Sci. Technol., 17(4), 227-231.

Chiou, C.T., Shoup, T.D., and Porter, P.E. (1985). "Mechanistic roles of soil humus and minerals in the sorption of nonionic organic compounds from aqueous and organic solutions." Org. Geochem., 8(1), 9-14.

Chu, K.H., and Alvarez-Cohen, L. (1999). "Evaluation of toxic effects of aeration and trichloroethylene oxidation on methanotrophic bacteria grown with different nitrogen sources." Appl. Environ. Microbiol., 65, 766-772.

Clement, T.P., Sun, Y., Hooker, B.S. and Petersen, J.N. (1998). "Modeling multi-species reactive transport in ground water." Ground Water Monitoring and Remediation, Spring, 79-92.

Clement, T.P. A Modular Computer Code for Simulating Reactive Multispecies Transport in 3-Dimensional Groundwater Systems (1997). Prepared for the US Department of Energy by Pacific Northwest National Laboratory, Richland, Washington.

Cirpka, O.A., Frind, E.O., and Helmig, R. (1999). "Numerical simulation of biodegradation controlled by transverse mixing." J. Contam. Hydrol., 40, 159-182.

Coates, J.T., Elzerman, A.W. (1986). "Desorption kinetics for selected PCB congeners from river sediments." J. Contam. Hydrol., 1(1/2), 191-210.

Coates, J.D., Anderson, R.T., and Lovley, D.R. (1996). "Oxidation of polycyclic aromatic hydrocarbons under sulfate-reducing conditions." Appl. Environ. Microbiol., 62, 1099-1101.

Coates, J.D., Woodward, J., Allen, J., Philp, P., and Lovley, D.R. (1997). "Anaerobic degradation of polycyclic aromatic hydrocarbons and alkanes in petroleum-contaminated marine harbor sediments." Appl. Environ. Microbiol., 63, 3589-3593.

Colby, J., Stirling, D.I. and Dalton, H. (1977). "The soluble methane monooxygenase of *Methylococcus capsulatus* (Bath)." Biochem. J., 165, 395-402.

Commandeur, L.C.M., Van Eyseren, H.E., Opmeer, M.R., Govers, H.A.J., and Parsons, J.R. (1995). "Biodegradation kinetics of highly chlorinated biphenyls by *Alcaligenes* sp. JB1 in an aerobic continuous culture system." Environ. Sci. Technol., 29, 3038-3043.

Corapcioglu, M.Y., and Baehr, A.L. (1987). "A compositional multiphase model for groundwater contamination by petroleum products, 1. Theoretical considerations." Water Resour. Res. 23(1), 191-200.

Criddle, C.S. (1993). "The kinetics of cometabolism." Biotechnol. Bioengin., 41, 1048-1056.

Dabrock, B., Riedel, J., Bertram, J., and Gottschalk, G. (1992). "Isopropylbenzene (cumene)—a new substrate for the isolation of trichloroethene-degrading bacteria." Arch. Microbiol., 158, 9-13.

Dalton, H. and Stirling, D.I. (1982). "Cometabolism." Phil. Trans. Roy. Soc. Lond., B. Biol. Sci., 297, 481-496.

Dean-Raymond, D. and Bartha, R. (1975). "Biodegradation of some polynuclear aromatic petroleum components by marine bacteria." Dev. Ind. Microbiol., 16, 97-110.

Devinny, J.S. (1990). "Chemical and physical alteration of wastes and leachates." Subsurface migration of hazardous wastes, J.S. Devinny, L.G. Everett, J.C.S. Lu, and R.L. Stollar, eds., Van Nostrand Reinhold, New York, NY, 142- 168.

Dolan, M.E., and McCarty, P.L. (1995). "Methanotrophic chloroethene transformation capacities and 1,1-dichloroethene transformation product toxicity." Environ. Sci. Technol., 29, 2741-2747.

Domenico, P.A., and Schwartz, F.W. (1998). Physical and chemical hydrogeology, 2nd edition, John Wiley & Sons, Inc., New York, NY.

Domenico, P.A. (1987). "An analytical model for multidimensional transport of a decaying contaminant species." Journal of Hydrology, 91, 49-58.

Drzyzga, O., Bruns-Nagel, D., Gorontzy, T., Blotevogel, K.H., and von Low, E. (1999). "Anaerobic incorporation of the radiolabeled explosive TNT and metabolites into the organic soil matrix of contaminated soil after different treatment procedures." Chemosphere, 38, 2081-2095.

Edwards, E.A., Wills, L.E., Reinhard, M., and Grbic-Galic, D. (1992). "Anaerobic degradation of toluene and xylene by aquifer microorganisms under sulfate-reducing conditions." Appl. Environ. Microbiol., 58, 794-800.

Edwards, E.A., and Grbic-Galic, D. (1994). "Anaerobic degradation of toluene and o-xylene by a methanogenic consortium." Appl. Environ. Microbiol., 60, 313-322.

Edwards, E.A., Edwards, A.M., and Grbic-Galic, D. (1994). "A method for detection of aromatic metabolites at very low concentrations: application to detection of metabolites of anaerobic toluene degradation." Appl. Environ. Microbiol., 60, 323-327.

Egli, T.W. (1992). "General strategies in the biodegradation of pollutants." *Metal Ions in Biological Systems, Vol. 28*, H. Sigel and A. Sigel eds., Marcel Dekker, Inc., New York, NY., 1-39.

Ehrhardt, H.M., and Rehm, H.J. (1985). "Phenol degradation by microorganisms adsorbed on activated carbon." Applied Microbiol. Technol., 21(1/2), 32-36.

Ensley, B.D. (1991). "Biochemical diversity of trichloroethylene metabolism." Annu. Rev. Microbiol., 45, 283-99.

Essaid, H.I., Bekins, B.A., Godsy, E.M., Warren, E., Baedecker, M.J. and Cozzarelli, I.M. (1995). "Simulation of aerobic and anaerobic biodegradation processes at a crude oil spill site." Water Resources Research, 31(12), 3309-3327.

Evans, P.J., Mang, D.T., and Young, L.Y. (1991). "Degradation of toluene and m-xylene and transformation of o-xylene by denitrifying enrichment cultures." Appl. Environ. Microbiol., 57, 450-454.

Fennell, D.E., Stover, M.A., Zinder, S.H., and Gossett, J.M. (1995). "Comparison of alternative electron donors to sustain PCE anaerobic reductive dechlorination." *Bioremediation of Chlorinated Solvents*, R.E. Hinchee, A. Leeson, and L. Semprini, eds., Battelle Press, Columbus, OH, 9-16.

Fennell, D.E., Gossett, J.M., and Zinder, S.H. (1997). "Comparison of butyric acid, ethanol, lactic acid, and propionic acid as hydrogen donors for the reductive dechlorination of tetrachloroethene." Environ. Sci. Technol., 31, 918-926.

Fewson, C.A. (1988). "Biodegradation of xenobiotic and other persistent compounds: the causes of recalcitrance." Trends Biotechnol., 6, 148-153.

Foght, J.M., Fedorak, P.M., and Westlake, D.W.S. (1989). "Mineralization of [^{14}C]hexadecane and [^{14}C]phenanthrene in crude oil: specificity among bacterial isolates." Can. J. Microbiol., 36, 169-175.

Folsom, B.R., Chapman, P.J., and Pritchard, P.H. (1990). "Phenol and trichloroethylene degradation by *Pseudomonas cepacia*G4: kinetics and interactions between substrates." Appl. Environ. Microbiol., 56, 1279-1285.

Fox, B.G., Borneman, J.G., Wackett, L.P., and Lipscomb, J.D. (1990). "Haloalkene oxidation by the soluble methane monooxygenase from *Methylosinus trichosporium* OB3b: Mechanistic and environmental implications." Biochemistry, 29, 6419-6427.

Freedman, D.L., and Gossett, J.M. (1989). "Biological reductive dechlorination of tetrachlorethylene and trichloroethylene to ethylene under methanogenic conditions." Appl. Environ. Microbiol., 55, 2144-2151.

Freeze, R.A., Cherry, J.A. (1979). Groundwater, Prentice-Hall, Inc., Englewood Cliffs, NJ.

Freyberg, D.L. (1986). "A natural gradient experiment on solute transport in a sand aquifer. 2. Spatial moments and the advection and dispersion of nonreactive tracers." Water Resour. Res., 22(13), 2031-2046.

Fried, J.J., and Combarnous, M.A. (1971). "Dispersion in porous media." Advances in hydroscience, Vol. 7, V.T. Chow, ed., Academic Press, Inc., New York, NY, 169-282.

Fries, M.R., Hopkins, G.D., McCarty, P.L., Forney, L.J., and Tiedje, J.M. (1997a). "Microbial succession during a field evaluation of phenol and toluene as the primary substrates for TCE cometabolism." Appl. Environ. Microbiol., 63, 1515-1522.

Fries, M.R., Forney, L.J., and Tiedje, J.M. (1997b). "Phenol- and toluene-degrading microbial populations from an aquifer in which successful TCE cometabolism occurred." Appl. Environ. Microbiol., 63, 1523-1530.

Frind, E.O., and Hokkanen, G.E. (1987). "Simulation of the Borden plume using alternating direction Galerkin technique." Water Resour. Res., 23(5), 918-930.

Fry, V.A., Istok, J.D., Semprini, L., O'Reilly, K.T., and Buscheck, T.E. (1995). "Retardation of dissolved oxygen due to a trapped gas phase in porous media." Ground Water, 33(3), 391-398.

Gallo, Y.L., Bildstein, O., and Brosse, E. (1998). "Coupled reaction-flow modeling of diagenetic changes in reservoir permeability, porosity and mineral compositions" J. Hydrol., 209:366–388.

Gao, J., Skeen, R.S., Hooker, B.S., and Quesenberry, R.D. (1997). "Effects of several electron donors on tetrachloroethene dechlorination in anaerobic soil microcosms." Wat. Res., 31, 2479-2486.

Garbarini, D.R., and Lion, L.W. (1986). "Influence of the nature of soil organics on the sorption of toluene and trichloroethylene." Environ. Sci. Technol., 20(12), 1263-1269.

Gelhar, L.W., Gutjahr, A.L., and Naff, R.L. (1979). "Stochastic analysis of macrodispersion in a stratified aquifer." Water Resour. Res., 15(6), 1387-1397.

Gelhar, L.W., and Axness, C.L. (1983). "Three-dimensional stochastic analysis of macrodispersion in aquifers." Water Resour. Res., 19(1), 161-180.

Gelhar, L.W., Welty, C., and Rehfeldt, K.R. (1992). "A critical review of data on field-scale dispersion in aquifers." Water Resour. Res. 28(7), 1955-1974.

Gelhar, L.W., Welty, C., and Rehfeldt, K.R. (1993). "Reply to 'Comment on "A critical review of data on field-scale dispersion in aquifers",' by S.P. Neuman." Water Resour. Res., 29(6), 1867-1869.

Gerritse, J., Renard, V., Pedro Gomes, T.M., Lawson, P.A., Collins, M.D., and Gottschal, J.C. (1996). "*Desulfitobacterium* sp. strain PCE1, an anaerobic bacterium that can grow by reductive dechlorination of tetrachloroethene or *ortho*-chlorinated phenols." Arch. Microbiol., 165, 132-140.

Gerritse, J., Drzyzga, O., Kloetstra, G., Keijmel, M., Wiersum, L.P. Hutson, R., Collins, M.D., and Gottschal, J.C. (1999). "Influence of different electron donors and acceptors on dehalorespiration of tetrachloroethene by *Desulfitobacterium frappieri* TCE1." Appl. Environ. Microbiol., 65, 5212-5221.

Gibson, S.A., Roberson, D.S., Russell, H.H., and Sewell, G.W. (1994). "Effects of three concentrations of mixed fatty-acids on dechlorination of tetrachloroethene in aquifer microcosms." Environ. Toxicol. Chem., 13, 453-460.

Gillham, R.W., and Cherry, J.A. (1982). "Contaminant migration in saturated unconsolidated geologic deposits." Recent trends in hydrogeology, T.N. Narasimhan, ed., Special Paper 189, The Geological Society of America, Boulder, CO, 31-62.

Glotz, M.N., and Roberts, P.V. (1986). "Interpreting organic solute transport data from a field experiment using physical nonequilibrium models." J. Contam. Hydrol., 1(1/2), 77-93.

Goblet, P. (1989). "Programme METIS: Simulation d'ecoulement et de transport miscible en milieu poreux et fracture" Technical Report LHM /RD /89/23, CIG, Ecole des Mines de Paris, Fontainebleau, France.

Grady, C.P.L., Jr. (1985). "Biodegradation: Its measurement and microbiological basis." Biotechnol. Bioengin., 27, 660-674.

Hamaker, J.W., and Thompson, J.M. (1972). "Adsorption." Organic chemicals in the soil environment, Vol. 1, C.A.I. Goring and J.W. Hamaker, eds. Marcel Dekker, Inc., New York, NY, 49-143.

Hamamura, N., Page, C., Long, T., Semprini, L. and Arp, D.J. (1997) "Chloroform cometabolism by butane-grown CF8, *Pseudomonas butanovora*, and *Mycobacterium vaccae* JOB5 and methane-grown *Methylosinus trichoporium* OB3b." Appl. Environ. Microbiol., 63, 3607-3613.

Haner, A., Hohener, P., and Zeyer, J. (1997). "Degradation of trimethylbenzene isomers by an enrichment culture under N_2O-reducing conditions." Appl. Environ. Microbiol., 63, 1171-1174.

Hardison, L.K., Curry, S.S., Ciuffetti, L.M., and Hyman, M.R. (1997) "Metabolism of diethyl ether and cometabolism of methyl tert-butyl ether by a filamentous fungus, a *Graphium* sp." Appl. Environ. Microbiol., 63, 3059-3067.

Hartmans, S. and Debont, J.A.M. (1992). "Aerobic vinyl-chloride metabolism in *Mycobacterium aurum* L1." Appl. Environ. Microbiol. 58(4): 1220-1226.

Hauschild, I., Schroer, A., Siedersleben, M., and Starnick, J. (1994). "The microbial growth of *Mycobacterium aurum* L1 on vinyl chloride with respect to inhibitory and limiting influence of substrate and oxygen." Wat. Sci. Technol. 30(9): 125-132.

Hayduk, W., and Laudie, H. (1974). "Prediction of diffusion coefficients for nonelectrolytes in dilute aqueous solutions." AIChE J., 20(3), 611-615.

Henry, S.M. and Grbic-Galic, D. (1991) "Influence of endogenous and exogenous electron donors and trichloroethylene oxidation toxicity on trichloroethylene oxidation by methanotrophic cultures from a groundwater aquifer." Appl. Environ. Microbiol. 57: 236-244.

Holliger, C., Schraa, G., Stams, A.J.M., and Zehnder, A.J.B. (1993). "A highly purified enrichment culture couples the reductive dechlorination of tetrachloroethene to growth." Appl. Environ. Microbiol., 59, 2991-2997.

Hopkins, G.D., Munakata, J., Semprini, L., and McCarty, P.L. (1993a). "Trichloroethylene concentration effects on pilot field-scale in situ groundwater bioremediation by phenol-oxidizing microorganisms." Environ. Sci. Technol., 27, 2542-2547.

Hopkins, G.D., Semprini, L., and McCarty, P.L. (1993b). "Microcosm and in situ field studies of enhanced biotransformation of trichloroethylene by phenol-utilizing microorganisms." Appl. Environ. Microbiol., 59, 2277-2285.

Hopkins, G.D. and McCarty, P.L. (1995). "Field evaluation of *in situ* aerobic cometabolism of trichloroethylene and three DCE isomers using phenol and toluene as primary substrates." Environ. Sci. Technol., 29, 1628-1637.

Horvath, R.S. (1972). "Microbial co-metabolism and the degradation of organic compounds in nature." Bacteriol. Rev., 36, 146-155.

Hunkeler, D., Jorger, D., Haberli, K., Hohener, P., and Zeyer, J. (1998). "Petroleum hydrocarbon mineralization in anaerobic laboratory aquifer columns." J. Contam. Hydrol., 32, 41-61.

Hur, H-G., Newman, L.M., Wackett, L.P., and Sadowsky, M.J. (1997) "Toluene 2-monooxygenase-dependent growth of *Burkholderia cepacia* G4/PR1 on diethyl ether." Appl. Environ. Microbiol., 63, 1606-1609.

Hutchins, S.R., Sewell, G.W., Kovacs, D.A., and Smith, G.A. (1991). "Biodegradation of aromatic hydrocarbons by aquifer microorganisms under denitrifying conditions." Environ. Sci. Technol., 25, 68-76.

Janssen, D.B. and Witholt, B. (1992) "Aerobic and anaerobic degradation of halogenated aliphatics, *Metal Ions in Biological Systems, Vol. 28*, H. Sigel and A. Sigel, eds., Marcel Dekker, New York, NY, 299-327.

Johnson, R.L., Palmer, C.D., and Fish, W. (1989). "Subsurface Chemical processes." Transport and fate of contaminants in the subsurface, United States Environmental Protection Agency, Center for Environmental Research Information, Robert S. Kerr Environmental Research Laboratory, EPA/625/4-89/019, Washington, D.C., 41-56.

Karickhoff, S.W., and Morris, K.R. (1985). "Sorption dynamics of hydrophobic pollutants in sediment suspensions." Environ. Toxicol. Chem., 4(4), 469-479.

Kavanaugh, M.C., and Trussell, R.R. (1980). "Design of aeration towers to strip volatile contaminants from drinking water." Jour. AWWA, 72(12), 684-692.

Kazumi, J., Caldwell, M.E., Suflita, J.M., Lovley, D.R., and Young, L.Y. (1997). "Anaerobic degradation of benzene in diverse anoxic environments." Environ. Sci. Technol., 31, 813-818.

Kearl, P.M., Korte, N.E., Gleason, T.A., and Beale, J.S. (1991). "Vapor extraction experiments with laboratory soil columns: Implications for field programs." Waste Manage., 11(4), 231-239.

Kim, Y., Semprini, L., and Arp, D.J. (1997). "Aerobic cometabolism of chloroform and 1,1,1-trichloroethane by butane-grown microorganisms." Biorem. J., 1, 135-148.

Kitanidis, P.K. (1994). "The concept of the dilution index." Water Resour. Res., 30(7), 2011-2026.

Klotz, D., and Moser, H. (1974). "Hydrodynamic dispersion as aquifer characteristic: Model experiments with radioactive tracers." Isotope techniques in groundwater

hydrology, Vol. 12, International Atomic Energy Agency, Vienna, Austria, 341-354.
Krainov, S.R.. (1999). "Geochemical models for prediction of groundwater quality: A review of potentialities and restrictions" Water Resources, 26 (3), 286-296.
Krone, U.E., Thauer, R.K., and Hogenkamp, H.P.C. (1989a). "Reductive dehalogenation of chlorinated C_1-hydrocarbons mediated by corrinoids." Biochemistry, 28, 4908-4914.
Krone, U.E., Laufer, K., Thauer, R.K., and Hogenkamp, H.P.C. (1989b). "Coenzyme F_{430} as a possible catalyst for the reductive dehalogenation of chlorinated C_1 hydrocarbons in methanogenic bacteria." Biochemistry, 28, 10061-10065.
Lagenhoff, A.A.M., Zehnder, A.J.B., and Schraa, G. (1996). "Behavior of toluene, benzene, and naphthalene under anaerobic conditions in sediment columns." Biodegradation, 7, 267-274.
LaGrega, M.D., Buckingham, P.L., and Evans, J.C. (1994). Hazardous waste management, McGraw-Hill, Inc., New York, NY.
Langworthy, D.E., Stapleton, R.D., Sayler, G.S., and Findlay, R.H. (1998). "Genotypic and phenotypic responses of a riverine microbial community to polycyclic aromatic hydrocarbon contamination." Appl. Environ. Microbiol., 64, 3422-3428.
Lapidus, L., and Amundson, N.R. (1957). "Mathematics of adsorption in beds. VI. The effect of longitudinal diffusion in ion exchange and chromatographic columns." J. Phys. Chem., 56(8), 984-988.
Larsen, T., Christensen, T.H., and Brusseau, M. (1992). "Predicting nonequilibrium transport of naphthalene through aquifer materials using batch determined sorption parameters." Chemosphere, 24(2), 141-153.
Li, Y.-H., Gregory, S. (1974). "Diffusion of ions in sea water and in deep-sea sediments." Geochim. Cosmochim. Acta, 38(5), 703-714.
Little, C.D., Palumbo, A.V., Herbes, S.E., Lidstrom, M.E., Tyndall, R.L., and Gilmer, P.J. (1988). "Trichloroethylene biodegradation by a methane-oxidizing bacterium." Appl. Environ. Microbiol., 54, 951-956.
Lovley, D.R. and Goodwin, S. (1988). "Hydrogen concentrations as an indicator of the predominant terminal electron-accepting reactions in aquatic sediments." Geochim. Cosmochim. Acta, 52, 2993-3003.
Lovley, D.R., Woodward, J.C., and Chapelle, F.H. (1994). "Stimulated anoxic biodegradation of aromatic hydrocarbons using Fe(III) ligands." Nature, 370, 128-131.
Lovley, D.R., Coates, J.D., Woodward, J.C., and Phillips, E.J. (1995). "Benzene oxidation coupled to sulfate reduction." Appl. Environ. Microbiol., 61, 953-958.
Lovley, D.R., Woodward, J.C., and Chapelle, F.H. (1996). "Rapid anaerobic benzene oxidation with a variety of chelated Fe(III) forms." Appl. Environ. Microbiol., 62, 288-291.
Lyman, W. J., Reehl, W. F., and Rosenblatt, D. H., (1990). "Handbook of Chemical Property Estimation Methods" American Chemical Society, Washington, DC
Lyman, W.J., Reidy, P.J., and Levy, B. (1992). Mobility and degradation of organic contaminants in subsurface environments, C.K. Smoley, Inc., Chelsea, MI.

Mabey, W.R., Barich, V., and Mill, T., (1980). "Hydrolysis of polychlorinated ethanes" Extended Abstracts, 186th National Meeting of the American Scoiety, Washington, DC.

MacFarlane, D.S., Cherry, J.A., Gillham, R.W., and Sudicky, E.A. (1983). "Migration of contaminants in groundwater at a landfill: A case study, 1. Groundwater flow and plume delineation." J. Hydrol., 63, 1-29.

MacIntyre, W.G., Stauffer, T.B., and Antworth, C.P. (1991). "A comparison of sorption coefficients determined by batch, column, and box methods on a low organic carbon aquifer material." Ground Water, 29(6), 908-913.

MacQuarrie, K.T.B., and Sudicky, E.A. (1990). "Simulation of biodegradable organic contaminants in groundwater. 2. Plume behavior in uniform and random flow fields." Water Resour. Res., 26(2), 223-239.

Malachowsky, K.J., Phelps, T.J., Teboli, A.B., Minnikin, D.E., and White, D.C. (1994). "Aerobic mineralization of trichloroethylene, vinyl chloride, and aromatic compounds by *Rhodococcus* species." Appl. Environ. Microbiol., 60, 542-548.

Mars, A.E., Prins, G.T., Wietzes, P., deKoning, W., and Janssen, D.B. (1998). "Effect of TCE on the competitive behavior of toluene-degrading bacteria." Appl. Environ. Microbiol., 64, 208-215.

McCarthy, K.A., and Johnson, R.L. (1993). "Transport of volatile organic compounds across the capillary fringe." Water Resour. Res., 29(6), 1675-1683.

McCarty, P.L., Reinhard, M., and Rittmann, B.E. (1981). "Trace organics in groundwater." Environ. Sci. Technol., 15(1), 40-51.

McCarty, P.L., Goltz, M.N., Hopkins, G.D., Dolan, M.E., Allan, J.P., Kawakami, B.T., and Carruthers, T.J. (1998). "Full-scale evaluation of *in situ* cometabolic degradation of TCE in groundwater through toluene injection." Environ. Sci. Technol., 32, 88-100.

McMahon, P.B., and Chapelle, F.H. (1991). "Microbial production of organic acids in aquitard sediments and its role in aquifer geochemistry." Nature (London), 349, 233-235.

Means, J.C., Wood, S.G., Hassett, J.J., and Banwart, W.L. (1982). "Sorption of amino- and carboxy-substituted polynuclear aromatic hydrocarbons by sediments and soils." Environ. Sci. Technol., 16(2), 93-98.

Meckenstock, R.U., Annweiler, E., Michaelis, W., Richnow, H.H., and Schink, B. (2000). "Anaerobic naphthalene degradation by a sulfate-reducing enrichment culture." Appl. Environ. Microbiol., 66, 2743-2747.

Mercer, J.W., and Cohen, R.M. (1990). "A review of immiscible fluids in the subsurface: Properties, models, characterization, and remediation." J. Contam. Hydrol. 6, 107-164.

Mercurio, J.W.; Beljin, M.S.; and Maynard, J.B. (1999). "Groundwater models and wellfield management: a case study" Environmental Engineering and Policy, 1 (3),155-164.

Mihelcic, J.R., and Luthy, R.G. (1988a). "Degradation of polycyclic aromatic hydrocarbon compounds under various redox conditions in soil-water systems." Appl. Environ. Microbiol., 54, 1182-1187.

Mihelcic, J.R., and Luthy, R.G. (1988b). "Microbial degradation of acenphthene and naphthalene under denitrification conditions in soil-water systems." Appl. Environ. Microbiol., 54, 1188-1198.

Minsker, B.S., and Shoemaker, C.A. (1998). "Quantifying the effects of uncertainty on optimal groundwater bioremediation policies." Water Resources Research, 34(12), 3615-3625.

Miralles-Wilhelm, F., Gelhar, L.W. and Kapoor, V. (1994) Modeling oxygen-transport limited biodegradation in three dimensionally heterogeneous aquifers, Health and Environmental Sciences Departmental Report No. DR192, American Petroleum Institute.

Mo, K., Lora, C.O., Wanken, A.E., Javanmardian, M., Yang, X., and Kulpa, C.F. (1997) "Biodegradation of methyl-t-butyl ether by pure bacterial cultures." Appl. Microbiol. Biotechnol., 47, 69-72.

Molz, F.J., Widdowson, M.A. and Benefield, L.D. (1986). "Simulation of microbial growth dynamics couple to nutrient and oxygen transport in porous media." Water Resources Research, 22(8), 1207-1216.

Molz, F.J., and Widdowson, M.A. (1988). "Internal inconsistencies in dispersion-dominated models that incorporate chemical and microbial kinetics." Water Resour. Res. 24(4), 615-619.

Morsen, A., and Rehm, H.J. (1987). "Degradation of phenol by a mixed culture of *Pseudomonas putida* and *Cryptococcus elinovii* adsorbed on activated carbon." Applied Microbiol. Biotechnol., 26(3), 283-288.

Moulin, C. (1996). "Logiciel de qualite de l'eau SUBIEF, version 3.1, manuel de l'utilisateur" Technical Report HE-43/95/054/B, Electricite de France EDF, Paris.

Mulkins-Phillips, G.J., and Stewart, J.E. (1974). "Distribution of hydrocarbon-utilizing bacteria in northwestern Atlantic waters and coastal sediments." Can. J. Microbiol., 20, 955-962.

Murphy, E.M., Ginn, T.R., Chilakapati, A., Resch, C.T., Phillips, J.L., Wietsma, T.W., and Spadoni, C.M. (1997). "The influence of physical heterogeneity on microbial degradation and distribution in porous media." Water Resour. Res., 33(5), 1087-1103.

Neuman, S.P. (1990). "Universal scaling of hydraulic conductivities and dispersivities in geologic media." Water Resour. Res., 26(8), 1749-1758.

Newell, C.J., McLeod, R.K. and Gonzales, J.R. (1996) BIOSCREEN: Natural Attenuation Decision Support System User's Manual, Version 1.3 EPA/600/R-96/087. National Risk Management Research Laboratory, Office of Research and Development, US Environmental Protection Agency, Cincinnati.

Odencrantz, J.E. (1992). "Modeling the biodegradation kinetics of dissolved organic contaminants in a heterogeneous two-dimensional aquifer." PhD dissertation, University of Illinois at Urbana-Champaign, IL.

Ogram, A.V., Jessup, R.E., Ou, L.T., and Rao, P.S.C. (1985). "Effects of sorption on biological degradation rates of (2,4-dichlorophenoxy) acetic acid in soils." Appl. Environ. Microbiol., 49(3), 582-587.

Ollar, P., Lucille, P., and Burnol, A. (1997). "Architecture du code couple chimie-transport CHEMTRAP. Exercises d'application et programme de travail" Technical Report HT-45/97/027/B, EDF-LNH, Chatou, France.

Oya, S., and Valocchi, A.J. (1997). "Characterization of traveling waves and analytical estimation of pollutant removal in one-dimensional subsurface bioremediation modeling." Water Resour. Res., 33(5), 1117-1127.

Palmer C.D., and Johnson, R.L. (1989). "Physical processes controlling the transport of contaminants in the aqueous phase." Transport and fate of contaminants in the subsurface, United States Environmental Protection Agency, Center for Environmental Research Information, Robert S. Kerr Environmental Research Laboratory, EPA/625/4-89/019, Washington, D.C., 29-40.

Parkhurst, D.L., Thorstenson, D.C. and Plummer, L.N. (1980). "PHREEQE—A computer program for geochemical calculations" U.S. Geological Survey Water Resources Investigations Report 80-96, p.195. (Revised and reprinted August, 1990). (Available from U.S. Geological Survey, 703/648-6977).

Parkhurst, D.L. (1995). "User's guide to PHREEQC--A computer program for speciation, reaction-path, advective-transport, and inverse geochemical calculations" U.S. Geological Survey Water Resources Investigations Report 95-4227, p.143. (Available from U.S. Geological Survey, 703/648-6977).

Parkhurst, D., Thorstenson, D., and Plummer, L. (1980). "PHREEQE, a computer program for geochemical calculations" Water Resources Investigations. U.S. Geological Survey, pp. 80–96.

Patel, R.N., Hou, C.T., Laskin, A.I., and Felix, A. (1982). "Microbial oxidation of hydrocarbons: Properties of a soluble methane monooxygenase from a facultative methane-utilizing organism, *Methylobacterium* sp. strain CRL-26." Appl. Environ. Microbiol., 44, 1130-1137.

Perkins, T.K., and Johnston, O.C. (1963). "A review of diffusion and dispersion in porous media." Soc. Pet. Eng. Jour., 3(1), 70-84.

Picardal, F.W., Arnold, R.G., Couch, H., Little, A.M., and Smith, M.E. (1993). "Involvement of cytochromes in the anaerobic biotransformation of tetrachloromethane by *Shewanella putrefaciens* 200." Appl. Environ. Microbiol., 59, 3763-3770.

Powell, R.M., Bledsoe, B.E., Curtis, G.P., and Johnson, R.L. (1989). "Interlaboratory methods comparison for the total organic carbon analysis of aquifer materials." Environ. Sci. Technol., 23(10), 1246-1249.

Pruess, K., (1991). "TOUGH2: a general numerical simulator for multiphase fluid and heat flow" Technical Report LBL-29400, Lawrence Berkeley Laboratory, CA, USA

Preuss, A., Fimpel, J., and Diekert, G. (1993). "Anaerobic transformation of 2,4,6-trinitrotoluene (TNT)." Arch. Microbiol., 159, 345-353.

Rabus, R. and Widdel, F. (1995). "Anaerobic degradation of ethylbenzene and other aromatic hydrocarbons by new denitrifying bacteria." Arch. Microbiol., 163, 96-103.

Rabus, R., Wilkes, H., Schramm, A., Harms, G., Behrends, A., Amann, R., and Widdel, F. (1999). "Anaerobic utilization of alkylbenzenes and n-alkanes from crude oil in an enrichment culture of denitrifying bacteria affiliating with the β-subclass of *Proteobacteria*." Environ. Microbiol., 1, 145-157.

Reieke, W. (1984). "Microbial degradation of halogenated aromatic compounds." *Microbial Degradation of Organic Compounds*, D.T. Gibson, ed., Marcel Dekker, Inc., New York, NY, pp. 319-359.

Rifai, H.S., Newell, C.J., Gonzales, J.R., Dendrou, S., Kennedy, L. and Wilson, J.T. (1998). BIOPLUME III: Natural Attenuation Decision Support System User's Manual, Version 1.0 EPA/600/R-98/010. National Risk Management Research Laboratory, Office of Research and Development, US Environmental Protection Agency, Cincinnati.

Rijnaarts, H.H.M., Bachmann, A., Jumelet, J.C., and Zehnder, A.J.B. (1990). "Effect of desorption and intraparticle mass transfer on the aerobic biomineralization of α-hexachlorocyclohexane in a contaminated calcareous soil." Environ. Sci. Technol., 24(9), 1349-1354.

Rittmann, B.E., Seagren, E., Wrenn, B.A., Valocchi, A.J., Ray, C., and Raskin, L. (1994). In Situ Bioremediation, 2nd edition, Noyes Publishers, Inc., Park Ridge, NJ.

Rivett, M.O. (1995). "Soil-gas signatures from volatile chlorinated solvents: Borden field experiments." Ground Water, 33(1), 84-98.

Robinson, J.A. and Tiedje, J.M. (1983). "Nonlinear estimation of Monod growth kinetic parameters from a single substrate depletion curve." Appl. Environ. Microbiol., 45, 1453-1458.

Rogers, R.D., McFarlane, J.C., and Cross, A.J. (1980). "Adsorption and desorption of benzene in two soils and montmorillonite clay." Environ. Sci. Technol., 14(4), 457-460.

Roubal, G. and Atlas, R.M. (1978). "Distribution of hydrocarbon-utilizing microorganisms and hydrocarbon biodegradation potentials in Alaskan continental shelf areas." Appl. Environ. Microbiol., 35, 897-905.

Roy, W.R., Krapac, I.G., Chou, S.F.J., and Griffin, R.A. (1992). "Batch-type procedures for estimating soil adsorption of chemicals." United States Environmental Protection Agency, EPA/530-SW-87-006-F, Washington, D.C.

Rueter, P., Rabus, R., Wilkes, H., Aeckersberg, F., Rainey, F.A., Jannasch, H.W., and Widdel, F. (1994). "Anaerobic oxidation of hydrocarbons in crude oil by new types of sulphate-reducing bacteria." Nature, 372, 455-458.

Sawyer, C.N., and McCarty, P.L. (1978). Chemistry for environmental engineering, 3rd edition, McGraw-Hill Book Company, New York, NY.

Schäfer, W., and Kinzelbach, W. (1992). "Stochastic modeling of in situ bioremediation in heterogeneous aquifers." J. Contam. Hydrol., 10(1), 47-73.

Schackmann, A. and Muller, R. (1991). "Reduction of nitroaromatic compounds by different *Pseudomonas* species under aerobic conditions." Appl. Microbiol. Biotechnol., 34, 809-813.

Schneider, J., Grosser, R., Jayasimhulu, K., Xue, W., and Warshawsky, D. (1996). "Degradation of pyrene, benzo[a]anthracene, and benzo[a]pyrene by *Mycobacterium* sp. strain RJGII-135, isolated from a former coal gasification site." Appl. Environ. Microbiol., 62, 13-19.

Scholz-Murumatsu, H., Neumann, A., Meßmer, M., Moore, E., and Diekert, G. (1995). "Isolation and characterization of *Dehalospirillum multivorans* gen. nov., sp. nov., a tetrachloroethene-utilizing, strictly anaerobic bacterium." Arch. Microbiol., 163, 48-56.

Schwarzenbach, R.P., and Westall, J. (1981). "Transport of nonpolar organic compounds from surface water to groundwater. Laboratory sorption studies." Environ. Sci. Technol., 15(11), 1360-1367.

Schwarzenbach, R.P., Gschwend, P.M., and Imboden, D.M. (1993). Environmental organic chemistry, John Wiley & Sons, Inc., New York, NY.

Schwille, F. (1967). "Petroleum contamination of the subsoil—A hydrological problem." The joint problems of the oil and water industries, P. Hepple, ed., The Institute of Petroleum, London, 23-54.

Schwille, F. (1984). "Migration of organic fluids immiscible with water in the unsaturated zone." Pollutants in porous media, the unsaturated zone between soil surface and groundwater, Ecological Studies, Vol. 47, B. Yaron, G. Dagan, and J. Goldshmid, eds., Springer-Verlag, Berlin, 27-48.

Selifonov, S.A., Chapman, P.J., Akkerman, S.B., Gurst, J.E., Bortiatynski, J.M., Nanny, M.A., and Hutcher, P.G. (1998). "Use of ^{13}C NMR to assess fossil fuel biodegradation: Fate of [1-^{13}C] acenaphthene in creosote polycyclic aromatic compound mixtures degraded by bacteria." Appl. Environ. Microbiol., 64, 1447-1453.

Semprini, L, Roberts, P.V., Hopkins, G.D., and McCarty, P.L. (1990). "A field evaluation of *in-situ* biodegradation of chlorinated ethenes: Part 2, Results of biostimulation and biotransformation experiments." Ground Water, 28, 715-727.

Semprini, L, Hopkins, G.D., Roberts, P.V., Grbic-Galic, D., and McCarty, P.L. (1991). "A field evaluation of *in-situ* biodegradation of chlorinated ethenes: Part 3, Studies of competitive inhibition." Ground Water, 29, 239-250.

Seto, M., Masai, E., Ida, M., Hatta, T., Kimbara, K., Fukuda, M., Yano, K. (1995). "Multiple polychlorinated biphenyl transformation systems in the Gram-positive bacterium *Rhodococcus* sp., strain RHA1." Appl. Environ. Microbiol., 61, 4510-4513.

Shackelford, C.D. (1991). "Laboratory diffusion testing for waste disposal—a review." J. Contam. Hydrol., 7(3), 177-217.

Shackelford, C.D., and Daniel, D.E. (1991). "Diffusion in saturated soil. I: Background." J. Geotech. Engrg., ASCE, 117(3), 467-484

Sleep, B.E., and Sykes, J.F. (1989). "Modeling the transport of volatile organics in variably saturated media." Water Resour. Res., 25(1), 81-92.

Seth, R., MacKay, D., and Muncke, J. (1999). "Estimating the organic carbon partition coefficient and its variability for hydrophobic chemicals." Environ. Sci. Technol., 33, 2390-2394.

Smatlak, C.R., Gossett, J.M., and Zinder, S.H. (1996). "Comparative kinetics of hydrogen utilization for reductive dechlorination of tetrachloroethene and methanogenesis in an anaerobic enrichment culture." Environ. Sci. Technol., 30, 2850-2858.

Spitz, K., and Moreno, J. (1996). A Practical Guide to Groundwater and Solute Transport Modeling. John Wiley & Sons, Inc., New York.

Smith, M.R. (1990). "The biodegradation of aromatic hydrocarbons by bacteria." Biodegradation, 1, 191-206.

Smith, R.L., Harvey, R.W., and LeBlanc, D.R. (1991a). "Importance of closely spaced vertical sampling in delineating chemical and microbiological gradients in groundwater studies." J. Contam. Hydrol., 7(3), 285-300.

Smith, R.L., Howes, B.L., and Duff, J.H. (1991b). "Denitrification in nitrate-contaminated groundwater: Occurrence in steep vertical geochemical gradients." Geochim. Cosmochim. Acta, 55(7), 1815-1825.

Snoeyink, V.L. (1988). "Adsorption of organic compounds." Water quality and treatment: A handbook of community water supplies, 4th Edition, American Water Works Association, McGraw-Hill, Inc., New York, NY, 781-876.

So, C.M. and Young, L.Y. (1999). "Initial reactions in anaerobic alkane degradation by a sulfate reducer, strain AK-01." Appl. Environ. Microbiol., 65, 5532-5540.

Soetaert, K., Herman, P., and Middleburg, J. (1996). "A model of early diagenetic processes from shelf to abyssal depths" Geochim. Cosmochim. Acta, 60,1019–1040.

Steffan, R.J., McClay, K., Vainber, S., Condee, C.W., and Zhang, D.L. (1997). "Biodegradation of the gasoline oxygenates methyl *tert*-butyl ether, ethyl *tert*-butyl ether, and *tert*-amyl methyl ether by propane-oxidizing bacteria." Appl. Environ. Microbiol., 63, 4216-4222.

Stoessell, R., 1988. 258C and 1 atm dissolution experiments for sepiolite and kerolite. Geochim. Cosmochim. Acta 52, 365–374.

Sturman, P.J., Stewart, P.S., Cunningham, A.B., Bouwer, E.J., and Wolfram, J.H. (1995). "Engineering scale-up of in situ bioremediation processes: A review." J. Contam. Hydrol. 19, 171-203.

Sudicky, E.A. (1986). "A natural gradient experiment on solute transport in a sand aquifer: Spatial variability of hydraulic conductivity and its role in the dispersion process. Water Resour. Res., 22(13), 2069-2082.

Sudicky, E.A., Gillham, R.W., and Frind, E.O. (1985). "Experimental investigation of solute transport in stratified porous media. 1. The nonreactive case." Water Resour. Res. 21(7), 1035-1041.

Sutfin, J.A. and Ramey, D. (1997). "*In situ* biological treatment of TCE-impacted soil and groundwater: Demonstration results." Environ. Prog., 16, 287-296.

Szecsody, J.E. and Bales, R.C. (1989). "Sorption kinetics of low-molecular-weight hydrophobic organic compounds on surface-modified silica." J. Contam. Hydrol., 4, 181-203.

Szecsody, J.E., Brockman, F.J., Wood, B.D., Streile, G.P., and Truex, M.J. (1994). "Transport and biodegradation of quinoline in horizontally stratified porous media." J. Contam. Hydrol., 15(4), 277-304.

Tharakan, J.P. and Gordon, J.A. (1999). "Cometabolic biotransformation of trinitrotoluene (TNT) supported by aromatic and non-aromatic cosubstrates." Chemosphere, 38, 1323-1330.

Thierrin, J., Davis, G.B., and Barber, C. (1995). "A ground-water tracer test with deuterated compounds for monitoring *in situ* biodegradation and retardation of aromatic hydrocarbons." Ground Water, 33, 469-475.

Thierrin, J., and Kitanidis, P.K. (1994). "Solute dilution at the Borden and Cape Cod groundwater tracer tests." Water Resour. Res., 30(11), 2883-2890.

Thiez, P.L., and Lemonnier, P. (1990). "An in-situ combustion reservoir simulator with a new representation of chemical reactions" SPE Reservoir Eng., 9, 285–292.

Theis, T.L. and Iyer, R. (1994). "Trace metal chemical reactions in groundwater: Parameterizing coupled chemistry-transport models" In: Metal Speciation and Contamination of Soil, Allen, Huang, Bailey and Bowers eds., Lewis Publishers, Boca Raton, FL, 207-225.

U. S. EPA. (1987). "Groundwater." Office of Research and Development, Center for Environmental Research Information, Robert S. Kerr Environmental Research Laboratory, EPA/625/6-87/016, Ada, OK.

U.S. EPA (1998) Technical Protocol for Evaluating Natural Attenuation of Chlorinated Solvents in Ground. Water EPA/600/R-98/128., US Environmental Protection Agency, Washington, D.C.

van der Lee, J., (1997). "Mechanistic modeling of heavy metal mobility in natural sand" Technical Report LHM/RD/97/02, CIG, Ecole des Mines de Paris, Fontainebleau, France.

van der Lee, J. (1998). "Thermodynamic and mathematical concepts of CHESS" Technical Report LHM/RD/98/39, CIG, Ecole des Mines de Paris, Fontainebleau, France.

van der Lee, J., and De Windt, L. (2001). "Present state and future directions of modeling of geochemistry in hydrogeological systems" J. Contaminant Hydrology, 47, 265–282.

Vogel, T.M., and McCarthy, P.L., (1987). "Abiotic and biotic transformations of 1,1,1-trichloroethane under methanogenic conditions" Environ. Sci. Technol., 21(12), 1208-13.

Vanderloop, S.L., Suidan, M.T., Moteleb, M.A., and Maloney, S.W. (1999). "Biotransformation of 2,4-dinitrotoluene under different electron acceptor conditions." Water Res., 33, 1287-1295.

van Hylckama Vlieg, J.E.T., deKoning, W., and Janssen, D.B. (1997) "Effect of chlorinated ethene conversion on viability and activity of *Methylosinus trichosporium* OB3b." Appl. Environ. Microbiol., 63, 4961-4964.

Verce, M.F., Ulrich, R.L., and Freedman, D.L. (2000). "Characterization of an isolate that uses vinyl chloride as a growth substrate under aerobic conditions." Appl. Environ. Microbiol., 66, 3535-3542.

Vroblesky, D.A., and Chapelle, F.H. (1994). "Temporal and spatial changes of terminal electron-accepting processes in a petroleum hydrocarbon-contaminated aquifer and the significance for contaminant biodegradation." Water Resour. Res., 30(5), 1561-1570.

Wackett, L.P. and Schanke, C.A. (1992). "Mechanisms if reductive dehalogenation by transition metal cofactors found in anaerobic bacteria." *Metal Ions in Biological Systems, Vol. 28*, H. Sigel and A. Sigel, eds., Marcel Dekker, Inc., New York, NY, 329-356.

Wade, R.S. and Castro, C.E. (1973). "Oxidation of heme proteins by alkyl halides." J. Am. Chem. Soc., 95, 231-234.

Watkinson, R.J. and Morgan, P. (1990). "Physiology of aliphatic hydrocarbon-degrading microorganisms." Biodegradation, 1, 79-92.
Weber, J.B., and Coble, H.D. (1968). "Microbial decomposition of diquat adsorbed on montmorillonite and kaolinite clays." J. Agr. Food Chem., 16(3),475-478.
Weber, W.J., Jr. (1972). Physicochemical processes for water quality control, John Wiley & Sons, New York, NY.
Weber, W.J. Jr., and Smith, E.H. (1987). "Simulation and design models for adsorption processes." Environ. Sci. Technol., 21, 1040-1050.
Weber, W.J., Jr., and DiGiano, F.A. (1996). Process dynamics in environmental systems, John Wiley & Sons, Inc., New York, NY.
Weber, W.J., Jr., McGinley, P.M., and Katz, L.E. (1991). "Sorption phenomena in subsurface systems: Concepts, models and effects on contaminant fate and transport." Wat. Res., 25(5), 499-528.
Weber, W.J., Jr., McGinley, P.M., and Katz, L.E. (1992). "A distributed reactivity model for sorption by soils and sediments. 1. Conceptual basis and equilibrium assessments." Environ. Sci. Technol., 26(10), 1955-1962.
Westall, J., (1986). "MINEQL: a computer program for the calculation of the chemical equilibrium composition of aqueous systems" Technical Report 86-01, Department of Chemistry, Oregon State University.
Weiner, J.M. and Lovley, D.R. (1998a). "Anaerobic benzene degradation in petroleum-contaminated aquifer sediments after inoculation with a benzene-oxidizing enrichment." Appl. Environ. Microbiol., 64, 775-778.
Weiner, J.M. and Lovley, D.R. (1998b). "Rapid benzene degradation in methanogenic sediments from a petroleum-contaminated aquifer." Appl. Environ. Microbiol., 64, 1937-1939.
Wheeler, M.F. and Dawson, C.N. (1988) An operator-splitting method for advection diffusion reaction problems, MAFELAP Proceedings VI, J.R. Whiteman ed., 463-482.
Wiedemeier, T. H., Wilson, J. T., Kampbell, D. H., Miller, R. N., and Hansen, J. E. (1995). "Technical protocol for implementing intrinsic remediation with long-term monitoring for natural attenuation of fuel contamination dissolved in groundwater." Volume I, Air Force Center for Environmental Excellence, Technology Transfer Division, Brooks AFB, San Antonio, TX.
Wiedemeier, T. H., Swanson, M. A., Moutoux, D. E., Gordon, E. K., Wilson, J. T., Wilson, B. H., Kampbell, D. H., Hansen, J. E., Haas, P., and Chapelle, F. H. (1996). "Technical protocol for evaluating natural attenuation of chlorinated solvents in groundwater." Air Force Center for Environmental Excellence, Technology Transfer Division, Brooks AFB, San Antonio, TX.
Wiedemeier, T.H., Rifai, H.S., Newell, C.J., and Wilson, J.T. (1999). *Natural Attenuation of Fuels and Chlorinated Solvents in the Subsurface*, John Wiley and Sons, Inc., New York, NY.
Wiedemeier, T.H., Swanson, M.A., Wilson, J.T., Kampbell, D.H., Miller, R.N., and Hansen, J.E. (1996). "Approximation of biodegradation rate constants for monoaromatic hydrocarbons (BTEX) in ground water." Ground Water Monitoring and Remediation, 16(3): 186-194.

Wild, A., Hermann, R., and Leisinger, T. (1996). "Isolation of an anaerobic bacterium which reductively dechlorinates tetrachloroethene and trichloroethene." Biodegradation, 7, 507-511.
Wilkins, M.D., Abriola, L.M., and Pennell, K.D. (1995). "An experimental investigation of rate-limited nonaqueous phase liquid volatilization in unsaturated porous media: Steady state mass transfer." Water Resour. Res., 31(9), 2159-2172.
Wilson, B.H., Smith, G.B., and Rees, J.F. (1986). "Biotransformations of selected alkylbenzenes and halogenated aliphatic hydrocarbons in methanogenic aquifer material." Environ. Sci. Technol., 20, 992-1002.
Wilson, J.L., and Miller, P.J. (1978). "Two-dimensional plume in uniform groundwater flow." Journal of the Hydraulics Division, ASCE, (104) HY4, 503-512.
Wilson, J.T., Pfeffer, F.M., Weaver, J.W. and Kampbell, D.H., (1994). "Intrinsic bioremediation of JP4 jet fuel. Symposium on Intrinsic Bioremediation of Ground Water", EPA/540/R-94/515, US EPA, Washington, D.C: 60-72.
Witherspoon, P.A., and Bonoli, L. (1969). "Correlation of diffusion coefficients for paraffin, aromatic, and cycloparaffin hydrocarbons in water." Ind. Eng. Chem., 8(3), 589-591.
Wood, B.D., Dawson, C.N., Szecsody, J.E., and Streile, G.P. (1994). "Modeling contaminant transport and biodegradation in a layered porous media system." Water Resour. Res., 30(6), 1833-1845.
Wood, J.M., Kennedy, F.S., and Wolfe, R.S. (1968). "The reaction of multihalogenated hydrocarbons with free and bound reduced vitamin B_{12}." Biochemistry, 7, 1707-1713.
Wood, B.D., Dawson, C.N., Szecsody, J.E. and Streile, G.P., (1994). "Modeling contaminant transport and biodegradation in a layered porous system." Water Resources Research, (30) 6, 1833-1845.
Wrenn, B.A. and Rittmann, B.E. (1996). "Evaluation of a model for the effects of substrate interactions on the kinetics of reductive dehalogenation." Biodegradation, 7, 49-64.
Wu, S.-C. and Gschwend, P.M. (1986). "Sorption kinetics of hydrophobic organic compounds to natural sediments and soils." Environ. Sci. Technol., 20(7), 717-725.
Xu, M., and Eckstein, Y. (1995). "Use of weighted least-squares method in evaluation of the relationship between dispersivity and field scale." Ground Water, 33(6), 905-908.
Yang, X., Erickson, L.E., and Fan, L.T. (1994). "Dispersive-convective characteristics in the biorestoration of contaminated soil with a heterogeneous formation." J. Hazard. Mater., 38(1), 163-185.
Young, D.F., and Ball, W.P. (1994). "A priori simulation of tetrachloroethene transport through aquifer material using an intraparticle diffusion model." Environ. Progress, 13(1), 9-20.
Zachara, J.M., Ainsworth, C.C., Cowan, C.E., and Schmidt, R.L. (1990). "Sorption of aminonaphthalene and quinoline on amorphous silica." Environ. Sci. Technol., 24(1), 118-126.

Zehnder, A.J.B. and Stumm, W. (1988). "Geochemistry and biogeochemistry of anaerobic habitats." *Biology of Anaerobic Microorganisms*, A.J.B. Zehnder, ed., John Wiley and Sons, Inc., 1-38.

Zheng, C. (1990). MT3D, A Modular Three-dimensional Transport Model for Simulation of Advection, Dispersion, and Chemical Reactions of Contaminants in Groundwater Systems, Report to the Kerr Environmental Research Laboratory, US Environmental Protection Agency, Ada, OK.

Zheng, C. and Wang, P.P. (1999). MT3DMS: A Modular Three-dimensional Multispecies Model for Simulation of Advection, Dispersion and Chemical Reactions of Contaminants in Groundwater Systems; Documentation and User's Guide, Contract Report SERDP-99-1, U.S. Army Engineer Research and Development Center, Vicksburg, MS.

CHAPTER 4

Long-Term Monitoring

4.1 Introduction

Long-term performance monitoring (LTPM) is an important element in evaluating the effectiveness of Natural Attenuation (NA) to ensure protection of human health and the environment. LTPM is more important in NA than in other remediation methods because of potentially longer remediation time required, potential for ongoing contaminant migration, and other uncertainties associated with natural attenuation. The LTPM generally requires locating monitoring wells and developing a sampling and analysis plan. The LTPM plan is used to monitor migration of groundwater plume over time, and to determine the extent of natural attenuation that is occurring at rates adequate to protect the potential down-gradient receptors. The LTPM plan should be developed based on data collected during site characterization, the analysis of solute-fate and transport modeling, and the analysis of exposure-pathway results. The LTPM program developed for each site should specify the location and depth of monitoring wells, frequency and type of samples, and measurements necessary to evaluate the performance of natural attenuation that is expected to achieve the intended remediation objectives.

The United States Environmental Protection Agency final policy directive (US EPA, 1999) suggests that all monitoring programs should be designed to accomplish the following:
- demonstrate that natural attenuation is occurring according to expectations;
- detect changes in hydrologic, geochemical, microbiological, and other changes that may reduce the efficiency of any of the natural attenuation processes;
- identify any potentially toxic or other transformation products;
- verify that groundwater plumes are not expanding, either down-gradient, and or laterally or vertically;
- verify no unacceptable impact to down-gradient receptors;
- detect new releases of contaminants to the environment that could impact the effectiveness of the natural attenuation remedy;
- demonstrate the efficiency of institutional controls (state, federal, etc.) that were put in place to protect potential receptors; and
- verify achievement of remediation objectives.

The LTPM should continue until the NA objectives are achieved, and should be continued longer, if necessary, such that the site poses no threat to human health or

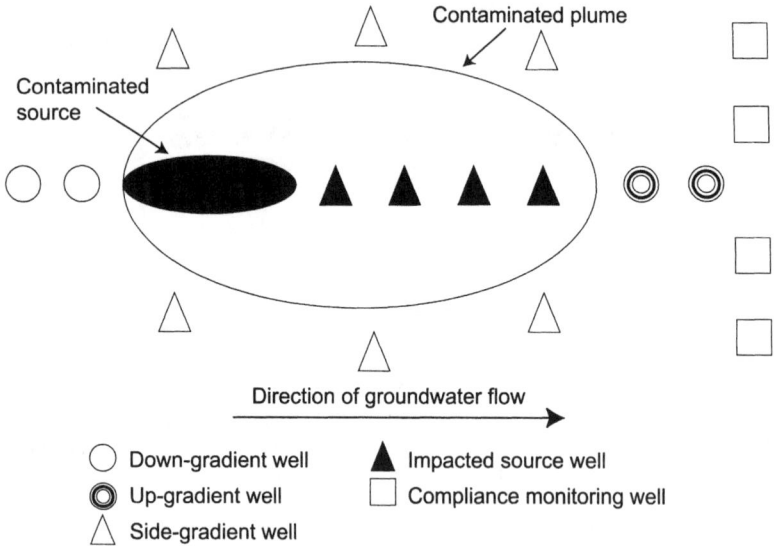

Figure 4-1. Long-term monitoring for natural attenuation

the environment. Generally, the institutional (state and federal) guidance or negotiations establish the performance monitoring requirements.

Further information on long-term performance monitoring can be found in US EPA and Department of Defense publications (US EPA, 1989; US EPA, 1992; US EPA, 1994a; US EPA, 1994b; US EPA, 1997; US EPA, 1998; and Wiedmeier et al, 1995).

4.2 Monitoring Wells

Developing long-term monitoring include: siting monitoring wells to evaluate plume migration over time and to verify that natural attenuation is occurring to protect the health and environment of the down-gradient receptors. Design of monitoring wells should be based on site survey data, groundwater modeling data, and present and future exposure pathways analysis. Geochemical and hydrogeological data should also be taken into consideration in siting monitoring wells. Geochemical parameters such as: dissolved oxygen, pH, oxidation-reduction potential, sulfate, nitrate, methane, and Fe(II) are used, in conjunction with contaminant data, to assess the natural attenuation of given site. Hydrogeological data, such as seepage velocity of groundwater is used in determining the frequency of sampling.

There are four types of monitoring wells: up-gradient wells, side-gradient wells, impacted source wells, and down-gradient wells (Figure 4-1). Up-gradient wells, impacted source wells, and down-gradient wells are used to measure performance. Side-gradient wells, and down-gradient wells located most downstream from the plume are used for compliance monitoring to ensure that the plume is not migrating past the pre-established boundaries. Figure 4-1 shows: a) up-gradient wells, located in uncontaminated groundwater area; b) impacted source wells, located in the contaminated groundwater area; c) side-gradient wells, located along the periphery of the plume; and d) down-gradient wells, located just past and most further from the plume. In certain sites, multi-level monitoring wells may be needed to correctly assess the extent and changes in the plume throughout the aquifer.

The final monitoring plan for a natural attenuation site should be based on: aquifer geological and stratigraphic characteristics; type of contaminant; distance to potential receptors; groundwater velocity; and the impacts of natural attenuation. The most important factors to consider are distance to potential receptors and the groundwater seepage velocity. These two factors determine the well spacing and frequency of sampling.

4.3 Sampling and Analysis

The long-term performance monitoring (LTPM) plan also includes sampling and analysis strategy. Frequency of sampling should be such that it yields useful data to predict plume migration and also provides contaminant concentrations over time to measure the performance of natural attenuation. At some sites, seasonal variations in groundwater flow and level may require more frequent sampling.

The sample analytical plan for a natural attenuation site depends mainly on the type of contaminant. Contamination with chlorinated solvents requires analysis of different geochemical parameters than needed for fuel hydrocarbons because different biodegradation mechanisms or pathways are involved. In case of metals, the sorption seems to be the major removal pathway.

Monitoring degradation of fuel hydrocarbons in natural attenuation is based on the following observations:
- disappearance of contaminant compounds in down-gradient wells;
- loss of electron acceptors, such as dissolved oxygen, nitrate, and sulfate in the impacted source wells;
- biodegradation products, Fe(II), hydrogen sulfide, carbon dioxide and methane in the impacted source wells and down-gradient wells;
- monitoring of specific microorganisms responsible for biodegradation of compounds.

Monitoring degradation of chlorinated solvents in natural attenuation requires the following observations:
- reduction of contaminant concentration in the down-gradient wells;
- depletion of electron donors and acceptors, and accumulation of metabolic by products in the down-gradient wells; and
- additional microbiological data to support biodegradation.

For metals, sorption to iron hydroxides, carbonate minerals and organic matter, and formation of insoluble sulfides are the major processes of remediation.

The geochemical parameters that should be sampled and analyzed during biodegradation of fuel hydrocarbons and chlorinated solvents are: volatile organic compounds, total organic carbon, dissolved organic carbon, carbon dioxide, methane, volatile fatty acids, PAHs, vinyl chloride, TCE, PCE, dissolved oxygen, nitrate, sulfate, iron (II), chloride, hydrogen, alkalinity, oxidation reduction potential, conductivity, pH, temperature, and other necessary byproducts and parameters related to source contaminant. For metals, appropriate metal byproducts should be analyzed. The final selection of geochemical parameters should be determined for each site in conjuction with the appropriate agencies (state and federal). Selection may also depend on the availability of historical contaminant data, the site complexity, and the types of contaminant present. Table 4-1 lists the purpose of geochemical parameters analysis for natural attenuation in soil and groundwater. Table 4-2 shows the geochemical parameters that need to be analyzed for long-term monitoring in natural attenuation.

Sampling frequency should be determined based on placement of monitoring wells and the groundwater plume velocity. The frequency of sampling should be adequate to detect, in a timely manner, the potential changes in plume degradation or migration. At minimum, the sampling program should be sufficient to determine the rates of attenuation and how the rate is changing over time. The sampling program should be flexible. Sampling frequency could be decreased when it is determined that natural attenuation is progressing as expected and very little change is noticed from one sample to the next. The sampling frequency could be increased if plume migration is observed.

Generally, the site sampling and analysis should be continued until the natural attenuation objective is achieved, and longer if necessary, to prove that the site is no longer threat to human health and the environment. Typically, the monitoring is continued for two to three years, after natural attenuation objectives have been met, to ensure that concentration levels are stable and remain below the targeted levels.

4.4 Contingency Plan

A contingency plan is a back-up plan, in the event of plume behavior change over time. Therefore, a contingency plan should be an integral part of the long-term monitoring plan. Contingency plans are implemented when a plume migrates further or faster than predicted, to protect the health and the environment of receptors. Generally, contingency plans involve some form of engineered remediation. The following criteria are suggested (US EPA, 1999):
- contaminant concentrations in soil or groundwater at specified locations exhibit an increasing trend not originally predicted during remedy selection;
- near-source wells exhibit large concentration increases indicative of a new or renewed release;
- contaminants are identified in monitoring wells located outside of the original plume boundary;
- contaminant concentrations are not decreasing at a sufficiently rapid rate to meet the remediation objectives; or
- changes in land or groundwater use will adversely affect the protectiveness of the natural attenuation remedy.

Table 4-1. Purpose of geochemical parameters analysis for natural attenuation in soil and groundwater

Geochemical parameter	Media	
	Water	Soil
Dissolved oxygen	Determines metabolic pathway and bioactivity	% oxygen is useful to determine bioactivity in the vadose zone
Iron (III)		Predicts iron reduction
Volatile organic compounds	Determines extent of groundwater contamination	Determines extent of soil contamination
Total organic carbon	Indicates contaminant migration and biodegradation	Indicates contaminant migration and biodegradation
Carbon dioxide		Determines bioactivity in the vadose zone
Nitrate	Indicates respiration in the absence of oxygen	Electron acceptor for organic compounds oxidation under some conditions
Iron (II)	Indicates anaerobic degradation	Electron donor
PAHs	Fuel components	Fuel components
Chloride	Chlorinated solvent reduction product	Chlorinated solvent reduction product
Sulfate	Indicates anaerobic microbial respiration	Electron acceptor for organic compounds oxidation under some conditions
Oxidation reduction potential	Indicates nature of degradation	Indicates nature of degradation
Alkalinity	Indicates buffering capacity	Indicates buffering capacity
Methane	Methanogenesis	—
pH	Metabolic processes are pH sensitive	Important condition for some metabolic processes to occur
Temperature	Useful in indicating stabilization of low flow monitoring	Selection of particular microbial species and degradation rate
Conductivity	Water quality parameter	—

Table 4-2. Analysis of geochemical parameters for long-term monitoring in natural attenuation

Well location	Sampling purpose	Geochemical parameters Initial samples	Subsequent samples
Up-gradient well	Background water quality	Contaminants, byproducts, and full suite of geochemical parameters	Pertinent geochemical parameters
Side-gradient well	Background water quality	Contaminants, byproducts, and full suite of geochemical parameters	Pertinent geochemical parameters
Source impact well	Changing source strength	Contaminants, byproducts, and full suite of geochemical parameters	Contaminants, byproducts, and pertinent geochemical parameters
Down-gradient source impact well	Contaminant/plume behavior over time	Contaminants, byproducts, and full suite of geochemical parameters	Contaminants and byproducts.
Down-gradient well just past the plume	Detect plume migration	Contaminants, byproducts, and full suite of geochemical parameters	Contaminants and byproducts.
Most down-gradient well	Detect plume migration	Contaminants, byproducts, and full suite of geochemical parameters	Contaminants and byproducts.
Farthest down-gradient well	Compliance monitoring	Contaminants, byproducts, and full suite of geochemical parameters	Contaminants and byproducts.

Some of the back-up remediation systems suggested for natural attenuation sites are: placing oxygen releasing materials in the plume; addition of chemical oxidation such as permanganate; ozone and hydrogen peroxide; phytoremediation; bioslurping; air sparging if light nonaqueous phase liquids are found to be present; pump and treat; and in-well circulation or stripping.

References

U.S. EPA, (1999). Use of monitored natural attenuation at superfund, RCRA corrective action, and underground storage tank sites, Directive 9200.4-17P,

Office of Solid Waste and Emergency Response, U.S. Environmental Protection Agency, Washington, D.C..

U.S. EPA, (1998). Monitored natural attenuation for groundwater: EPA/625/K-98/001, U.S. Environmental Protection Agency, Washington, D.C.

U.S. EPA, (1997). Proceedings of the symposium on natural attenuation of chlorinated organics in groundwater: EPA/540/R-97/504, U.S. Environmental Protection Agency, Washington, D.C.

U.S. EPA, (1994a). Proceedings of the symposium on natural attenuation of groundwater: EPA/600/R-94/162, U.S. Environmental Protection Agency, Washington, D.C.

U.S. EPA, (1994b). Methods for monitoring pump-and-treat performance: EPA/600/R-94/123, U.S. Environmental Protection Agency, Washington, D.C.

U.S. EPA, (1992). Methods for evaluating attainment of cleanup standards, Volume 2, Groundwater: EPA/230-R-92-014, U.S. Environmental Protection Agency, Washington, D.C.

U.S. EPA, (1989). Methods for evaluating attainment of cleanup standards, Volume 1, Soils and solid media: EPA/230/02-89-042, U.S. Environmental Protection Agency, Washington, D.C.

Wiedmeier, T. H., Wilson, J.T., Kampbell, D.H., Miller, R.N., and Hansen, J.E., (1995). Technical protocol for implementing intrinsic remediation with long-term monitoring for natural attenuation of fuel contamination dissolved in groundwater, U.S. Air Force Center for Environmental Excellence, Technology Transfer Division, Brooks Air Force Base, San Antonio, Texas.

CHAPTER 5

Natural Attenuation in Soils, Sediments and Groundwater

Natural attenuation is being considered as a remedy, alone or as part of a cleanup strategy, for an increasing variety of contaminants; however, the current level of understanding of the processes involved and field evidence of success are variable (NRC, 2000). The goal of this chapter is to review the current application of natural attenuation to several of the different contaminant classes of interest, including: Petroleum Hydrocarbons and MTBE (Section 5.1); Chlorinated Solvents (Section 5.2); Polycyclic Aromatic Hydrocarbons (Section 5.3); Metals (Section 5.4); and Radioactive Contaminants (Section 5.5).

5.1   Petroleum Hydrocarbons and MTBE

5.1.1   Background

Several developments over the last three decades have contributed to the evolution of natural attenuation as a remedial approach for petroleum hydrocarbon releases. By the 1970s, national attention had been drawn to a significant groundwater pollution problem in the United States, e.g., (Council on Environmental Quality, 1981), including petroleum hydrocarbon contamination resulting from leaking underground storage tanks and transfer lines, e.g., (Williams and Wilder, 1971). At about the same time, some researchers recognized that microorganisms present in the subsurface could be utilized to cleanup aquifers contaminated by petroleum hydrocarbons, (e.g., API, 1972; Jamison et al., 1975; Jamison et al., 1976; McKee et al., 1972; McNabb and Dunlap, 1975; Williams and Wilder, 1971). During the 1980s, these developments were the impetus for intense studies of potential microbial activity in the subsurface and its impact on groundwater quality, (e.g., see reviews by Chapelle, 1999; Ghiorse and Wilson, 1988), including studies focused on the biodegradation of petroleum hydrocarbons, (e.g., Barker et al., 1987; Ehrlich et al., 1985; Wilson et al., 1983a; Wilson et al., 1983b).

The 1980s were also a period during which a significant effort was expended to assess and monitor the contamination of groundwater by petroleum hydrocarbon releases in the United States and Europe (Chapelle, 1999). Numerous case studies of actual field sites and field research studies indicated that contaminant plumes resulting from releases of petroleum hydrocarbons in groundwater migrated more slowly than expected, reached a steady state, or decreased in extent and concentration over time. Plume migration was inhibited due to naturally occurring attenuation mechanisms, especially biodegradation, but also by dilution, dispersion, and sorption, (e.g., Baedecker et al., 1993; Barbaro et al., 1992; Barker et al., 1987; Chiang et al., 1989; Kemblowski et al., 1987; McAllister and Chiang, 1994; Wiedemeier et al., 1996b).

Currently, these natural attenuation processes are used as a remedial approach in the managed strategy of intrinsic remediation or monitored natural attenuation at petroleum release sites. Several factors contributed significantly to the acceptance of monitored natural attenuation as an appropriate "technology" for treating hydrocarbon-contaminated groundwater. A key factor was a lack of efficient, cost-effective engineered remedial technologies for cleanup of the large number of contaminated sites (Wiedemeier et al., 1999). Also important was the seminal report, "In situ Bioremediation: When does it work?," (NRC, 1993), which clearly distinguished intrinsic bioremediation from engineered bioremediation as an approach that, "manages the innate capabilities of naturally occurring microbial communities to degrade environmental pollutants without taking any engineering steps to enhance the process." Another factor that has contributed to the increased acceptance and use of natural attenuation as a treatment strategy for benzene, toluene, ethylbenzene, and xylenes (BTEX) contaminated groundwater was the development and utilization of risk-based corrective action (RBCA) programs as a means for categorizing and prioritizing leaking underground storage tank (LUST) site cleanups, in the face of large numbers of contaminated sites and limited cleanup funds (Small, 1998). For example, as part of RBCA, estimates of natural attenuation are used to predict the migration of petroleum hydrocarbon releases and to estimate the potential impact to the receptors of concern.

The goal of this section is to review the state of the art of remediation of petroleum hydrocarbon releases using natural attenuation. After discussing the contaminants of concern, we review the key attenuation mechanisms, emphasizing groundwater impacted by petroleum products, discuss how to demonstrate natural attenuation of petroleum hydrocarbons, and summarize case studies demonstrating the natural attenuation of petroleum hydrocarbon releases. This review does not address specific technical details of how to implement remediation by natural attenuation for petroleum hydrocarbons, nor does this review cover remedial technologies requiring human intervention other than monitoring.

5.1.2 Petroleum hydrocarbons and additives

The composition of petroleum or crude oil is highly variable and depends on the source of petroleum (Speight, 1991). On a molecular basis, petroleum is a complex mixture containing hydrocarbons plus organic compounds of sulfur, oxygen, and nitrogen and minor fractions of metallic constituents. The hydrocarbon components of petroleum can be divided into five classes: (1) paraffins (alkanes), (2) naphthenes (cycloalkanes), (3) olefins (alkenes), (4) acetylenes (alkynes), and (4) aromatics (monocyclic and polycyclic aromatic hydrocarbons). The relative proportions vary with the type of crude oil, but the alkene and alkyne hydrocarbons are usually not present.

Crude oil is refined into petroleum products, the exact composition of which is very variable and depends on several factors (Speight, 1991; Watts, 1998). This complexity is exemplified by gasoline, which is the primary refinery product. Gasoline is also one of the most frequent causes of groundwater pollution in many parts of the United States (Tangley, 1984), having been released in the majority of incidents involving petroleum products (Hartley and Englande, 1992). Gasoline is a complex mixture of hydrocarbons that boils below 180-200°C (Speight, 1991). The hydrocarbon constituents in this boiling range have molecular structures with 4-12 carbon atoms, and fall into three general categories: alkanes (including cycloalkanes and branched alkanes), alkenes, and aromatics. There are about 500 members of the alkane, alkene, and aromatic series that boil below 200°C, and the majority of these have been found in the gasoline fraction of petroleum. For example, 233 hydrocarbons were identified in an analysis of a premium-grade gasoline (Maynard and Sanders, 1969).

The highly complex composition of gasoline is a function of three variables (Speight 1991): (1) the source of the crude oil from which the gasoline is refined, (2) the processes used in the gasoline production, and (3) the blending of the hydrocarbon streams produced by the various production processes with special-purpose additives. The special-purpose additives can be grouped into eight classes (Hill and Moxey, 1960): (1) anti-knock compounds, (2) dyes, (3) anti-oxidants, (4) metal deactivators, (5) anti-rust agents, (6) anti-stall agents, (7) anti-preignition agents, and (8) upper-cylinder lubricants. Of the additives that have been used, the focus here is on methyl *tert*-butyl ether (MTBE). MTBE was originally added to gasoline as a high octane replacement for organometallic antiknock compounds, such as tetraethyl lead (Lane, 1980). Other high-octane oxygenated enhancers of gasoline include methanol and a number of other alcohols (e.g., *tert*-butyl alcohol (TBA)) and ethers (Speight, 1991).

The final compositions of example leaded and unleaded commercial gasolines are presented in Table 5-1. Unleaded brands of gasoline have ≤0.05 g lead/gal and generally have a higher aromatic hydrocarbon fraction than leaded gasolines (Cline et al., 1991). The differences in the hydrocarbon compositions of gasoline are important because they influence the equilibrium hydrocarbon concentrations in water in contact with the gasoline (Table 5-2).

The Clean Air Act Amendments of 1990 require that "reformulated" fuel, which has a minimum oxygen content of 2 wt. % (equivalent to 11% MTBE by vol.), be used year round or seasonally in some areas of the U.S. to reduce atmospheric concentrations of CO and ozone, and "oxygenated" gasolines containing ≥ 2.7 wt. % of oxygen (equivalent to 15% MTBE or 8% ethanol by volume), may be required in winter months for the most problematic areas (Delzer et al., 1996; Grisham, 1999; Haggin, 1992). This law also restricted the benzene concentration in gasoline to a maximum of 1.0% by volume, and aromatic hydrocarbons to 25% by volume (Squillace et al., 1996). In comparison, gasolines formulated before 1990 contained higher benzene concentrations (ca. 2–5% by volume) (Table 5-1).

Addition of polar organic fuel oxygenates (e.g., methanol, ethanol, MTBE), which are completely or highly soluble in water, to a mixture of petroleum hydrocarbons and water may have a cosolvent effect, (e.g., Heermann and Powers, 1998; Poulsen et al., 1992). This might increase the aqueous phase concentrations of

Table 5-1. Example compositions of leaded and unleaded gasolines

Compound	Unleaded[1]	Leaded[1]
Normal/iso-hydrocarbons	55	59
Isopentane	9–11	9–11
n-butane	4–5	4–5
n-pentane	2.6–2.7	2.6–2.7
Aromatic hydrocarbons	34	26
Xylenes	6–7	6–7
Toluene	6–7	6–7
Ethylbenzene	5	5
Benzene	2–5	2–5
Naphthalene	0.2–0.5	0.2–0.5
Benzo(b)fluoranthene	3.9 mg/L	3.9 mg/L
Anthracene	1.8 mg/L	1.8 mg/L
Olefins	5	10
Cyclic hydrocarbons	5	5
Additives		
Tetraethyllead		600 mg/L
Tetramethyllead		5 mg/L
Dichloroethane		210 mg/L
Dibromoethane		190 mg/L

[1]Approximate percent volume, unless otherwise noted.
Source: from Table 1, p.914, (Cline et al., 1991).

hydrocarbons at petroleum release sites. Nevertheless, studies have shown a minimal cosolvent effect of MTBE in gasoline on hydrocarbon partitioning at the current levels of addition, (e.g., Cline et al., 1991; Poulsen et al., 1992).

Although many hydrocarbons are present in gasoline and other refined petroleum products, this section focuses primarily on the natural attenuation of the monoaromatic BTEX compounds and the additive MTBE, because the BTEX compounds and oxygenated additives are commonly the major constituents of regulatory importance at petroleum release sites (ASTM, 1998; McAllister and Chiang, 1994).

Table 5-2. Gasoline composition and equilibrium aqueous phase concentration of hydrocarbons in 31 gasolines

Compound	Gasoline composition (wt. %)		Aqueous-phase composition (mg/L)	
	Average (min–max)	SD	Average (min–max)	SD
Benzene	1.73 (0.7–3.8)	0.68	42.6 (12.3–130)	18.9
Toluene	9.51 (4.5–21.0)	3.59	69.4 (23–185)	25.4
Ethyl-benzene	1.61 (0.7–2.8)	0.48	3.2 (1.3–5.7)	0.8
m-, p-Xylene	5.95 (3.7–14.5)	2.07	11.4 (2.6–22.9)	3.8
o-Xylene	2.33 (1.1–3.7)	0.72	5.6 (2.6–9.7)	1.8
n-Propyl-benzene	0.57 (0.13–0.85)	0.14	0.4 (0.1–3)	0.1
3-,4-ethyl-toluene	2.20 (1.5–3.2)	0.40	1.7 (0.8–3.8)	0.3
1,2,3-trimethyl-benzene	0.8 (0.6–1.1)	0.12	0.7 (0.2–2)	0.2

Note: SD = standard deviation.
Source: from Table II, p.917, Cline et al., 1991.

5.1.3 Attenuation Mechanisms for BTEX & MTBE

Once petroleum hydrocarbons are released into the environment, the non-aqueous phase liquid (NAPL) source material may adsorbed onto soil, volatilize into the soil gas and/or dissolve into the aqueous phase. Hydrocarbons that have dissolved into the aqueous phase may be subsequently transported with the groundwater and are also subject to several possible natural attenuation mechanisms, including physical phenomena, chemical reactions, and biological processes as summarized in Table 5-3.

5.1.3.1 Physical Mechanisms

There are several hydrologic processes (e.g., dispersion, infiltration/recharge) and interphase partitioning processes (e.g., sorption, volatilization) that may contribute to the natural attenuation of petroleum hydrocarbons. These are all nondestructive attenuation mechanisms that may result in a decrease in contaminant concentration, but do not reduce the total contaminant mass.

5.1.3.1.1 *Diffusion and Dispersion*

Diffusion and dispersion are the only mixing processes for solutes in the deep subsurface and so play a key role in mixing of the contaminant plume with electron acceptors and donors present in the surrounding aquifer. For example, at a JP-4 jet fuel release site, the transport of BTEX and MTBE from the residual light NAPL (LNAPL) source was controlled by dissolution from the fuel spill into the

Table 5-3. Summary of key trends in natural attenuation mechanisms for BTEX and MTBE

Mechanism	Description	Potential for BTEX and MTBE Attenuation
Physical		
Dispersion	Mixing due to molecular diffusion and mechanical dispersion resulting from aquifer heterogeneities.	May result in: (1) mixing of the contaminant plume with electron acceptors and donors present in the surrounding pristine portions of the aquifer, and (2) dilution of the contaminant concentrations, but no net loss of mass.
Infiltration/ recharge	Entry of water into the soil from the ground surface, and subsequent flow within the unsaturated zone and into the saturated zone.	May result in: (1) dilution, and (2) transfer of materials (e.g., electron donors and acceptors, or contaminants) from the unsaturated zone to the saturated zone
Sorption	Partitioning of contaminant between groundwater and mineral or organic aquifer solids.	Reduces rate of contaminant migration and dissolved concentrations, but does not result in net loss of mass or permanently remove BTEX from groundwater.
Volatilization	Partitioning of contaminants from groundwater into the soil gas of the unsaturated zone.	Typically minor contribution compared to biodegradation, although may be more important for shallow contamination or cases with a highly fluctuating water table.
Chemical		
Chemical reaction	Reactions such as hydrolysis, which occur in groundwater and reduce contaminant mass.	Not observed for BTEX or MTBE.
Biological		
Aerobic	O_2 is used as an e^- acceptor by microbes to degrade contaminants via growth-related processes (e.g., destruction by mineralization to CO_2, H_2O, and biomass) or cometabolic processes.	Growth-related degradation most important destructive attenuation mechanism for BTEX, if sufficient DO present; MTBE may be transformed via growth-related processes or cometabolism, if at all.
Anaerobic	Alternative e^- acceptors (e.g., NO_3^-, SO_4^{2-}, Fe(III), and CO_2) are used by microbes to degrade contaminants	Growth-related degradation most important destructive attenuation mechanism for BTEX, although benzene is not always degraded and rates typically slower than aerobic biodegradation; MTBE probably transformed via growth-related processes, if at all.

Source: adapted from Knox et al., 1993; McAllister and Chiang, 1994.

groundwater, diffusion in the groundwater from the spill to permeable layers in the aquifer, and advective transport in the permeable layers, where groundwater flow supplied the electron acceptors for biodegradation (Cho et al., 1997).

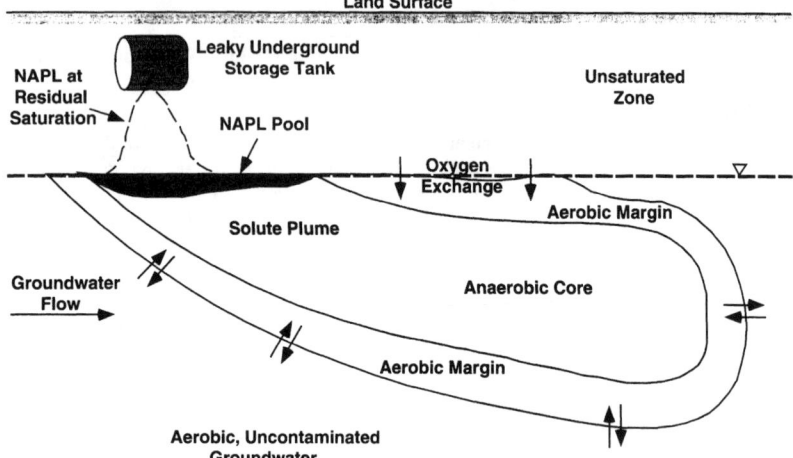

Figure 5-1. Oxygen supply to a petroleum-hydrocarbon release undergoing natural attenuation
Source: Adapted from Figures 9.1, 9.2, p.184, in Borden (1994).

Diffusion and dispersion can also play a key role in the supply of electron acceptors, such as oxygen, to the contaminated zone. There are three sources of oxygen in the subsurface (Fig. 5-1; Kemblowski et al., 1987): (1) oxygen in the groundwater flow from upgradient of the contaminated zone, (2) oxygen transport via dispersion from the surrounding groundwater into the dissolved hydrocarbon plume with low DO, and (3) oxygen diffusion down from the unsaturated zone soil gas into the groundwater. The oxygen in the upgradient groundwater is transported by advection into the contaminated zone. Because the consumption of oxygen is relatively fast, and the supply of oxygen is slow due to the low oxygen solubility in water, microorganisms near the source area may consume the available oxygen to biodegrade a portion of the hydrocarbons, in which case the remaining hydrocarbons are transported down gradient in an oxygen-limited plume (Borden, 1994). If this occurs, diffusion and dispersion become important modes of oxygen transport for hydrocarbon degradation downstream of the source, (e.g., Borden et al. 1997; Kemblowski et al., 1987), resulting in aerobic biodegradation occurring primarily at the plume fringes (Borden, 1994). Because dispersion is a relatively limited process, the overall aerobic biodegradation rate may be slow, and oxygen availability may play an important role in determining reaction rates and hydrocarbon persistence, (e.g., Barker et al., 1987).

Transfer of oxygen across the capillary fringe/water table may be especially critical for LNAPLs at or near the water table. Theoretical and field studies, e.g., (Borden and Bedient, 1986; Borden et al., 1986; Molz and Widdowson, 1988) indicate that transverse dispersion can have a significant effect on oxygen exchange and the aerobic biodegradation rate. However, vertical dispersion is a weak process, and the microbial oxygen demand is high, resulting in steep vertical oxygen gradients and limiting the impact of vertical dispersion to relatively shallow contaminated plumes (less than roughly 2-3 m below the water table), e.g., (MacQuarrie and Sudicky, 1990; Smith et al., 1991a; Smith et al., 1991b).

Hydrodynamic dispersion may also result in dilution of the contaminant concentrations (Kitandis, 1994). This may be a particularly important attenuation mechanism for MTBE, because of its apparent recalcitrance to biodegradation under some conditions and its tendency to remain in the aqueous phase and move unretarded with the groundwater flow (Anthony et al., 1999). For example, based on the occurrence of recharge events, and a low biodegradation potential, Landmeyer et al. (1998) concluded that the observed decreases in MTBE concentrations relative to benzene in a shallow, gasoline-contaminated aquifer were the result of dilution and dispersion with less-contaminated groundwater, rather than biodegradation. However, a modeling analysis using measured dispersivity values and pseudo-first-order biodegradation rates based on the field mass loss of MTBE and BTEX compounds suggested that dispersion alone is not a very effective natural attenuation process (Schirmer et al., 1999). In contrast, inclusion of biodegradation dramatically reduced the predicted number of years required to lower the initial benzene and MTBE to acceptable levels. The relatively high, near-source concentrations of MTBE also indicate that dispersion may be insufficient for receptor protection in many cases (Schirmer and Barker, 1998).

5.1.3.1.2 Infiltration and Recharge

The combined processes of infiltration and recharge can potentially contribute to natural attenuation in groundwater via: (1) dilution, and (2) transfer of materials from the unsaturated zone to the saturated zone. Observations of vertical "sinking" of BTEX and MTBE plumes with distance, (e.g., Borden et al., 1997; Schirmer and Barker, 1998) or of higher dissolved contaminant concentrations at greater depths (Landmeyer et al. 1998) demonstrate the impact of recharge to natural attenuation of BTEX and MTBE. However, given the weak nature of vertical transverse dispersion, there may be little mixing between the overlaying zone of recharge water and the contaminant plume (e.g., MacFarlane et al., 1983).

Infiltration may also dissolve electron acceptors, which then may enter the groundwater system. This effect may increase the overall electron-accepting capacity within the contaminated plume, (e.g., Borden et al., 1997), and/or change the predominant terminal electron accepting process (TEAP), which may affect the hydrocarbon biodegradation potential, (e.g., Vroblesky and Chapelle, 1994).

5.1.3.1.3 Sorption

Based on batch sorption isotherms, the BTEX compounds have different sorption tendencies (e.g., Chiou et al., 1983; Kemblowski et al., 1987; Stuart et al., 1991; Zytner, 1994). Correspondingly, the individual BTEX compounds have been observed to migrate at different velocities in laboratory column studies (e.g., Angley et al., 1992; Chen et al., 1992; Larsen et al., 1992; Stuart et al., 1991). This same "chromatographic separation" of BTEX is seen in petroleum-hydrocarbon impacted

Table 5-4. Physical/chemical properties of BTEX and MTBE (values at 25°C)[1]

Compound	Water solubility (mg/L)	Low K_{OW}	Log K_{OC}	Vapor pressure (mm Hg)	Henry's constant, K_H (atm·m³/mol)	Henry's constant, K_H (dimensionless)
Benzene	1,791[2]	2.13[2]	1.09–2.53[5]	95.19[2]	5.43(10⁻³)[2] 5.28(10⁻³)[8]	2.22(10⁻¹)[7] 2.16(10⁻¹)[8]
Toluene	534.8[2]	2.73[2]	1.12–2.85[5]	28.4[2]	5.94(10⁻³)[2] 6.43(10⁻³)[8]	2.43(10⁻¹)[7] 2.63(10⁻¹)[8]
Ethylbenzene	161[3]	3.15[3]	1.98–3.04[5]	9.53[3]	8.44(10⁻³)[3] 7.78(10⁻³)[8]	3.45(10⁻¹)[7] 3.18(10⁻¹)[8]
o-Xylene	175[2]	3.12[2]	1.68–2.73[5]	6.6[2]	5.1(10⁻³)[2] 4.99(10⁻³)[8]	2.084(10⁻¹)[7] 2.04(10⁻¹)[8]
m-Xylene	146[2]	3.20[2]	2.04–3.15[5]	8.3[2]	7.68(10⁻³)[2]	3.139(10⁻¹)[7]
p-Xylene	156[2]	3.15[2]	2.05–3.08[5]	8.7[2]	7.68(10⁻³)[2]	3.139(10⁻¹)[7]
MTBE	51,000[4] 48,000–54,300[6]	1.24[4]	1.05[7]	249[4]	5.87(10⁻⁴)[4] 5.28(10⁻⁴)[8]	2.399(10⁻²)[7] 2.16(10⁻²)[8]

[1] More extensive compilations of measured and calculated values are available elsewhere, e.g., Mackay et al., 1992; Mackay et al., 1993.
[2] Sage et al., 1990.
[3] Jarvis et al., 1989.
[4] Michalenko et al., 1993.
[5] Mackay et al., 1992.
[6] Mackay et al., 1993.
[7] Squillace et al., 1997. Dimensionless Henry's constants generally were calculated from documented references.
[8] Robbins et al., 1993).

groundwater aquifers (e.g., Odermatt, 1994), thereby affecting the relative spatial and temporal proportions of these compounds in the solute plume.

These observed effects can be analyzed by applying the linear equilibrium sorption model to calculate retardation factors, R. Values for water solubility, and octanol-water (K_{ow}) and organic-carbon (K_{oc}) partition coefficients determined for the BTEX species are presented in Table 5-4. Assuming an aquifer material with bulk density, $\rho_b = 1.86$ g·cm⁻³, porosity, $n = 0.3$, and organic content, $f_{oc} = 0.01$, and using the log K_{oc} values from Table 5-4, the following ranges of R values for the BTEX compounds are calculated: benzene, $1.76 < R < 22.0$; toluene, $1.82 < R < 45.9$; ethylbenzene, $6.92 < R < 69.0$; o-xylene, $2.97 < R < 34.3$; m-xylene, $7.8 < R < 88.6$; and p-xylene, $7.96 < R < 75.5$. These differences in mobility (benzene > toluene > ethylbenzene and xylenes) could explain the observed chromatographic separation of the BTEX compounds, and the resulting changes in the BTEX ratios at different locations in the plume (Alvarez et al., 1998).

A number of R values determined for the BTEX compounds via different techniques and for different solid phases are summarized in Table 5-5. The majority of these data were collected for aquifer solids with relatively low f_{oc} values and, as a result, most of the R values are significantly below those predicted in the previous paragraph. Consistent with the above predictions, these R values indicate that benzene has the fastest, and ethylbenzene and the xylenes have the slowest, relative mobilities, respectively. In particular, the sorption tendency of benzene, which has a lower log K_{ow} value, is different from the TEX compounds, which have relatively similar log K_{ow} values (Table 5-4.) and more similar sorption tendencies (Zytner, 1994).

These values also indicate that, in general, sorption of the BTEX compounds in groundwater may be assumed to be relatively limited in aquifer materials with low organic carbon content, (e.g., Chiang et al., 1989). This is supported by field experiments in an uncontaminated, unconfined sand aquifer indicating that sorption caused BTX components to migrate only slightly slower than the groundwater (Barker et al., 1987). Of course, BTEX sorption could be more significant in higher organic content soils, such as peat moss or organic top soil, (e.g., Zytner, 1994). Furthermore, dramatic spatial variation in sorptive properties may be observed at a given site, even at sites geologically defined as being relatively homogeneous (Kemblowski et al., 1987). Sorption of BTEX onto inorganic surfaces can also occur, (e.g., Rogers et al., 1980), and may be as important as partitioning into a natural organic phase in some situations (Ball et al., 1997).

The partition coefficients for MTBE are even lower than those for the BTEX compounds (see Table 5-4) (Squillace et al., 1997). For example, using the same assumptions as above for the BTEX compounds, the estimated R value for MTBE is 1.70. Thus, as demonstrated by the estimated R values from field studies summarized in Table 5-5, which are near 1, sorption is generally an insignificant natural attenuation process for MTBE. As a result, at field sites, MTBE migrates more rapidly than the BTEX compounds, causing a chromatographic-like separation of MTBE and the BTEX compounds in contaminant plumes (e.g., Landmeyer et al., 1998). The greater susceptibility of BTEX compounds to biodegradation compared to MTBE (discussed below) further amplifies this separation effect. Therefore, the leading edge of a plume emanating from an MTBE-oxygenated gasoline spill may contain substantial levels of MTBE with very little or no BTEX (Squillace et al., 1997).

Table 5-5. Retardation factors for BTEX compounds, measured using various techniques

Compound	Soil	Retardation factor, R, estimation technique				
		Empirical correlation	Batch isotherm	Laboratory column	Field "in situ" column	Natural gradient
Benzene	Sandy aquifer material					
	($f_{oc} \approx 0.02\%$)[1]	1.1	1.2	1.0	1.4	1.1
	($f_{oc} = 0.01–0.08\%$)[2]		1.4	1.6		
	($f_{oc} = 0.015\%$)[3]			1.4		
	($f_{oc} = 0.007–0.025\%$)[4]			1.02–1.33		
	($f_{oc} = 0.12–1.08\%$)[5]		1.35–4.48			
	Clayey sand aquifer underlain by course sand/silty aquifer (average $f_{oc} = 0.05\%$)[7]	1.03				
Toluene	Sandy aquifer material					
	($f_{oc} \approx 0.02\%$)[1]	1.2	1.3	1.2	1.5	1.2
	($f_{oc} = 0.01–0.08\%$)[2]		1.6	2.1		
	($f_{oc} = 0.015\%$)[3]			1.7		
	($f_{oc} = 0.007–0.025\%$)[4]			1.01–1.52		
	($f_{oc} = 0.12–1.08\%$)[5]		2.08–8.10			
	Clayey sand aquifer underlain by course sand/silty aquifer (average $f_{oc} = 0.05\%$)[7]	1.08				
Ethylbenzene	Sandy aquifer material ($f_{oc} = 0.015\%$)[3]			1.7		
	Clayey sand aquifer underlain by course sand/silty aquifer (average $f_{oc} = 0.05\%$)[7]	1.18				
o-Xylene	Sandy aquifer material					
	($f_{oc} \approx 0.02\%$)[1]	1.4	1.7	1.3	1.7	1.4
	($f_{oc} = 0.015\%$)[3]			1.7		
	($f_{oc} = 0.007–0.025\%$)[4]			1.10–2.18		
	($f_{oc} = 0.12–1.08\%$)[5]		3.0–16.0			
	Clayey sand aquifer underlain by course sand/silty aquifer (average $f_{oc} = 0.05\%$)[7]	1.09				

continued on next page

Table 5-5. Retardation factors for BTEX compounds, measured using various techniques (continued)

Compound	Soil	Retardation factor, R, estimation technique				
		Empirical correlation	Batch isotherm	Laboratory column	Field "in situ" column	Natural gradient
m-Xylene	Sandy aquifer material					
	($f_{oc} \approx 0.02\%$)[1]	1.5	2.0	1.5	1.8	1.4
	($f_{oc} = 0.015\%$)[3]			2.0		
	Clayey sand aquifer underlain by course sand/silty aquifer (average $f_{oc} = 0.05\%$)[7]	1.19				
p-Xylene	Sandy aquifer material					
	($f_{oc} \approx 0.02\%$)[1]	1.4	2.0	1.4	1.8	1.4
	($f_{oc} = 0.015\%$)[3]			2.0		
	Clayey sand aquifer underlain by course sand/silty aquifer (average $f_{oc} = 0.05\%$)[7]	1.19				
MTBE	Sand aquifer[6]	1.02				
	Clayey sand aquifer underlain by course sand/silty aquifer (average $f_{oc} = 0.05\%$)[7]	1.003				

[1]Ptacek et al., 1987.
[2]Chen et al., 1992.
[3]Angley et al., 1992.
[4]Larsen et al., 1992.
[5]Kemblowski et al., 1987.
[6]Schirmer and Barker, 1998.
[7]Borden et al., 1997.

5.1.3.1.4 Volatilization

Equilibrium partitioning of a volatile chemical from dilute aqueous solutions into the soil gas is governed by Henry's Law, while the ideal equilibrium partitioning between a NAPL and the soil gas can be quantified using Raoult's Law and the vapor pressure. Henry's Law constants and vapor pressures for the BTEX compounds and MTBE are provided in Table 5-4. Although partitioning can occur from the groundwater phase and from the petroleum product NAPL, the focus here is on volatilization from the dissolved contaminant plume. This is not a destructive mechanism, although it does result in the removal of contaminants from the groundwater phase (ASTM, 1998).

Henry's Law constants, K_H (atm·m^3·mole^{-1}) for hydrocarbons range over three orders of magnitude (Lyman et al., 1992). Most volatile are the saturated aliphatics ($K_H \approx 1-10$ atm·m^3·mole^{-1} at 25°C), slightly less volatile are the unsaturated and cyclo-aliphatics ($K_H \approx 0.1-1$ atm·m^3·mole^{-1} at 25°C), and least volatile are the aromatics ($K_H \approx 0.005-0.02$ atm·m^3·mole^{-1} at 25°C) (Table 5-4). The Henry's constants for the BTEX compounds are relatively similar, suggesting similar levels of volatilization. The combined effects of relatively low volatilization and sorption lead to an enrichment in the dissolved-phase of the BTEX compounds relative to other petroleum hydrocarbons (Lyman et al., 1992). Although the tendency to volatilize from water is relatively similar among the BTEX compounds, the gas-phase contaminant concentrations that result from equilibration with an aqueous solution are also a function of the aqueous contaminant concentration and, thus, may vary from compound to compound. For example, for benzene and o-xylene present at the average aqueous equilibrium concentrations for gasoline in contact with water provided in Table 5-2, the equilibrium gas-phase concentrations are 9.46 mg/L$_{gas}$ and 1.2 mg/L$_{gas}$, respectively. These values, which represent the maximum achievable concentration for this situation, demonstrate that partitioning from water contaminated by petroleum products results in relatively low equilibrium gas-phase concentrations.

Given the relatively low Henry's constants, and the small surface area of the groundwater flow system exposed to soil gas, the volatilization of BTEX compounds from groundwater is a relatively slow process and its impact on dissolved contaminant concentrations can generally be neglected (Wiedemeier et al., 1995). For example, based on a mass balance analysis, only 5% or less of the total mass loss of benzene at a petroleum-product impacted sandy aquifer could be attributed to volatilization (Chiang et al., 1989; McAllister and Chiang, 1994). Nevertheless, volatilization may account for a larger fraction of the total mass loss if: the water table is shallower, or fluctuates significantly (ASTM, 1998); the groundwater temperature is relatively high, which increases the value of the Henry's constant (e.g., Robbins et al., 1993); or rates of biodegradation are relatively slow (McAllister and Chiang, 1994).

Dimensionless Henry's constants, K_H, for MTBE (Table 5-4) are an order of magnitude lower than the K_H of the BTEX compounds. The dimensionless Henry's constants for MTBE and other alkyl ether oxygenates are similar in magnitude; therefore, these additives partition to a high degree into water (Squillace et al., 1997). However, given the higher aqueous solubility of MTBE compared to the petroleum hydrocarbons, the gas-phase concentration could be higher than those for the BTEX compounds. For example, for gasoline spiked with MTBE at 10% (by volume), Poulsen et al. (1992) measured an equilibrium aqueous-phase concentration for MTBE of 3,650 mg/L, which would result in an equilibrium gas phase concentration of 87.6 mg/L$_{gas}$.

Mass losses of BTEX and MTBE from groundwater via volatilization may be further reduced for portions of plumes that are > 1 m below the water table (Schirmer and Barker, 1998; Wiedemeier et al., 1996a). This is due to the weakness of vertical transverse dispersion in the groundwater and the downward groundwater velocity in the vicinity of the water table during recharge (Rivett, 1995).

5.1.3.2 Chemical Mechanisms

BTEX compounds are rarely transformed by naturally-occurring chemical reactions under typical groundwater conditions (McAllister and Chiang, 1994). Although benzene has a molecular formula indicating a high degree of unsaturation, benzene is highly stable and resistant to the reactions characteristic of other unsaturated compounds, e.g., alkenes (Morrison and Boyd, 1973). The stability of benzene and other aromatic compounds is due to several factors, in particular, the delocalization of the π electrons, and the manner in which the various orbitals that make up the π cloud are filled (e.g., Morrison and Boyd, 1973). The stability and low reactivity of benzene and its derivatives is borne out by observations of little or no BTEX mass loss in aerobic and anaerobic batch (e.g., Baedecker et al., 1993; Barker et al., 1987; Chiang et al., 1989; Kemblowski et al., 1987) and column experiments (e.g., Angley et al., 1992; Larsen et al., 1992) in which biological activity was prevented.

Similarly, transformation of MTBE by naturally-occurring chemical reactions under typical groundwater conditions is not expected (Schirmer and Barker, 1998; Squillace et al., 1997). In general, ethers are relatively unreactive compounds (Morrison and Boyd, 1973). This is due to the chemically-unreactive ether linkage, which is stable toward bases, oxidizing agents, and reducing agents, and only undergoes one known type of chemical reaction, i.e., cleavage by acids under very vigorous conditions. For example, Yeh and Novak (1995), observed chemical oxidation of MTBE to TBA and acetone in soil microcosms and in solution when H_2O_2 was added in the presence of catalytic ferrous iron. However, the reaction was favored at low pH (pH 4) and did not occur after oxidation and precipitation of the iron occurred. Therefore, the reaction is unlikely to occur in aerobic and near-neutral to alkaline groundwater environments (Squillace et al., 1997). Indeed, MTBE has been shown to be chemically unreactive under aerobic and anaerobic conditions in the sterile controls used in numerous batch experiments, (e.g., Borden et al., 1997; Bradley et al., 1999; Suflita and Mormile, 1993; Yeh and Novak, 1994; Yeh and Novak, 1995).

5.1.3.3 Biological Mechanisms

5.1.3.3.1 BTEX

BTEX-degrading bacteria appear to be ubiquitous in the environment and are estimated to be present at approximately 99% of contaminated sites (Norris, 1994). As summarized by Rittmann et al. (1994), cometabolic biotransformation of BTEX compounds can be mediated by organisms, e.g., *Nocardia* strains, that are growing on aliphatic hydrocarbons. However, BTEX compounds can also be used as growth substrates, specifically, as sole sources of electrons and carbon, by a wide variety of bacterial strains (Rittmann et al., 1994), and it is generally assumed that growth-related transformations are much more important than cometabolic reactions in bringing about the removal of subsurface BTEX contamination.

The flow of electrons from a BTEX compound electron donor to an electron acceptor releases free energy (ΔG_R). Some of this free energy is used by the bacteria for cell synthesis (McCarty, 1972). As ΔG_R values become less negative, less energy is available for cell synthesis. The paradigm for BTEX biodegradation in subsurface environments is that contaminant removal initially occurs via aerobic respiration because oxidation of a BTEX compound coupled to the reduction of oxygen releases the maximum free energy (most negative ΔG_R) giving microorganisms that use the reaction an ecological competitive advantage. After oxygen is depleted, oxidation of the BTEX compounds is coupled to the reduction of electron acceptors that yield progressively less free energy (less negative ΔG_R values). According to this paradigm, electron acceptors are utilized for the oxidation of BTEX compounds and depleted in the following order: oxygen, ferric iron, nitrate, sulfate, and carbon dioxide.

However, utilization of electron acceptors for the oxidation of BTEX compounds may not always proceed as predicted on the basis of thermodynamic considerations. The dominant TEAP may vary both temporally *and* spatially in a contaminated aquifer (Vroblesky and Chapelle, 1994), and in some cases, TEAPs may not be mutually exclusive. Furthermore, at a given site, biodegradation of certain BTEX compounds may not occur under some redox conditions, e.g., (Cho et al., 1997), even though the potential for biodegradation has been demonstrated in laboratory studies. Biodegradation potential may not be realized in the field due to the absence of the appropriate electron acceptor, organisms, or genetic information, or due to inhibitory conditions (e.g., toxic levels of co-contaminants, high levels of preferred growth substrates, or unfavorable pH). Nevertheless, the pool of anaerobic electron acceptors in subsurface environments contaminated with petroleum hydrocarbons is typically much larger than the supply of oxygen, which is rapidly depleted by oxidation of readily biodegradable contaminants near the heart of a plume, as discussed above (Fig. 5-1). Therefore, greater amounts of dissolved BTEX compounds may be removed from contaminated groundwater plumes via anaerobic metabolic processes, compared to aerobic respiration (Wiedemeier et al., 1999).

Electron acceptor and nutrient requirements

The dominant TEAP will influence both the rate of biodegradation and the geochemical changes that occur as a result of contaminant biodegradation. Therefore, estimates of the electron acceptor requirements required to sustain BTEX oxidation are required in order to accurately model the environmental fate of the contaminants and to develop and interpret evaluations of natural attenuation.

Two different approaches can be used to estimate the amounts of electron acceptors required to oxidize dissolved BTEX compounds and the products of contaminant metabolism. One method involves the construction of balanced stoichiometric reactions of electron donor utilization for energy generation and cell synthesis, and is based on thermodynamic considerations (McCarty, 1972). Importantly, this method accounts for the fact that not all of the electrons derived from an electron donor (in this case, a BTEX compound) are transferred by growing microorganisms to an electron acceptor for energy generation. Instead, a fraction of electrons are directed to cell synthesis in reactions that consume some of the energy generated by the reduction of the electron acceptor. The relative fractions of electrons consumed in the energy reaction and in cell synthesis are determined by the free energy changes of the energy (ΔG_R) and synthesis reactions (ΔG_S), which in turn are functions of the electron donor, electron acceptor, carbon source and nitrogen source used.

Using aerobic respiration as an example, balanced stoichiometric reactions of benzene, toluene, ethylbenzene, and *m*-xylene utilization for energy generation and cell synthesis were determined using the method of McCarty (1972) (see Table 5-6).

Table 5-6. Stoichiometric requirement for O_2, including synthesis[1]

Compound	Reaction	Mass ratio of oxygen to BTEX compound (g/g)[2]
B	$C_6H_6 + 3.066\ O_2 + 0.888\ HCO_3^- + 0.888\ NH_4^+ \rightarrow$ $0.888\ C_5H_7O_2N + 2.454\ CO_2 + 2.115\ H_2O$	1.26:1 (2.08:1)
T	$C_6H_5CH_3 + 3.737\ O_2 + 1.055\ HCO_3^- + 1.055\ NH_4^+ \rightarrow$ $1.055\ C_5H_7O_2N + 2.786\ CO_2 + 2.948\ H_2O$	1.30:1 $(1.99–2.34:1)^{[3]}$
E	$C_6H_5C_2H_5 + 4.200\ O_2 + 1.260\ HCO_3^- + 1.260\ NH_4^+ \rightarrow$ $1.260\ C_5H_7O_2N + 2.961\ CO_2 + 3.738\ H_2O$	1.27:1 $(1.87–2.17:1)^{[3]}$
m-X	$C_6H_4(CH_3)_2 + 4.234\ O_2 + 1.256\ HCO_3^- + 1.256\ NH_4^+ \rightarrow$ $1.256\ C_5H_7O_2N + 2.986\ CO_2 + 3.746\ H_2O$	1.28:1 (2.18:1)
	Average	1.28:1

[1]McCarty's (1972) assumptions regarding the efficiency of energy capture and transfer are implicit, and the electron acceptor requirements of cell maintenance were neglected, in these determinations.

[2]Value in parentheses includes co-reactant requirements for O^2.

[3]Different O_2 requirements are feasible because more than one pathway for the aerobic biodegradation of this compound exists, and the distinct pathways include different numbers of reactions that involve O_2 as a co-reactant.

In determining these and comparable stoichiometric equations for BTEX oxidation under other TEAPs (discussed below), it is assumed that the BTEX compound serves as the electron donor and carbon source and NH_4^+ is the nitrogen source. The assumptions made by McCarty (1972) are also implicit in these determinations. The mass requirements of each BTEX compound for O_2, which were obtained from the molar relationships defined by the stoichiometric equations, are also shown in Table 5-6.

In reality, BTEX-biodegrading organisms also consume electron acceptors through endogenous respiration. However, it is difficult to estimate the amount of electron acceptor required to meet the demands of endogenous respiration without having any information about the rate of biomass decay. Therefore, the electron acceptor requirements needed to sustain the endogenous decay of BTEX-degrading organisms is neglected here. It is important to note that the effects of cell synthesis and endogenous decay on electron acceptor requirement cannot be disregarded simply on the basis of a contaminant plume being at steady-state, as has been suggested (Wiedemeier et al., 1999). Although the steady-state condition may indicate that, overall, the growth of BTEX-degrading organisms in the plume is being balanced by biomass losses due to decay and other mechanisms, both synthesis and decay exert a demand for electron acceptors and do not cancel each other out.

As discussed further below, aerobic BTEX biodegradation involves reactions mediated by oxygenases. Reactions catalyzed by oxygenases utilize molecular oxygen as an obligatory cosubstrate, which represents an additional electron acceptor requirement for aerobic BTEX oxidation. This cosubstrate requirement must be

Table 5-7. Stoichiometric requirement for O_2, neglecting synthesis and co-reactant requirements

Compound	Reaction	Mass ratio of oxygen to BTEX compound (g/g)
B	$C_6H_6 + 7.5\ O_2 \rightarrow 6\ CO_2 + 3\ H_2O$	3.07:1
T	$C_6H_5CH_3 + 9\ O_2 \rightarrow 7\ CO_2 + 4\ H_2O$	3.13:1
E	$C_6H_5C_2H_5 + 10.5\ O_2 \rightarrow 8\ CO_2 + 5\ H_2O$	3.16:1
m-X	$C_6H_4(CH_3)_2 + 10.5\ O_2 \rightarrow 8\ CO_2 + 5\ H_2O$	3.16:1
	Average	3.13:1

quantified along with the demand for oxygen as a terminal electron acceptor. The known metabolic pathways for aerobic BTEX biodegradation (Ellis and Wackett, 2000; Gibson and Subramanian, 1984; Ribbons and Eaton, 1982; Rittmann et al., 1994) involve a total of two or three oxygenase-mediated reactions and thus consume two or three moles of molecular oxygen per mole BTEX. These co-reactant needs were added to the demands for oxygen as a terminal electron acceptor (Table 5-6).

The second method that can be used to estimate electron acceptor requirements simply involves balancing two redox half-reactions, i.e., the oxidation of a BTEX compound to CO_2, and the reduction of an electron acceptor to its reduced form. This method neglects the utilization of some of the electron donor for synthesis. Balanced reactions for the mineralization of benzene, toluene, ethylbenzene, and m-xylene coupled to the reduction of O_2 were constructed using this approach and are summarized in Table 5-7, along with the mass requirements for oxygen calculated from these reactions. Comparison of the mass ratios of oxygen to BTEX compound in Tables 5-6 and 5-7 reveals that by neglecting cell synthesis, the amount of oxygen required to achieve complete oxidation of the BTEX compounds is overestimated by a factor of more than two. As shown below, the magnitude of the errors introduced into estimates of electron acceptor mass requirements by neglecting cell synthesis diminishes as the electron acceptors become energetically less favorable. This occurs because as electron acceptors become less favorable, and consequently, ΔG_R values become less negative, microorganisms have to direct a greater fraction of the electrons derived from the electron donor to the electron acceptor for energy generation. Correspondingly, a smaller fraction of electrons is consumed for cell synthesis. Similarly, the bioenergetic model predictions of electron acceptor requirements will approach the predictions of the redox half-reactions under conditions that significantly decrease the efficiency of growth (summarized by McCarty (1972)).

In addition to an electron donor/carbon source and appropriate electron acceptor supply, BTEX-degrading populations, like all bacteria, also require macro- and micro-nutrients. However, endogenous sources of nutrients appear to be adequate to support biodegradation of BTEX compounds in most subsurface environments (Wiedemeier et al., 1999).

Biodegradation of BTEX compounds under aerobic conditions

A common feature of all the known peripheral and central pathways involved in the aerobic biodegradation of the BTEX compounds is that they include reactions mediated by monooxygenases and/or dioxygenases. The oxygenases introduce hydroxyl groups into the BTEX compounds, making them susceptible to ring cleavage, which is mediated by dioxygenases (Heider and Fuchs, 1997).

As discussed in Chap. 3.3, catechol and substituted catechols are commonly formed aromatic intermediates in the aerobic biodegradation of the BTEX compounds and are subject to ring cleavage (Gibson and Subramanian, 1984; Ribbons and Eaton, 1982; Rittmann et al., 1994). Biodegradation of benzene occurs via initial attack on the aromatic nucleus, forming catechol, whereas oxidation of the alkyl-substituted benzenes may be initiated via either attack on the methyl group or on the aromatic nucleus. For example, biodegradation of toluene may proceed via an initial side-chain oxidation that leads to the formation of benzoate, which is converted to catechol through dioxygenase-mediated insertion of oxygen into the aromatic ring. Several different mechanisms involving an initial attack on the aromatic nucleus of toluene that result in the formation of a methyl-substituted catechol or protocatechuate are also possible (Whited and Gibson, 1991). Ethylbenzene biodegradation may also be initiated by ring hydroxylation, producing an ethyl-substituted catechol (Gibson and Subramanian, 1984; Ribbons and Eaton, 1982; Rittmann et al., 1994). Most bacteria that grow on m- or p-xylene first oxidize a methyl group, resulting in the formation of a methyl-substituted catechol. Organisms that are able to grow on o-xylene appear to be much less common than those that use m- or p-xylene as growth substrates, but growth on o-xylene has been demonstrated and occurred via ring hydroxylation leading to a dimethylcatechol (Gibson and Subramanian, 1984; Rittmann et al., 1994). Cleavage of the catechol and substituted catechols formed during the aerobic oxidation of BTEX compounds occurs via either the *ortho-* or *meta-*cleavage pathway and leads to the formation of compounds that can be funneled into the citric acid cycle or other central metabolic pathways.

Biodegradation of the BTEX compounds under iron-reducing conditions

When subsurface environments become anaerobic, Fe(III) is often one of the most abundant electron acceptors available for microbial metabolism (Lovley, 1991; Lovley and Lonergan, 1990). Geochemical data that show elevated levels of dissolved Fe(II) in groundwater, formation of extracellular magnetite, leaching of Fe(III) oxides from sediment, and/or production of organic acids and reduced electron acceptors occurring concomitantly with the oxidation of BTEX compounds, suggest that dissimilatory iron reduction may contribute significantly to the natural attenuation of some aquifers polluted with BTEX and other organic contaminants (e.g., Borden et al., 1995; Cozzarelli et al., 1990; Lovley, 1997; Lovley et al., 1989; Lyngkilde and Christensen, 1992; Rugge et al., 1995). Although benzene often appears to be recalcitrant under denitrifying conditions, oxidation of benzene under Fe(III)-reducing conditions has frequently been demonstrated in both laboratory and field studies, (e.g., Kazumi et al., 1997; Lyngkilde and Christensen, 1992; Rugge et al., 1995).

Despite the importance of anaerobic metabolic processes in the removal of BTEX contaminants from the subsurface, relatively little is known about the pathways of BTEX oxidation under Fe(III)-reducing and other anaerobic conditions. However, complete oxidation of benzene and toluene to carbon dioxide has been observed in Fe(III)-reducing aquifer materials and a Fe(III)-reducing bacterium, *Geobacter metallireducens*, is able to mineralize toluene in pure culture (Lovley and Lonergan, 1990). Although no intermediates were detected in the oxidation of toluene to carbon dioxide by *G. metallireducens*, two pathways for the oxidation of toluene under Fe(III)-reducing conditions have been proposed. One involves sequential oxidations of the methyl group to form benzyl alcohol, benzaldehyde, and finally benzoate, which are all substrates that are readily metabolized by *G. metallireducens* (Lovley and Lonergan, 1990). An alternative proposed pathway involves the hydroxylation of the aromatic ring to form *p*-cresol, which is subsequently oxidized and dehydroxylated to form *p*-hydroxybenzoate and benzoate, respectively. Mixed culture studies indicate that phenol and benzoate may be intermediates in the biodegradation of benzene under Fe(III)-reducing conditions (Caldwell and Suflita, 2000). Hydroxylation of benzene presumably explains the detection of phenol, which could have been converted to benzoate via well-established sequential carboxylation and dehydroxylation reactions (Heider and Fuchs, 1997). Alternatively, benzoate could have been produced via direct carboxylation of benzene. Benzoate is formed as an intermediate in BTEX oxidation under other TEAP conditions as well, and its potential use as an indicator of anaerobic BTEX metabolism is discussed below.

Balanced stoichiometric reactions of the oxidation of benzene, toluene, ethylbenzene, and *m*-xylene coupled to dissimilatory Fe(III) reduction for energy generation and cell synthesis are summarized in Table 5-8, with the mass requirements of each BTEX compound for Fe(III). Neglecting the utilization of a fraction of the BTEX compounds for cell synthesis increases the estimates of Fe(III) requirements by a factor of over two (Table 5-9) because thermodynamically, Fe(III) is nearly as good an electron acceptor as oxygen. Under Fe(III)-reducing and other anaerobic conditions, oxygenase-mediated reactions are not possible so oxygen co-reactant requirements do not exist.

Biodegradation of the BTEX compounds under denitrifying conditions

Denitrification may be an important mechanism for removing some BTEX compounds in anaerobic subsurface environments that are contaminated not only with petroleum hydrocarbons, but also with nutrients (e.g., from agricultural runoff), and therefore contain relatively high groundwater nitrate concentrations. In general, under denitrifying conditions, oxidation of toluene, ethylbenzene, and *m*- and *p*-xylene is reported to occur rather rapidly, (e.g., Borden et al., 1997; Hutchins, 1991; Zeyer et al., 1986), while removal of *o*-xylene occurs relatively slowly (e.g., Borden et al., 1997; Hutchins, 1991; Kuhn et al., 1985). Oxidation of benzene to carbon dioxide by two *Dechloromonas* isolates using nitrate as the terminal electron acceptor has been observed (Coates et al., 2001), and oxidation of benzene to carbon dioxide has been linked to nitrate reduction in enrichment cultures developed from soil and groundwater (Burland and Edwards, 1999). However, in most other laboratory studies (e.g., Hutchins et al., 1991; Kazumi et al., 1997; Kuhn et al., 1988) and field

NATURAL ATTENUATION OF HAZARDOUS WASTES 139

Table 5-8. Stoichiometric requirement for Fe^{3+}, including synthesis[1]

Compound	Reaction	Mass ratio of Fe^{3+} to BTEX compound (g/g)
B	$C_6H_6 + 12.54\ Fe^{3+} + 0.87\ HCO_3^- + 0.87\ NH_4^+ + 4.14\ H_2O \rightarrow$ $0.87\ C_5H_7O_2N + 12.54\ Fe^{2+} + 2.52\ CO_2 + 12.54\ H^+$	8.97:1
T	$C_6H_5CH_3 + 15.264\ Fe^{3+} + 1.044\ HCO_3^- + 1.044\ NH_4^+ + 4.68\ H_2O \rightarrow$ $1.044\ C_5H_7O_2N + 15.264\ Fe^{2+} + 2.844\ CO_2 + 15.264\ H^+$	9.25:1
E	$C_6H_5C_2H_5 + 17.136\ Fe^{3+} + 1.26\ HCO_3^- + 1.26\ NH_4^+ + 4.83\ H_2O \rightarrow$ $1.26\ C_5H_7O_2N + 17.136\ Fe^{2+} + 3.024\ CO_2 + 17.136\ H^+$	9.01:1
m-X	$C_6H_4(CH_3)_2 + 17.304\ Fe^{3+} + 1.218\ HCO_3^- + 1.218\ NH_4^+ + 4.872\ H_2O \rightarrow$ $1.218\ C_5H_7O_2N + 17.304\ Fe^{2+} + 3.024\ CO_2 + 17.304\ H^+$	9.10:1
	Average	9.08:1

[1]McCarty's (1972) assumptions regarding the efficiency of energy capture and transfer are implicit, and the electron acceptor requirements of cell maintenance were neglected, in these determinations.

Table 5-9. Stoichiometric requirement for Fe^{3+}, neglecting synthesis

Compound	Reaction	Mass ratio of Fe^{3+} to BTEX compound (g/g)
B	$C_6H_6 + 30\ Fe^{3+} + 12\ H_2O \rightarrow 30\ Fe^{2+} + 6\ CO_2 + 30\ H^+$	21.45:1
T	$C_6H_5CH_3 + 36\ Fe^{3+} + 14\ H_2O \rightarrow 36\ Fe^{2+} + 7\ CO_2 + 36\ H^+$	21.82:1
E	$C_6H_5C_2H_5 + 42\ Fe^{3+} + 16\ H_2O \rightarrow 42\ Fe^{2+} + 8\ CO_2 + 42\ H^+$	22.09:1
m-X	$C_6H_4(CH_3)_2 + 42\ Fe^{3+} + 16\ H_2O \rightarrow 42\ Fe^{2+} + 8\ CO_2 + 42\ H^+$	22.09:1
	Average	21.86:1

investigations (e.g., Barbaro et al., 1992; Borden et al., 1997), benzene appeared to be biologically recalcitrant under denitrifying conditions.

Benzoate is a common intermediate during the biodegradation of toluene and ethylbenzene under denitrifying conditions (Ball et al., 1996; Beller and Spormann, 1997; Biegert et al., 1996; Heider et al., 1998). However, the reactions leading to the production of benzoate from toluene and ethylbenzene appear to be quite different. The first reaction in the toluene biodegradation pathway is enzymatically-mediated and involves the addition of toluene to fumarate, which results in the formation of benzylsuccinate and E-phenylitaconate (Beller and Spormann, 1997; Biegert et al., 1996). In contrast, biodegradation of ethylbenzene under denitrifying conditions is initiated by dehydrogenation of the benzylic carbon (Ball et al., 1996). Conversion of m-xylene to 3-methylbenzoate occurs via a series of reactions that are analogous to those involved in the conversion of toluene to benzoate (Krieger et al., 1999), and eventually leads to the formation of methyl homologs of benzylsuccinate and E-phenylitaconate.

The usefulness of benzoate, benzylsuccinate, and E-phenylitaconate and their methyl homologs as indicators of anaerobic BTEX oxidation was reviewed by Beller (2000). All of these compounds are relatively stable and the benzoates in particular have frequently been observed at petroleum-contaminated field sites. However, a potential shortcoming of benzoate is that it is a common intermediate in the anaerobic metabolism of a wide variety of substituted aromatic compounds (Heider and Fuchs, 1997). Therefore, its formation is not uniquely indicative of BTEX metabolism. Further, benzoate may be derived from commercial or industrial sources. The detection of benzylsuccinate and E-phenylitaconate or their substituted homologs can be definitely related to the anaerobic metabolism of specific alkylbenzenes, and these compounds are not used in commercial and industrial applications. Unfortunately, efforts to quantify these compounds may be limited by the lack of commercially-available standards. The proposed pathways discussed above were primarily derived from studies conducted with denitrifying bacteria from the genera *Thauera* and *Azoarcus*. However, the addition of methyl-substituted benzenes to fumarate has been observed in other metabolic and phylogenetic groups, including toluene-degrading SRB and methanogenic enrichment cultures (discussed below), and therefore may be relatively widespread (Spormann and Widdel, 2000).

Balanced stoichiometric reactions of benzene, toluene, ethylbenzene, and m-xylene utilization for energy generation and cell synthesis under denitrifying conditions were determined. The equations are summarized in Table 5-10, along with the mass requirements of each BTEX compound for NO_3^-. Because the theoretical free energy yield of BTEX oxidation coupled to nitrate reduction is relatively high, neglecting the utilization of a fraction of the BTEX compounds for cell synthesis increases the estimates of nitrate requirements by a factor of over two (Table 5-11).

Biodegradation of the BTEX compounds under sulfate-reducing conditions

In groundwater systems that naturally contain relatively high levels of sulfate, sulfate reduction may become an important anaerobic respiratory process in areas of contaminant plumes that have largely been depleted of more favorable electron acceptors. Sulfate reduction may be an especially important process in groundwaters that contain high levels of sulfate due to dissolution of sulfate-bearing minerals, or in aquifers that are proximal to marine or estuarine systems.

Biodegradation of all of the BTEX compounds has been shown to occur under sulfate-reducing conditions, although in some field studies and laboratory studies conducted with mixed cultures, coupling of BTEX compound oxidation to the reduction of sulfate has not been unequivocally demonstrated. For example, in some anaerobic BTEX-degrading systems, biogeochemical evidence suggests that in

Table 5-10. Stoichiometric requirement for NO_3^-, including synthesis.[1]

Compound	Reaction	Mass ratio of NO_3^- to BTEX compound (g/g)
B	$C_6H_6 + 2.544\ NO_3^- + 0.864\ HCO_3^- + 0.864\ NH_4^+ + 2.55\ H^+ \rightarrow$ $0.864\ C_5H_7O_2N + 1.272\ N_2 + 2.55\ CO_2 + 3.39\ H_2O$	2.02:1
T	$C_6H_5CH_3 + 3.096\ NO_3^- + 1.026\ HCO_3^- + 1.026\ NH_4^+ + 3.096\ H^+ \rightarrow$ $1.026\ C_5H_7O_2N + 1.548\ N_2 + 2.88\ CO_2 + 4.536\ H_2O$	2.08:1
E	$C_6H_5C_2H_5 + 3.486\ NO_3^- + 1.218\ HCO_3^- + 1.218\ NH_4^+ + 3.486\ H^+ \rightarrow$ $1.218\ C_5H_7O_2N + 1.743\ N_2 + 3.066\ CO_2 + 5.502\ H_2O$	2.04:1
m-X	$C_6H_4(CH_3)_2 + 3.528\ NO_3^- + 1.218\ HCO_3^- + 1.218\ NH_4^+ + 3.528\ H^+ \rightarrow$ $1.218\ C_5H_7O_2N + 1.756\ N_2 + 3.108\ CO_2 + 5.544\ H_2O$	2.06:1
	Average	2.05:1

[1] McCarty's (1972) assumptions regarding the efficiency of energy capture and transfer are implicit, and the electron acceptor requirements of cell maintenance were neglected, in these determinations.

Table 5-11. Stoichiometric requirement for NO_3^-, neglecting synthesis

Compound	Reaction	Mass ratio of NO_3^- to BTEX compound (g/g)
B	$C_6H_6 + 6\ NO_3^- + 6\ H^+ \rightarrow 6\ CO_2 + 3\ N_2 + 6\ H_2O$	4.76:1
T	$C_6H_5CH_3 + 7.2\ NO_3^- + 7.2\ H^+ \rightarrow 7\ CO_2 + 3.6\ N_2 + 7.6\ H_2O$	4.85:1
E	$C_6H_5C_2H_5 + 8.4\ NO_3^- + 8.4\ H^+ \rightarrow 8\ CO_2 + 4.2\ N_2 + 9.2\ H_2O$	4.91:1
m-X	$C_6H_4(CH_3)_2 + 8.4\ NO_3^- + 8.4\ H^+ \rightarrow$ $8\ CO_2 + 4.2\ N_2 + 9.2\ H_2O$	4.91:1
	Average	4.86:1

addition to sulfate reduction, another TEAP, such as Fe(III)-reduction (Borden et al., 1995; Gieg et al., 1999) or methanogenesis (Acton and Barker, 1992; Gieg et al., 1999) occurred to a significant extent and therefore may have contributed to the oxidation of BTEX compounds. However, results obtained with pure cultures have demonstrated that the potential for oxidation of at least some of the BTEX compounds coupled to the reduction of sulfate in contaminated subsurface environments does exist. For example, sulfate-reducing strains have been isolated using toluene or o- or m-xylene as organic substrates (Harms et al., 1999; Heider et al., 1998). Several studies conducted using enrichment cultures also provide strong evidence that mineralization of benzene (Lovley et al., 1995; Phelps et al., 1996) and toluene (Beller et al., 1992; Edwards et al., 1992) is dependent on sulfate reduction. In general, when sulfate reduction is a dominant TEAP, toluene is one of the most rapidly degraded BTEX compounds in many monitored field sites, (e.g., Acton and Barker, 1992; Beller et al., 1995; Borden et al., 1995; Chapelle et al., 1996a; Davis et al., 1999), as well as in some laboratory microcosms and column reactors, (e.g., Barlaz et al., 1995; Edwards et al., 1992). The xylene isomers are also degraded relatively rapidly in BTEX mixtures under sulfate-reducing conditions. In some cases, m-xylene is degraded more rapidly than o- and p-xylene (e.g., Acton and Barker, 1992; Barlaz et al., 1995; Edwards et al., 1992)), while in other studies, oxidation of o-xylene preceded that of m- and p-xylene (e.g., Borden et al., 1995; Davis et al., 1999). However, ethylbenzene (Barlaz et al., 1995; Beller et al., 1995; Borden et al., 1995; Davis et al., 1999; Edwards et al., 1992) and, to a lesser extent, benzene (Beller et al., 1995; Edwards et al., 1992; Thierrin et al., 1993), sometimes appear to be recalcitrant or only minimally biodegradable when present in BTEX mixtures under sulfate-reducing conditions. The biodegradation of benzene is also often preceded by extended lag periods when it is present as the sole organic growth substrate in sulfate-reducing systems (e.g., Gieg et al., 1999; Kazumi et al., 1997), perhaps because benzene-degrading organisms are absent or are present only in low numbers in the sulfate-reducing zones of some aquifers (Weiner and Lovley, 1998a).

At least one sulfate-reducing strain appears to initiate toluene metabolism through addition to fumarate to form the intermediates benzylsuccinate, E-phenylitaconate, and benzoate, a series of reactions analogous to those used by some denitrifying strains in the biodegradation of toluene and m-xylene (Beller and Spormann, 1997; Biegert et al., 1996; Heider et al., 1998). Oxidation of benzene to CO_2 by a sulfate-reducing enrichment culture without the formation of phenol, benzoate, or several other potential intermediates has been observed (Lovley et al., 1995). However, in a recent study conducted with a sulfate-reducing enrichment from a petroleum-contaminated aquifer, transient accumulation of low concentrations of phenol and benzoate were observed during benzene metabolism (Caldwell and Suflita, 2000). The formation of phenol and benzoate can be explained as discussed above. The potential usefulness of benzoate, benzylsuccinate, and E-phenylitaconate as indicators of BTEX metabolism was also discussed above.

Balanced stoichiometric reactions of the oxidation of benzene, toluene, ethylbenzene, and m-xylene coupled to dissimilatory sulfate reduction for energy generation and cell synthesis are summarized in Table 5-12, along with the mass requirements of each BTEX compound for sulfate. Because the theoretical free energy yield of BTEX oxidation coupled to sulfate reduction is relatively low compared to BTEX oxidation under aerobic, denitrifying, or Fe(III)-reducing conditions, neglecting the utilization of a fraction of the BTEX compounds for cell synthesis (Table 5-13) increases the estimates of sulfate requirements only slightly.

NATURAL ATTENUATION OF HAZARDOUS WASTES 143

Table 5-12. Stoichiometric requirement for SO_4^{2-}, including synthesis[1]

Compound	Reaction	Mass ratio of SO_4^{2-} to BTEX compound (g/g)
B	$C_6H_6 + 3.42\ SO_4^{2-} + 0.12\ HCO_3^- + 0.12\ NH_4^+ + 5.16\ H^+ \rightarrow$ $0.12\ C_5H_7O_2N + 1.71\ H_2S + 1.71\ HS^- + 5.49\ CO_2 + 2.88\ H_2O$	4.21:1
T	$C_6H_5CH_3 + 4.14\ SO_4^{2-} + 0.144\ HCO_3^- + 0.144\ NH_4^+ + 6.23\ H^+ \rightarrow$ $0.144\ C_5H_7O_2N + 2.09\ H_2S + 2.09\ HS^- + 6.41\ CO_2 + 3.85\ H_2O$	4.32:1
E	$C_6H_5C_2H_5 + 4.75\ SO_4^{2-} + 0.21\ HCO_3^- + 0.21\ NH_4^+ + 7.098\ H^+ \rightarrow$ $0.21\ C_5H_7O_2N + 2.352\ H_2S + 2.352\ HS^- + 7.182\ CO_2 + 4.83\ H_2O$	4.29:1
m-X	$C_6H_4(CH_3)_2 + 4.75\ SO_4^{2-} + 0.21\ HCO_3^- + 0.21\ NH_4^+ + 7.14\ H^+ \rightarrow$ $0.21\ C_5H_7O_2N + 2.394\ H_2S + 2.394\ HS^- + 7.182\ CO_2 + 4.83\ H_2O$	4.29:1
	Average	4.28:1

[1]McCarty's (1972) assumptions regarding the efficiency of energy capture and transfer are implicit, and the electron acceptor requirements of cell maintenance were neglected, in these determinations.

Table 5-13. Stoichiometric requirement for SO_4^{2-}, neglecting synthesis

Compound	Reaction	Mass ratio of SO_4^{2-} to BTEX compound (g/g)
B	$C_6H_6 + 3.75\ SO_4^{2-} + 7.5\ H^+ \rightarrow 6\ CO_2 + 3.75\ H_2S + 3\ H_2O$	4.61:1
T	$C_6H_5CH_3 + 4.5\ SO_4^{2-} + 9\ H^+ \rightarrow 7\ CO_2 + 4.5\ H_2S + 4\ H_2O$	4.69:1
E	$C_6H_5C_2H_5 + 5.25\ SO_4^{2-} + 10.5\ H^+ \rightarrow$ $8\ CO_2 + 5.25\ H_2S + 5\ H_2O$	4.75:1
m-X	$C_6H_4(CH_3)_2 + 5.25\ SO_4^{2-} + 10.5\ H^+ \rightarrow$ $8\ CO_2 + 5.25\ H_2S + 5\ H_2O$	4.75:1
	Average	4.70:1

Biodegradation of the BTEX compounds under methanogenic conditions

There have been a number of reports of removal of BTEX compounds under methanogenic conditions in sediment systems, as summarized below. In most studies, the potential for methanogenic BTEX biodegradation was demonstrated by monitoring substrate removal and, in some cases, metabolite production, in laboratory incubations. However, evidence collected in several field studies also suggests that at least some of the BTEX compounds may be transformed *in situ* under methanogenic conditions (e.g., Gieg et al., 1999; Reinhard et al., 1984; Wilson et al., 1990). Most convincingly, Gieg et al. (1999) used collective lines of evidence to demonstrate that biodegradation contributed to the *in situ* removal of toluene in gas condensate-contaminated groundwater sediments, even under anaerobic conditions, and that methanogenesis occurred in the contaminated aquifer.

Consistent with observations of BTEX biodegradation under other anaerobic electron-accepting conditions, toluene generally appeared to be the most readily degradable BTEX compound under methanogenic conditions. Biodegradation of toluene frequently preceded the removal of the other BTEX compounds tested (e.g., Wilson et al., 1986). In some cases, recalcitrance of benzene (Acton and Barker, 1992; Edwards and Grbic-Galic, 1994; Phelps and Young, 1999), and to a lesser extent, of the xylene isomers and ethylbenzene (Gieg et al., 1999; Johnston et al., 1996) has been observed in toluene-degrading methanogenic systems.

The characteristics and pollution history of aquifer materials may play a role in determining the fate of BTEX compounds under methanogenic conditions. For example, conditions that favor the growth of sulfate-reducing bacteria (SRB) over methanogens, e.g., as in marine environments, or that support low levels of microorganisms in general, e.g., due to low organic matter levels, may not be conducive to the biodegradation of the BTEX compounds under methanogenic conditions (Phelps and Young, 1999). However, in some cases, the presence of natural organic compounds and/or co-contaminants may inhibit the removal of BTEX compounds if they are utilized preferentially by BTEX-degrading populations (Edwards and Grbic-Galic, 1994).

Lengthy adaptation periods frequently precede the removal of BTEX compounds in methanogenic laboratory incubations. For example, adaptation periods preceding the methanogenic biodegradation of toluene ranged from approximately 40 to 60 days (Gieg et al., 1999; Phelps and Young, 1999) to 100 days (Edwards and Grbic-Galic, 1994). Even longer adaptation periods have frequently been reported for the other BTEX compounds, e.g., benzene (140 to 280 days (Wilson et al., 1986); 420 days (Kazumi et al., 1997)), ethylbenzene (161 days (Phelps and Young, 1999)) and *o*-xylene (200 (Edwards and Grbic-Galic, 1994) to 371 days (Phelps and Young, 1999)). As noted below in the discussion of MTBE degradation under methanogenic conditions, long lag times in laboratory biodegradation studies may indicate that indigenous microorganisms are not adapted to the contaminant(s) and may not be metabolizing these compounds *in situ*. On the other hand, in one instance, benzene was biodegraded without a significant lag period in methanogenic sediments from a petroleum-contaminated aquifer (Weiner and Lovley, 1998b).

The mineralization of aromatic compounds under anaerobic conditions is a multi-step process and requires the involvement of multiple populations. However, in

Table 5-14. Stoichiometric production of CH_4, including synthesis[1]

Compound	Reaction
B	$C_6H_6 + 0.090\ HCO_3^- + 0.090\ NH_4^+ + 4.200\ H_2O \rightarrow$ $0.090\ C_5H_7O_2N + 3.570\ CH_4 + 2.130\ CO_2$
T	$C_6H_5CH_3 + 0.072\ HCO_3^- + 0.072\ NH_4^+ + 4.680\ H_2O \rightarrow$ $0.072\ C_5H_7O_2N + 4.320\ CH_4 + 2.340\ CO_2$
E	$C_6H_5C_2H_5 + 0.126\ HCO_3^- + 0.126\ NH_4^+ + 4.998\ H_2O \rightarrow$ $0.126\ C_5H_7O_2N + 4.194\ CH_4 + 2.520\ CO_2$
m-X	$C_6H_4(CH_3)_2 + 0.126\ HCO_3^- + 0.126\ NH_4^+ + 4.998\ H_2O \rightarrow$ $0.126\ C_5H_7O_2N + 4.956\ CH_4 + 2.520\ CO_2$

[1] McCarty's (1972) assumptions regarding the efficiency of energy capture and transfer are implicit, and the electron acceptor requirements of cell maintenance were neglected, in these determinations.

Table 5-15. Stoichiometric production of CH_4, neglecting synthesis

Compound	Reaction
B	$C_6H_6 + 4.5\ H_2O \rightarrow 2.25\ CO_2 + 3.75\ CH_4$
T	$C_6H_5CH_3 + 5.0\ H_2O \rightarrow 2.5\ CO_2 + 4.5\ CH_4$
E	$C_6H_5C_2H_5 + 5.5\ H_2O \rightarrow 2.75\ CO_2 + 5.25\ CH_4$
m-X	$C_6H_4(CH_3)_2 + 5.5\ H_2O \rightarrow 2.75\ CO_2 + 5.25\ CH_4$

order to simplify the construction of balanced stoichiometric equations, the reactions mediated by individual populations involved in the mineralization of a BTEX compound were lumped together and the production of methane from a BTEX compound was treated as a single step. The stoichiometric equations for the net BTEX mineralization reactions are summarized in Table 5-14. Mineralization of a BTEX compound results in a net production of CO_2, which is not included in Table 5-14. Stoichiometric equations for the oxidation of BTEX compounds that neglect cell synthesis are summarized in Table 5-15.

As observed under other anaerobic electron-accepting conditions, benzoate was a frequently detected intermediate in the methanogenic biodegradation of BTEX compounds. For example, biodegradation of toluene led to the formation of benzoate in an undefined, mixed methanogenic culture (Edwards et al., 1994). The first step in this conversion may be the formation of benzylsuccinate through the addition of toluene to fumarate (Beller and Edwards, 2000), which has been described above for toluene-degrading denitrifiers and SRB. Benzoate was also a putative intermediate in the biodegradation of benzene by methanogenic enrichments derived from a petroleum-contaminated site (Caldwell and Suflita, 2000). Evidence that benzene was converted to phenol by these enrichments was also obtained. Possible pathways that could lead to the production of phenol and/or benzoate were discussed above for a

Fe(III)-reducing culture that degraded benzene. In addition, in some cases, phenol produced as an intermediate of methanogenic benzene degradation may undergo transformation via a ring reduction pathway (Bakker, 1977), rather than undergoing conversion to benzoate (Weiner and Lovley, 1998b).

5.1.3.3.2 MTBE

Two structural features of the MTBE molecule, the ether linkage and methyl branching, are known barriers to the biodegradation of natural and anthropogenic compounds (Alexander, 1999; White et al., 1996). Until fairly recently MTBE was thought to be recalcitrant (Odencrantz, 1998; Squillace et al., 1998). However, increasingly the results of laboratory and field studies indicate that under certain conditions, intrinsic biodegradation of MTBE can occur. Aerobic biotransformation of MTBE occurs either cometabolically or in the absence of a primary growth substrate. Biotransformation of MTBE under a variety of anaerobic TEAP conditions has also been observed.

Aerobic degradation of MTBE

A wide range of pure and mixed microbial cultures are able to utilize short (C3–C5) normal or branched alkanes with a single methyl substitution as primary growth substrates for the aerobic cometabolism of MTBE (Garnier et al., 1999b; Hardison et al., 1997; Hyman et al., 1998; Hyman and O'Reilly, 1999; Hyman et al., 2000; Steffan et al., 1997). Cometabolism of MTBE using alkanes with slightly longer chains as primary growth substrates has also been demonstrated (Garnier et al., 1999a; Hyman et al., 2000). Other growth substrates that have been shown to support MTBE in some cultures include alcohols (Hernandez-Perez et al., 2001; Steffan et al., 1997), cyclohexane (Corcho et al., 2000), camphor (Steffan et al., 1997), diethyl ether (Garnier et al., 1999a; Hardison et al., 1997), and benzene (Koenigsberg et al., 1999).

The widespread ability to use short chain alkanes as primary growth substrates for cometabolism of MTBE is significant because some of these compounds are major components of gasoline and, therefore, are likely to be initially present in MTBE-impacted environments as co-contaminants. However, the alkane compounds found in gasoline tend to be relatively immobile due to their low solubilities, and typically do not migrate significantly beyond the source. In contrast, MTBE has a high solubility and low retardation factor, as discussed above. As a result, short-chain alkanes are unlikely to be present and support cometabolic transformations throughout MTBE plumes.

The most frequently detected intermediate in the cometabolic oxidation of MTBE is TBA, which has been observed in studies involving fungal (Hardison et al., 1997) or bacterial (Hyman et al., 1998; Steffan et al., 1997) populations. Two different routes for the cometabolic production of TBA from MTBE have been proposed (Fig. 5-2). The production of TBA in *Pseudomonas putida* strain CAM and strain ENV425 is attributed to one of the most prevalent strategies for disrupting aliphatic ether bonds (White et al., 1996). In this pathway, an initial monooxygenase attack hydroxylates the methoxy group by inserting one oxygen atom, which leads to the formation of the unstable hemiacetal *tert*-butoxymethanol. This intermediate spontaneously dismutates to form formaldehyde and TBA. Rapid oxidation of *tert*-

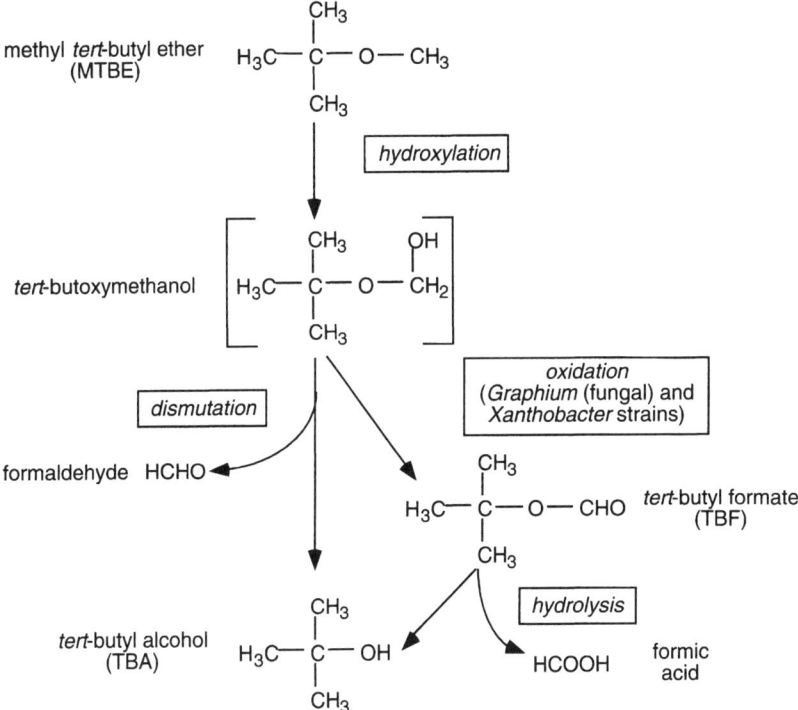

Figure 5-2. Proposed pathways for the aerobic biotransformation of MTBE to TBA.

Source: Modified from Hardison et al. (1997).

butoxymethanol to *tert*-butyl formate (TBF) by an alcohol dehydrogenase, followed by biotic and abiotic hydrolysis of TBF to formate and TBA (Fig. 5-2), has been proposed to explain the detection of TBF and TBA as intermediates in the cometabolism of MTBE by fungal (Hardison et al., 1997) and *Xanthobacter* strains (Hyman et al., 1998). In addition to TBA, 2-methyl-2-hydroxymethanol, 2-hydroxy isobutyric acid (HIBA), and formaldehyde were detected as intermediates of MTBE cometabolism by the propane-grown strain ENV425 (Steffan et al., 1997). Presumably, formaldehyde was produced, along with TBA, via the monooxygenation/dismutation mechanism shown in Fig. 5-2.

As summarized by Deeb et al. (2000), strain ENV425 and several other cultures oxidize TBA at a slower rate than MTBE, which may explain why it is detected as an intermediate in MTBE cometabolism in some cases. In other organisms, MTBE is not transformed beyond TBA (Hardison et al., 1997). The accumulation of TBA during aerobic MTBE biotransformation is of interest because of its demonstrated carcinogenicity in lab animals (Squillace et al., 1998) and potential use as an indicator of MTBE bioattenuation (Church et al., 1997). However, Piveteau et al. (2001) recently isolated a facultative methylotroph (strain CIP I-2052) that is able to transform TBA at a relatively high rate, using it as a sole carbon and energy source. These results, and other observations of TBA degradation (summarized in Deeb et al. (2000)) suggest that complete mineralization of MTBE is feasible through association of TBA-utilizing organisms like strain CIP I-2052 with populations that cometabolically transform MTBE.

The results of several studies suggest that cometabolism of MTBE may also ultimately lead to the production of at least some CO_2 in pure cultures of strain ENV425 and other organisms. Because there are five carbon atoms in a MTBE molecule, the conversion of greater than 20% of the radioactivity added as [U-^{14}C]MTBE to $^{14}CO_2$ indicates that the tertiary carbon structure was attacked. On the other hand, the production of approximately 20% $^{14}CO_2$ could result from the cleavage and subsequent oxidation of the methoxy methyl group alone, unless all of the remaining radioactivity is present as [U-^{14}C]MTBE. Thus, the conversion of >60% [U-^{14}C]-MTBE to $^{14}CO_2$ in less than 30 h by strain ENV425 (Steffan et al., 1997) indicates that in some organisms, the tertiary carbon structure of MTBE may be subject to cometabolic attack. On the other hand, in *P. putida* strain CAM, [U-^{14}C]-MTBE was converted to approximately 25% $^{14}CO_2$, which may be attributable to reactions involving only the methoxy methyl group.

Unlike the organisms described above, which transformed MTBE cometabolically, several isolates (Hanson et al., 1999; Mo et al., 1997) and a mixed culture (Salanitro et al., 1994) were able to transform MTBE in the absence of an alternative source of carbon and energy. However, the results obtained in these studies vary widely with respect to the percent removal and the extent of mineralization of MTBE. For example three bacterial strains, members of the genera *Methylobacterium*, *Rhodococcus*, and *Arthrobacter*, were able to transform only 28 to 29% of an initial MTBE dose of 200 ppm over a two-week period (Mo et al., 1997). In contrast, *Rubrivivax* sp. PM1 achieved complete removal of a 500 mg·l^{-1} dose of MTBE within ten days (Hanson et al., 1999). Complete removal of MTBE (120 mg·l^{-1}) was also observed in a batch experiment conducted with the mixed culture BC-1 (Salanitro et al., 1994).

Experiments conducted with [U-^{14}C]MTBE suggest that not only the percent removal, but the extent of MTBE mineralization mediated by the *Rubrivivax* strain (Hanson et al., 1999) was greater than that observed with the three MTBE-transforming isolates obtained by Mo et al. (1997). Only 7 to 8% of the [U-^{14}C]-MTBE added to the *Methylobacterium*, *Rhodococcus*, and *Arthrobacter* strains was recovered as $^{14}CO_2$ (Mo et al., 1997). Therefore, the results obtained by Mo et al. (1997) do not demonstrate that the entire MTBE molecule was mineralized. In contrast, transformation of [U-^{14}C]-MTBE by the *Rubrivivax* strain resulted in the production of 19% ^{14}C-biomass and 46% $^{14}CO_2$, which indicates that the tertiary carbon structure was attacked.

As noted above, TBA is a frequently detected intermediate in the cometabolic transformation of MTBE (Hardison et al., 1997; Hyman et al., 1998; Steffan et al., 1997). TBA was also detected as a transient intermediate in the transformation of MTBE by the mixed culture BC-1 (Salanitro et al., 1994). Therefore, the biodegradation of MTBE cannot be attributed to either a growth-related or cometabolic process based solely on the production of TBA. TBA was transformed by BC-1 at a slower rate than MTBE. However, as summarized by Deeb et al. (2000), growth of some cultures on MTBE does not result in accumulation of TBA.

Aerobic growth of a pure culture utilizing MTBE as the sole source of carbon and energy has been unequivocally demonstrated only for the *Rubrivivax* strain (Hanson et al., 1999) although increases in biomass associated with MTBE utilization by mixed cultures have also been reported, (e.g., Park and Cowan, 1997). In some studies, alternative electron donors were supplied to the MTBE-degrading cultures and may have contributed to the observed increases in cell numbers or biomass (Mo et al., 1997; Salanitro et al., 1994). Although utilization of MTBE as a sole growth substrate is clearly feasible, based on the laboratory studies conducted to date, it is difficult to predict how widely distributed this form of metabolism is and its importance in contaminated environments.

As noted by Garnier et al. (1999b), rates of MTBE transformation in the absence of an alternative growth substrate are often difficult to compare, especially with cometabolic MTBE transformation rates, due to differences in reporting and the methods used to determine these values. However, in general, it appears that pure cultures that grow on short *n*- and *iso*-alkanes can cometabolize MTBE at rates ranging from approximately 2 to 14 $nmol \cdot min^{-1} \cdot mg^{-1}$ of cell protein (Garnier et al., 1999b; Hyman et al., 2000; Steffan et al., 1997); although somewhat faster rates have been observed by Hyman et al. (1998). In comparison, a mixed culture (Salanitro et al., 1994) and a newly characterized isolate (Steffan et al., 2000) are able to transform MTBE at rates of 12 and 46 $nmol \cdot min^{-1} \cdot mg^{-1}$ of cell protein, respectively, in the absence of an alternative growth substrate, although transformation of MTBE in the absence of an alternative growth substrate sometimes proceeds at much slower rates. Based on the limited rate data available, it appears that organisms that cometabolize MTBE transform the fuel oxygenate at similar or slower rates than pure and mixed cultures that completely mineralize MTBE in the absence of an alternative growth substrate.

The studies of aerobic MTBE degradation described above were conducted with pure or enrichment cultures. Evaluations of field data and experiments involving undefined sediment samples suggest that indigenous organisms at a variety of sites can biodegrade MTBE if adequate O_2 is available. For example, MTBE removal was observed in a naturally oxic site where groundwater discharged to a ditch (Landmeyer et al., 2001). High MTBE levels were observed in anoxic groundwater unless an artificial source of O_2 was supplied. Significant conversion of [U-^{14}C]-MTBE to $^{14}CO_2$ was also observed in surface water sediments taken from sites throughout the United States (Bradley et al., 2001c). The amount of $^{14}CO_2$ produced was inversely related to the content of silt- and clay-sized grains in the sediment. Presumably, coarser, more permeable grains allowed greater diffusion of O_2 into the statically incubated sediments.

Degradation of MTBE under methanogenic conditions

The potential for intrinsic biodegradation of MTBE under anaerobic conditions is of interest because the source zone of petroleum-contaminated aquifers is typically anaerobic (Fig. 5-1). Biodegradation of MTBE under methanogenic conditions has been examined in a variety of systems including a contaminated groundwater plume (Wilson et al., 2000) and laboratory microcosms containing soil and water from uncontaminated sites (Yeh and Novak, 1994), contaminated river sediment (Mormile et al., 1994), or contaminated aquifer material (Wilson et al., 2000). In some of the batch laboratory experiments, the extent of MTBE transformation was limited (Bradley et al., 2001; Mormile et al., 1994) or required the addition of starch and nutrients (Yeh and Novak, 1994). However, extensive removal of the fuel oxygenate under methanogenic conditions has also been observed in the absence of added electron donors or nutrients (Wilson et al., 2000). Degradation of [U-^{14}C]-MTBE to $^{14}CO_2$ and $^{14}CH_4$ was also observed in microcosms constructed with aquatic sediment in which methane production, sulfate reduction, and iron reduction were simultaneously taking place (Finneran and Lovley, 2001).

It is important to note that in each of the laboratory batch experiments conducted under methanogenic conditions, transformation of MTBE was preceded by an adaptation period of at least 150 days. Based on the results of these studies, batch experiments designed to evaluate the potential for anaerobic biotransformation of MTBE in environmental samples should have a duration of at least 200 to 300 days. In addition, if lengthy adaptation periods are required in laboratory evaluations, the potential for removal of MTBE from methanogenic regions of contaminated aquifers may not be realized. As discussed below, the presence of other organic compounds may inhibit MTBE metabolism. In a methanogenic source area, other organic compounds will be continuously supplied. Given the high mobility of MTBE compared to other fuel components, MTBE is not likely to reside in methanogenic regions as long as alternative compounds. Therefore, it is feasible that MTBE will be washed out of methanogenic regions before adaptation can occur.

As previously noted, TBA has been observed as an intermediate of cometabolic and growth-related transformation of MTBE under aerobic conditions. Increased levels of TBA associated with decreases in MTBE concentrations under methanogenic conditions have also been observed in a contaminated groundwater plume (Wilson et al., 2000) and in a laboratory microcosm (Mormile et al., 1994), suggesting that TBA is also an intermediate of MTBE biodegradation under methanogenic conditions. O-Demethylation of MTBE has been suggested as an explanation for the production of TBA in these systems (Mormile et al., 1994). [U-^{14}C]TBA was rapidly converted to $^{14}CH_4$ and $^{14}CO_2$ by a variety of anaerobic TEAP, including methanogenesis, without a lag period, in aquatic sediment microcosms (Finneran and Lovley, 2001). These results suggest that if MTBE transformation is initiated under methanogenic conditions, complete mineralization may be feasible.

Potential for degradation of MTBE with other terminal electron acceptors

MTBE removal under other terminal electron-accepting conditions, e.g., nitrate-, sulfate-, and Fe(III)-reducing, has been evaluated in only a few studies (Bradley et al., 2001a, 2001b; Finneran and Lovley, 2001; Landmeyer et al., 1998; Mormile et al., 1994; Somsamak et al., 2001; Yeh and Novak, 1994). Bradley et al. (2001b) observed conversion of ~25% of the [U-^{14}C]-MTBE radioactivity to $^{14}CO_2$ in nitrate-amended anoxic surface water sediments. TBA did not accumulate and appeared to be biodegradable. Concomitant removal of MTBE and sulfate reduction have been observed in two studies involving sulfate-amended sediment microcosms (Bradley et al., 2001a; Somsamak et al., 2001). In other studies, MTBE was not transformed under nitrate- or sulfate-reducing conditions (Mormile et al., 1994; Yeh and Novak, 1994). MTBE was rapidly degraded in aquifer sediments that were amended with humic substances and Fe(III), but persisted in unamended sediments (Finneran and Lovley, 2001). Presumably, Fe(III)-reducers oxidized MTBE and transferred the electrons to humic substances, which can abiotically transfer electrons to Fe(III) and are thereby regenerated. In addition, conversion of a small percentage (~3%) of [U-^{14}C]-MTBE to $^{14}CO_2$ has been observed in microcosms constructed with naturally Fe(III)-reducing aquifer material and groundwater (Landmeyer et al., 1998).

Effect of different organic compounds on MTBE degradation

The presence of other organic compounds has been shown to either inhibit, stimulate, or have no effect on MTBE transformation in various systems. The presence of organic compounds may inhibit MTBE removal through toxicity effects, inhibition of enzymes involved in MTBE transformation, or their utilization as preferential growth substrates, which may also decrease the amounts of electron acceptors and nutrients available for MTBE biodegradation. Stimulatory effects of other organic compounds often may be indicative of removal of MTBE via a cometabolic mechanism. Therefore, the nature of the organic compound, the physiological properties of the organisms mediating MTBE biotransformation, and the availability of electron acceptors and nutrients also undoubtedly play a role in determining the effects of other organic compounds on MTBE removal.

As suggested above, the presence of a compound that can serve as primary growth substrate for MTBE-cometabolizing organisms (e.g., short chain n-alkanes) has been shown to enhance or be essential for MTBE removal via cometabolism, (e.g., Garnier et al., 1999a). However, short-chain n-alkanes may be toxic. Consequently, as observed in a study of MTBE cometabolism mediated by *Pseudomonas aeruginosa* growing on n-pentane (Garnier et al., 1999b), n-alkanes may serve as potential growth substrates and, at the same time, inhibit growth and cometabolic transformation of MTBE, if present at sufficiently high concentrations.

Cometabolism of MTBE with an alkane growth substrate has been observed in the presence of BTEX compounds (Steffan et al., 1997), although inhibition of MTBE cometabolism by relatively low concentrations of toluene and higher levels of benzene has also been noted (Corcho et al., 2000). Similarly, when MTBE was the sole contaminant supplied to model aquifer columns, conversion to TBA was detected (Church et al., 2000), but no removal of MTBE was detected when the BTEX compounds were supplied as co-contaminants.

The effect of the BTEX compounds on the transformation of MTBE by a pure culture in the absence of an alternative growth substrate has been studied extensively (Deeb et al., 2001). Growth of *Rubrivivax* sp. PM1 on either MTBE or benzene has been demonstrated, and both compounds were used simultaneously by the pure culture. However, benzene and MTBE degradation by strain PM1 appeared to occur via two independent and inducible pathways, and MTBE degradation by strain PM1 was inhibited by the BTEX compounds. Together with observations of inhibition of aerobic MTBE transformation in model aquifer columns by the BTEX compounds (Church et al., 2000), the results obtained with *Rubrivivax* sp. PM1 suggest that MTBE removal rates may be significantly reduced in the presence of BTEX co-contamination, even if organisms that utilize MTBE as a sole source of carbon and energy are present.

In addition to certain co-contaminants, high concentrations of endogenous organic matter may also inhibit MTBE removal in environmental samples under aerobic (Bradley et al., 1999) and methanogenic (Yeh and Novak, 1994) conditions. Under either condition, preferential utilization of the high levels of natural organic matter could explain the observed inhibition of MTBE transformation. Competitive consumption of electron acceptors could also play a role under aerobic conditions.

5.1.4 Demonstrating Natural Attenuation of Petroleum Hydrocarbons & MTBE

5.1.4.1 Protocols and Procedures

The application of natural attenuation is best developed for BTEX contamination, and several protocols and guidance documents on demonstrating natural attenuation of petroleum hydrocarbons have been developed (Chapter 1). As discussed in Chapter 1, many of these protocols for natural attenuation of petroleum hydrocarbons have used an independent lines-of-evidence approach to demonstrate that natural attenuation is occurring and at a sufficient rate to be protective of human health and the environment (NRC, 1993).

The focus here is on evidence documenting that the proposed natural attenuation process(es) actually occur(s) in the field. This is the most difficult evidence to obtain, and the most crucial, because it provides the linkage between the evidence of contaminant removal in the field and the natural attenuation mechanisms (NRC, 1993). It may be especially important if the historical contaminant concentration data are inconclusive, or if assessment efforts have only recently been initiated (ASTM, 1998). Because intrinsic biodegradation contributes most significantly to the natural attenuation of petroleum hydrocarbons, evidence that this process is occurring is especially important.

Although the attenuation mechanisms resulting in contaminant transformation/ degradation cannot typically themselves be measured directly, these processes often consume or result in production of other materials that can be measured, i.e., the processes leave "footprints" (NRC, 2000). The key measurable footprints of BTEX and MTBE biodegradation are reviewed in the following paragraphs, including: electron acceptors, inorganic carbon, alkalinity, metabolic intermediates (organic acids, TBA), redox potential, microbial numbers and activity, carbon stable isotopes, and the ratio of nondegradable to degradable substrates. Data sets from field sites demonstrating use of several different combinations of footprints as indicators of intrinsic bioremediation are summarized in Table 5-16.

Table 5-16. Summary of example data sets from field sites demonstrating use of various combinations of footprints as indicators of natural attenuation

Site	Contaminant(s)	Contaminants controlled?	Footprints
Bemidji, MN, U.S.A. (Cozzarelli et al. 1990; Baedecker et al., 1993)	BTEX (crude oil spill)	Yes	Loss of O_2; production of Fe^{2+}, Mn^{2+}, and CH_4; increase in DIC; increase in alkalinity; production of phenol plus aromatic, alicyclic, and straight- and branched-chain aliphatic organic acids in the anoxic plume; drop in E_h (probe); DIC enriched in ^{12}C under nonmethanogenic conditions; DIC enriched in 13C under methanogenic conditions; differential loss of alkylbenzenes
Aquifer near Rocky Point, NC, U.S.A. (Borden et al., 1995)	BTEX (gasoline contamination)	Yes	Loss of O_2, NO_3^-, SO_4^{2-}; production of Fe^{2+} and CH_4; increase in DIC; increase in alkalinity; drop in E_h (probe); preferential removal of certain BTEX components
Shallow aquifer underlying natural gas production site (Gieg et al., 1999)	Gas condensate hydrocarbons (96% w/w C_5–C_{15} compounds, including 18% w/w BTEX)	Yes	Loss of O_2, NO_3^-, SO_4^{2-}, and bioavailable Fe^{3+}; production of dissolved sulfides, Fe^{2+} and CH_4; increase in alkalinity; detected methylbenzylsuccinic acids, signature metabolites of anaerobic xylene biodegradation; dissolved H_2 levels indicative of iron reduction, sulfate reduction, methanogenesis; 10–1000 fold higher numbers of sulfate reducers and methanogens in contaminated sediments compared to background;
Aquifer in Sampson County, NC, U.S.A. (Borden et al., 1997)	MTBE (gasoline and diesel fuel spill)	Partially	Loss of O_2; high NO_3 availability; minor Fe^{2+} production; increase in DIC coinciding with BTEX plume; E_h (probe) > +200 mV indicating oxidizing conditions in all wells; mass flux analysis indicated MTBE degradation near contaminant source, but not in downgradient aquifer (supported by laboratory microcosm studies showing limited MTBE biodegradation and TBA production near the source but not downgradient)

5.1.4.2 "Footprints" for Petroleum Hydrocarbons

5.1.4.2.1 Electron Acceptors and Products of their Reduction

Consumption of electron-acceptor co-reactants can be a key qualitative indicator of intrinsic biodegradation. For example, depletion of oxygen correlated with contaminant loss provides evidence of intrinsic bioremediation (e.g., Chiang et al., 1989; McAllister and Chiang, 1994). In addition, given that aerobic and anaerobic biodegradation of the petroleum hydrocarbons may take place simultaneously in different parts of the plume (Fig. 5-1), the concentrations of alternative electron acceptors, especially NO_3^- and SO_4^{2-}, should also be monitored if they are initially present. For some TEAPs, the product of reduction of the electron acceptor is measured rather than the electron acceptor itself. For example, Fe(III) is reduced to Fe(II), which is much more soluble in water and can be used as an indicator for biodegradation via iron reduction (Wiedemeier et al., 1995). Similarly, the presence of methane at concentrations above background may be indicative of petroleum hydrocarbon biodegradation.

Changes in electron acceptors can also be evaluated quantitatively, based on the stoichiometry of BTEX biodegradation via various TEAPs summarized above. If the background concentrations of electron acceptors (or the product of reduction of the electron acceptor) are known, along with their concentrations in the contaminated zone, that information can be coupled with the stoichiometric relationships (e.g., Tables 5-6 through 5-15) and the volume of contaminated groundwater to estimate the mass of BTEX that has been lost due to biodegradation via the various TEAPs. These calculations provide an indication of the intrinsic capacity of the aquifer system to degrade BTEX. Such calculations are an important part of estimating the sustainability, i.e., the long-term viability, of natural attenuation (NRC, 2000). For example, the potential mass of benzene biodegraded by aerobic respiration is given by (Wiedemeier et al., 1995):

$$\Delta Benzene_{Bio,DO} = (1/\alpha)(O_B - O_M) \qquad (5.1.1)$$

Where:
- $\Delta Benzene_{Bio,DO}$ = the reduction in benzene concentration via aerobic respiration, mg/L;
- α = stoichiometric ratio, mg benzene degraded per mg dissolved oxygen consumed, e.g., α = 2.08 mg O_2/mg benzene, Table 5-6;
- O_B = background dissolved oxygen concentration, mg/L;
- O_M = lowest measured dissolved oxygen concentration, mg/L.

Thus, if the groundwater has a background dissolved oxygen concentration, O_B = 6.0 mg/L, and the concentration of dissolved oxygen in the core of the plume (i.e., where the highest BTEX concentration exists), O_M = 0.0 mg/L, then the groundwater has the capacity to assimilate 2.9 mg/L of benzene. Similar calculations can be performed for the other BTEX compounds, and for denitrifying, iron-reducing, sulfate-reducing, and methanogenic conditions (Wiedemeier et al., 1995). For an example of application of these calculations for calculating the assimilative capacity of site groundwater, see

Kampbell et al. (1996). One difficulty with this approach, however, is determining how to attribute the observed reduction in electron acceptors to the oxidation of several specific contaminants, such as BTEX, or other organic compounds present in the system. For example, which portion of the oxygen, nitrate, sulfate, etc. consumed in the contaminant plume can be attributed to benzene degradation versus toluene, ethylbenzene, or xylenes present in the system?

In addition to co-reactants, products and intermediates of petroleum hydrocarbon and MTBE biodegradation, and composite parameters that are influenced by these biodegradation products or intermediates may serve as measurable footprints. These potential footprints include: inorganic carbon, alkalinity, organic acids, TBA, redox potential, and microbial numbers and activity.

5.1.4.2.2 Dissolved Inorganic Carbon (DIC)

As shown in the stoichiometric reactions provided above, the oxidation of hydrocarbons in groundwater via the various TEAPs directly produces CO_2, which is absorbed by the water, forming mostly HCO_3^- and H_2CO_3. As a result, the DIC of the water should increase in proportion to the mineralization of the hydrocarbon. Therefore, estimating the electron acceptor supply rate, as discussed above, and correlating it with increases in DIC, can theoretically provide a quantitative estimate of the rate of petroleum hydrocarbon biodegradation (NRC, 2000). In reality, accurate measurement of DIC resulting from biodegradation can be difficult because of interactions with the carbonate-buffering system in groundwater (ASTM, 1998).

5.1.4.2.3 Total Alkalinity (TA)

TA is a parameter that may be influenced by consumption of microbial substrates and the products of microbial metabolism. Several protocols for evaluating natural attenuation of fuel hydrocarbons include measurement of TA in an array of analyses to be performed on field samples (e.g., ASTM, 1998; Wiedemeier et al., 1995). CO_2 produced during biodegradation forms carbonic acid (Table 5-17), which may dissolve carbonate minerals, if present in the aquifer, and increase the alkalinity (e.g., Chapelle and Bradley, 1997). Therefore, the Air Force Center for Environmental Excellence protocols, (e.g., Wiedemeier et al., 1995), suggest that increases in TA provide a qualitative indicator of CO_2 production and biodegradation (Table 5-17).

However, in addition to carbonate dissolution, many other chemical and microbially-mediated reactions relevant to bioremediation can impact the TA and its interpretation (Seagren et al., 1998). Some of these effects are summarized as reactions (ignoring biomass formation) in Table 5-17. Importantly, aerobic respiration alone should not alter the TA. In this case, the addition of dissolved CO_2 increases the acidity of the system and the total concentration of the dissolved carbonic species, but does not affect the alkalinity, because dissociation of H_2CO_3 produces equal numbers of acid and base ions (Stumm and Morgan, 1981). Similarly, hydrocarbon mineralization under methanogenic conditions does not directly affect the TA; however, organic acids produced as intermediates during anaerobic degradation, as discussed above, may contribute to the TA measured (e.g., Willey et al., 1975). On the other hand, an increase in alkalinity is expected under denitrifying, sulfate reducing, and dissimilatory Fe(III)-reducing conditions due to proton consumption (Table 5-17)

Table 5-17. Summary of chemical and microbially-mediated phenomena relevant to bioremediation and their effects on TA, using toluene as the example compound

Process with Example Reaction	TA change for forward reaction
CaCO₃ Dissolution:	
$CaCO_3 + CO_2 + H_2O \rightarrow Ca^{2+} + 2HCO_3^-$	Increase
Aerobic Respiration:	
$9O_2 + C_7H_8 \rightarrow 7CO_2 + 4H_2O$	No change
Denitrifying, Sulfate Reducing, and Fe(III) Reducing Conditions:	
Denitrification:	
$7.2NO_3^- + C_7H_8 + 7.2H^+ \rightarrow 7CO_2 + 3.6N_2 + 7.6H_2O$	Increase
Sulfate Reduction:	
$4.5SO_4^{2-} + C_7H_8 + 9H^+ \rightarrow 7CO_2 + 4.5H_2S + 4H_2O$	Increase
Fe(III) Reduction:	
$36Fe(OH)_3 + C_7H_8 + 72H^+ \rightarrow 7CO_2 + 36Fe^{2+} + 94H_2O$	Increase
Methanogenic Conditions (not considering organic acids):	
$5H_2O + C_7H_8 \rightarrow 2.5CO_2 + 4.5CH_4$	No change
Abiotic Oxidation of Inorganic Ions:	
Abiotic Fe(II) Oxidation Reaction:	
$Fe^{2+} + 0.25O_2 + 2OH^- + 0.5H_2O \rightarrow Fe(OH)_3 (s)$	Decrease

Source: adapted from Seagren et al., 1998.

(Seagren et al., 1998). Abiotic oxidation of inorganic ions can cause a decrease in TA. Based on these examples, although TA can be a useful monitoring tool, interpreting changes in TA, or the lack thereof, is not trivial.

5.1.4.2.4 Metabolic Intermediates

Organic acids

As discussed above, organic acids are known metabolic intermediates in the anaerobic microbial oxidation of petroleum hydrocarbons and, thus, may provide a footprint of intrinsic biodegradation. In fact, the detection of a variety of aliphatic and aromatic acids in the anoxic contaminant plumes at a number of petroleum hydrocarbon-impacted sites has been reported (e.g., Cozzarelli et al., 1994; Cozzarelli et al., 1990; Kampbell et al., 1996; Schmitt et al., 1996), including the "signature metabolites" of anaerobic biodegradation of BTEX compounds, which were introduced above (e.g., Beller, 2000; Beller et al., 1995; Gieg et al., 1999).

However, there are some complications in using organic acids as a footprint. For example, the transport and fate of organic acids in groundwater environments may be complex. As reviewed elsewhere (Barcelona et al., 1993), not only are organic acids biologically reactive, but they are also geochemically reactive, impacting processes such as mineral dissolution and metal complexation. In addition, analysis of organic acids may be complicated by two other factors: (1) significant amounts of the organic acids in background and contaminated samples may be associated with the aquifer solids, and (2) advanced acid derivatization techniques may be required for quantitative analysis of organic acids in environmental samples (e.g., Barcelona et al., 1995; Barcelona et al., 1993). Finally, any oxygenated metabolic intermediates produced during aerobic aromatic hydrocarbon degradation are apparently metabolized rapidly (Cozzarelli et al., 1990).

TBA

As noted above, the principal daughter product observed during MTBE biodegradation is TBA. TBA is sufficiently resistant to further degradation to accumulate as an intermediate before being degraded itself (Church et al., 1997), which suggests that TBA may be a useful indicator of MTBE biodegradation.

However, use of TBA as a footprint of MTBE biodegradation in practice is complicated by the fact that TBA itself is sometimes used as a fuel additive, as discussed earlier; therefore, its appearance does not provide conclusive evidence of MTBE biodegradation (Anthony et al., 1999). In addition, it is difficult to measure low concentrations of TBA in water (Church et al., 1997). Finally, to date, there is very little conclusive field evidence of TBA resulting from intrinsic bioremediation of MTBE. For example, groundwater samples at eight MTBE-contaminated sites were analyzed for the presence of TBA (Church et al., 1997; Landmeyer et al., 1998; Schirmer and Barker, 1998). At six sites, an improved analytical method with a TBA detection limit of 0.1 µg/L was used (Church et al., 1997). TBA was detected in the groundwater samples taken from only two sites (Church et al., 1997; Landmeyer et al., 1998). In both cases, it appeared that TBA was a constituent of the initial gasoline contaminant and, therefore, was not necessarily indicative of MTBE biodegradation. In one case, TBA was detected in a laboratory column study (but not in laboratory microcosms) (Schirmer et al., 1999), even though no TBA was detected in the groundwater samples (Schirmer and Barker, 1998).

Similarly, Wilson et al. (2000) found no evidence of accumulation of TBA in the overall groundwater plume at a former fuel farm site, with concentrations of TBA generally less than 200 µg/L. However, at a location immediately downgradient of the source, a higher concentration of TBA (near 2,000 µg/L) with a corresponding reduction of MTBE concentration was observed, which suggests that the TBA at that location was probably a transformation product of MTBE. Therefore, if MTBE is a potential contaminant at a site, the distribution and concentrations of MTBE and TBA should be evaluated to determine whether stoichiometric increases in TBA concentrations are associated with decreases in MTBE (Anthony et al., 1999).

Table 5-18. Dissolved hydrogen concentration ranges characteristic of various dominant terminal electron accepting processes in natural systems

Dominant terminal electron accepting process	Dissolved hydrogen concentration (nM)
Denitrification	<0.1
Iron Reduction	0.1 to 0.8
Sulfate Reduction	1 to 4
Methanogenesis	5 to 25

Source: Vroblesky and Chapelle, 1994.

5.1.4.2.5 Oxidation-Reduction (Redox) Potential

The biodegradation of organic contaminants in the subsurface typically results in the sequential depletion of increasingly less favorable electron acceptors. Thus, the redox potential (E_h) of contaminated groundwater should theoretically decrease as biodegradation proceeds and may be a useful indicator of bioremediation. Knowing the distribution of microbially-mediated redox processes in groundwater systems is also key to predicting the fate and transport of contaminants in groundwater systems because they affect the rate and extent of the biodegradation of organic contaminants, as discussed above (Chapelle et al., 1996b). The classic geochemical indicator of redox processes is platinum electrode measurement of redox potential. However, the problems associated with this measurement in complex systems are well documented (e.g., Chapelle et al., 1996b; Lovley and Goodwin, 1988). As a result of these limitations, E_h measurements at best may be able to delineate oxic, transitional, and anoxic zones (e.g., Chapelle et al., 1996b; Harkness and Braccom 1998).

Measurement of H_2 concentrations is an alternative and potentially more useful indicator of anoxic redox processes than the platinum electrode measurement (Chapelle et al., 1996b; Lovley and Goodwin, 1988). Based on several studies, Vroblesky and Chapelle (1994) identified ranges of H_2 concentrations that are indicative of various terminal electron acceptor processes (Table 5-18). These H_2 concentration ranges are increasingly being used to characterize microbially-mediated redox processes in contaminated groundwater. However, H_2 is more reliable if interpreted in the context of electron acceptor availability and the presence of microbial metabolic byproducts (Chapelle et al., 1996b). In some cases, it may not be possible to discern a pattern based on the geochemical indicators and H_2 measurements. Possible interferences include transport of soluble geochemical indicators (e.g., methane), the occurrence of different anaerobic processes simultaneously in heterogeneous microenvironments, volumetric averaging of H_2 concentrations during groundwater sampling, and H_2 detection limitations (Gieg et al., 1999). In addition, measured H_2 concentrations can be affected by a number of factors including the sampling and pumping methods, and the sample-well casing material (Chapelle et al., 1997).

Another product of microbial metabolism that can potentially provide an indication of the predominant electron-accepting reactions is organic acids (e.g., Barcelona, 1980; Barcelona et al., 1993; Cozzarelli et al., 1994; Lovley and Phillips, 1987). Seagren and Becker (1999) investigated the use of organic acid concentrations as indicators of redox conditions in natural and petroleum-hydrocarbon contaminated sites. Using examples from the literature, coupled with a theoretical evaluation, they concluded that as the terminal electron acceptor becomes more reduced, increasing concentrations of aliphatic acids, especially acetate, are observed. At sites contaminated with mono- and polycyclic aromatic hydrocarbons, variations in the composition of the acids are also observed, depending on the dominant TEAP. However, at one petroleum hydrocarbon-contaminated aquifer, a lack of correlation between the concentration of low molecular weight, aliphatic organic acids and dissolved H_2 was observed (Vroblesky et al., 1997). Thus, it appears that the relationship between microbially mediated redox reactions and the measurement of organic acid concentrations, like H_2 and platinum electrode measurements, may not be unambiguous at some sites.

5.1.4.2.6 Microbial Numbers and Activity

A correlation between an increase in the number and activity of contaminant-degrading bacteria and contaminant loss can be a key indication of successful bioremediation, because BTEX metabolism is often associated with microbial growth (NRC, 1993). At some sites, patterns in both the distribution of microbial numbers and activity have been used to assess the importance of intrinsic bioremediation. For example, at a fuel spill site near Barrow, AK, the proportion of heterotrophic bacteria that were able to use gasoline as a growth substrate and measured laboratory mineralization potentials for benzene (at 10°C) were greater in soil and groundwater samples collected near contaminated areas than in those taken from reference areas (Braddock and McCarthy, 1996). Differentiation of the relative amounts of eukaryotes, such as protozoans, and bacteria is also useful in evaluating the success of bioremediation (NRC, 1993), because if the number of bacteria increases as a result of active bioremediation, the number of protozoa preying on bacteria may also subsequently increase, (e.g., Madsen et al., 1991).

Nevertheless, little or no increase in bacterial numbers does not necessarily indicate that intrinsic bioremediation is not occurring. Increases in bacterial numbers above background levels may not be detected when biodegradation rates are low, due to sampling and measurement errors (NRC, 1993). Furthermore, increased numbers of specific types of microbes in the contaminated area relative to the background do not provide unequivocal evidence of ongoing activity, nor of biodegradative capabilities (Gieg et al., 1999). For example, Salanitro (1993) measured similar BTEX degradation rates, regardless of where the soil samples were collected from with respect to the hydrocarbon source, although the highest numbers of BTEX degraders were associated with samples taken nearer the source.

Adaptations of native microorganisms that lead to metabolism of initially persistent or slowly transformed site contaminants is another type of evidence of intrinsic bioremediation (NRC, 1993). To assess adaptation, the rate of contaminant transformation can be compared in samples from the contaminated plume and from an adjacent pristine location.

In addition to the co-reactants and intermediates or products of hydrocarbon degradation reviewed above, there are other innovative and potentially useful indicators of biodegradation, including carbon stable isotopes, and the ratio of nondegradable to degradable substrates.

5.1.4.2.7 Carbon-Stable Isotopes

When interpretation of changes in DIC concentration or soil gas CO_2 is confounded by abiotic processes (e.g., dissolution of atmospheric CO_2 or carbonate minerals), carbon stable isotope analyses may be applied to the inorganic carbon in liquid and gas samples. The ratio of the stable isotopes, $^{13}C:^{12}C$, is expressed as $\delta^{13}C$ in parts per thousand (per mil, ‰) relative to the PeeDee Belemnite standard, according to (Stumm and Morgan, 1981):

$$\delta^{13}C = \left[\frac{\left(^{13}C/^{12}C\right)_{sample} - \left(^{13}C/^{12}C\right)_{standard}}{\left(^{13}C/^{12}C\right)_{standard}}\right] 1,000$$

$$= \left[\left(\frac{\left(^{13}C/^{12}C\right)_{sample}}{\left(^{13}C/^{12}C\right)_{standard}}\right) - 1\right] 1,000 \quad (5.1.2)$$

A positive change in $\delta^{13}C$ values reflects a relative enrichment of ^{13}C with respect to ^{12}C, and a negative change reflects a relative enrichment of ^{12}C with respect to ^{13}C.

Most petroleum has $\delta^{13}C$ values between –21‰ and –32‰ (Deines 1980). As a result, inorganic carbon produced via microbial oxidation of petroleum is highly enriched in ^{12}C compared to inorganic carbon from the dissolution of atmospheric CO_2 ($\delta^{13}C \approx -7$ to –9‰) (Deines, 1980), soil carbonate minerals ($\delta^{13}C \approx 2$ to –10‰) or limestone ($\delta^{13}C \approx 2$‰) (Amundson et al. 1988). Indeed, several field studies of *in situ* petroleum hydrocarbon biodegradation under aerobic conditions have observed the $\delta^{13}C$ of the soil gas CO_2, (e.g., Aggarwal and Hinchee, 1991; Van de Velde et al., 1995) or groundwater DIC, (e.g., Baedecker et al., 1993), to be more enriched in ^{12}C than in uncontaminated locations.

Isotopic fractionation, i.e., shifting, of carbon isotopes also occurs during biologically-mediated reactions. This fractionation occurs because the bonds formed by light isotopes are more readily broken than the bonds involving heavy isotopes (Stumm and Morgan, 1981). As a result of this kinetic isotopic effect, the products of the reaction show a preferential enrichment in ^{12}C, while the remaining source material simultaneously becomes enriched in ^{13}C (e.g., Stehmeier et al., 1999). Varying degrees of fractionation of carbon isotopes occur during biodegradation of ^{13}C-depleted petroleum hydrocarbons, depending on the TEAP. In the case of aerobic hydrocarbon biodegradation, small amounts of fractionation have generally been observed, with $\delta^{13}C$ values of the soil gas CO_2 or DIC similar to that of the petroleum source material (e.g., Aggarwal and Hinchee, 1991; Baedecker et al., 1993; Landmeyer et al., 1996; Van de Velde et al., 1995). Somewhat larger fractionation has been observed during biodegradation of petroleum hydrocarbons under sulfate-reducing conditions, (e.g., Landmeyer et al., 1996). However, under methanogenic conditions, extensive fractionation of carbon isotopes occurs. Metabolically generated CH_4 has a very negative $\delta^{13}C$ ($\approx$ –100 to –40‰), while the CO_2 produced during methanogenesis has a more positive $\delta^{13}C$ ($\approx$ –30 to +10‰) (Whiticar et al. 1986). Thus, during methanogenic hydrocarbon biodegradation, the $\delta^{13}C$ value of the soil gas CO_2 or DIC may be higher than that of the petroleum source material (e.g., Baedecker et al., 1993; Landmeyer et al., 1996).

Compound-specific isotopic analysis may also provide footprints of biodegradation and recent developments in continuous-flow, compound-specific isotopic analysis have made it possible to analyze contaminants present at low concentrations in the vapor phase or dissolved in groundwater (Ahad et al., 2000). However, laboratory and field results to date are variable for BTEX, and relatively low levels of fractionation (ranging from no change to +5.9‰) have generally been observed with residual BTEX compounds (e.g., Ahad et al., 2000; Sherwood Lollar et al., 1999; Stehmeier et al., 1999). These results suggest that the observed isotopic fractionation effects may be a function of the specific compound, the TEAP, the metabolic pathways, or the microbial populations. Hunkeler et al. (2001) did observe greater carbon isotopic fractionation during degradation of MTBE as the sole substrate and with 3-methylpentane as the primary substrate in aerobic biodegradation experiments using aquifer material from the Borden site. The residual MTBE was enriched by +5.1 to 6.9‰ after 95-97% degradation.

Although carbon stable isotopic analyses is a promising technique, it may be complicated by several factors. For example, abiotic processes may also result in isotopic fractionation of petroleum hydrocarbons. However, physical fractionation of BTEX, e.g., due to sorption and volatilization, appears to be relatively minor (Harrington et al., 1999). Hunkeler et al.(2001) also observed relatively little fractionation during partitioning of MTBE between an organic phase, aqueous phase and gas phase. In addition, there may be a significant overlap between the $\delta^{13}C$ of the contaminant hydrocarbons and the natural organic matter (Conrad et al., 1997; Conrad et al., 1999). Furthermore, the large shifts between the $\delta^{13}C$ of the substrate and the product that occur during some microbially-mediated processes, such as methanogenesis, can lead to ambiguous results. To resolve these difficulties, Conrad et al. (1997; 1999) recommend that the ^{14}C content of the metabolic byproducts also be measured. This approach is recommended primarily because modern levels of ^{14}C, characteristic of natural organic matter in near-surface environments, are relatively high, while there is no measurable ^{14}C in fossil fuels. For example, in an analysis of a shallow plume of aviation gas, downgradient from the source, the concentrations of gasoline-derived compounds and CH_4 levels in soil gas were high (Conrad et al., 1999). The $\delta^{13}C$ of the CH_4 was indicative of methanogenesis via acetate fermentation ($\approx$ –50‰) and the $^{14}CH_4$ content, which was less than 10% of modern levels, provided confirmation that the CH_4 resulted from biodegradation of the aviation gas.

5.1.4.2.8 Ratio of Degradable to Nondegradable Substrates

If a mixture of contaminants is present at a site, then a decrease with time in the ratio of biodegradable to non-biodegradable (i.e., biomarker) organic contaminants in the field indicates microbiological activity (NRC, 1993). An ideal biomarker compound is non-biodegradable and has physical/chemical properties similar to those of the degradable compound being monitored. The ratios of heptadecane:pristane and octadecane:phytane have often been used to evaluate the biodegradation of crude oil (e.g., Pritchard and Costa, 1991); however, pristane and phytane have relatively high biodegradability under aerobic and anaerobic conditions (e.g., Bregnard et al., 1997; Cooney et al., 1985; Pirnik et al., 1974; Westlake et al., 1974). Therefore, hopanes have been used as an alternative biomarker for crude oils (e.g., Prince et al., 1994; Venosa et al., 1997). However, one difficulty in using hopanes as conservative tracers is that most crude oil components have different physical/chemical characteristics than hopanes. Alternatively, Moldowan et al. (1995) applied a biodegradation ranking scale (low to extensive biodegradation) based on the removal of a series of refractory biomarker hydrocarbons indigenous to, and ubiquitous in crude oil (n-paraffins, isoprenoids, steranes, hopanes, diasteranes, C26-C29 aromatics) to monitor bioremediation of refinery wastes. For refined petroleum products, e.g., diesel fuel and heating oil, which do not contain hopane, alkylated multi-ringed PAHs (e.g., C_3-chrysene) can be used as biomarkers for evaluating bioremediation, at least for a moderate period of time (Venosa et al., 1997). Diterpanes can be used as field biomarkers for prolonged periods in assessing bioremediation of diesel fuel.

Trimethylbenzene (TMB) isomers (1,2,3-TMB, 1,2,4-TMB, and 1,3,5-TMB) have been recommended (Wiedemeier et al., 1995) for use as conservative tracers to estimate the relative portions of observed decreases in BTEX concentrations that can be attributed to biodegradation or to dispersion, dilution, and sorption (e.g., Wiedemeier et al., 1996b). The TMB isomers and BTEX compounds have similar Henry's constants and soil sorption coefficients. In addition, the TMBs are generally present at detectable levels in the groundwater (Wiedemeier et al., 1995). However, Wiedemeier et al. note that, although the TMB isomers are relatively persistent under *anaerobic* conditions, the degree of recalcitrance is site-specific. Tetramethylbenzene is another compound that can potentially be used as a conservative tracer (Cozzarelli et al., 1990).

For compounds like MTBE, which may be recalcitrant at some sites and are not readily sorbed to the aquifer matrix, several techniques that do not require the use of a conservative tracer have been applied to correct for apparent mass loss (e.g., due to dispersion and dilution) (Anthony et al., 1999). These methods include: (1) performing a regression of the chemical concentration versus travel distance or time from the chemical source, assuming a steady-state plume (Buscheck and Alcantar, 1995); (2) comparing the changes in advective mass flux through lines of monitoring wells installed perpendicular to the groundwater flow direction (e.g., Borden et al., 1997); and (3) comparing the estimated total compound mass remaining in the aquifer to the estimated total compound mass in the original spill (e.g., Schirmer and Barker, 1998). The second and third approaches generally require more detailed site characterization and/or knowledge of the site history than non-reactive tracer methods and are difficult to apply.

5.1.4.3 Intrinsic Biodegradation Rates

Simulation of the fate and transport of BTEX compounds in groundwater and assessing the efficacy of intrinsic bioremediation requires reliable estimates of biodegradation rate constants. The literature is replete with aerobic and anaerobic BTEX biodegradation rate data obtained from laboratory and field studies. Many of these data have been compiled by Wiedemeier et al. (1999). The methods used to obtain biodegradation rates and the conditions under which these measurements are made may significantly impact the values that are obtained. Consequently, it is not surprising that in some cases, reported biodegradation rates for the BTEX compounds vary over three orders of magnitude. For example, first-order rate constants for BTEX biodegradation at selected Air Force sites ranged from 0.0002 to 0.08 day^{-1}. Because of the influence of methodology and site conditions on rate estimates, the biodegradation rate values obtained in one study may not be applicable in another system. Caution should be exercised in applying rate data from one study to another. In particular, values obtained from field measurements are typically influenced by the specific conditions at the site and reflect BTEX oxidation under a range of TEAPs.

There are several approaches available for estimating *in situ* intrinsic biodegradation rates. One approach is to monitor the removal of dissolved hydrocarbons or electron acceptors along the groundwater flow path (e.g., Barker et al., 1987; Kemblowski et al., 1987; MacIntyre et al., 1993; McAllister and Chiang, 1994). First-order kinetic constants are then estimated from concentration data along the groundwater flowpath by two methods: (1) using a one-dimensional steady-state analytical solution to the advection-dispersion equation (Eq. 3.5.10; Buscheck and Alcantar, 1995); or, (2) using a recalcitrant compound in the dissolved BTEX plume as a conservative tracer (Eq. 3.5.11; Wiedemeier et al., 1995; Wiedemeier et al., 1996b). For example, in an analysis of data from a jet fuel spill, the first-order BTEX biodegradation rate constants estimated using the first method usually were within 20% of the values obtained using the second method (Wiedemeier et al., 1996b). When the steady-state assumption is not appropriate, nonsteady-state numerical solutions can be used to estimate biodegradation rates.

A mass-balance approach can also be used for estimating first-order attenuation rate constants due to all attenuation mechanisms (e.g., Chiang et al., 1989; Schirmer et al., 1999). However, the cost of the extensive monitoring network that is required and the relatively complex data analyses preclude its application at most sites (McAllister and Chiang, 1994).

A third approach is to estimate biodegradation rates based on controlled field tracer studies. For example, Thierrin et al. (1993) estimated natural degradation rates with a groundwater tracer test by injecting deuterated organic compounds and bromide into the dissolved hydrocarbon plume down-gradient of a leaking underground storage tank. The first-order toluene biodegradation rate obtained using the reactive tracer test compared well to the model-derived field estimates, but the values obtained in the laboratory columns were 200-500 times higher.

The fourth key approach is to use laboratory microcosm studies (batch or continuous flow) to estimate intrinsic biodegradation rates based on the disappearance of target compounds (e.g., Barker et al., 1987; Chiang et al., 1989; Kemblowski et al.,

1987) or the production of $^{14}CO_2$ from radiolabeled compounds (e.g., Chapelle et al., 1996a; Landmeyer et al., 1998). Such studies have several advantages, including the capacity to construct a mass balance by performing the assay in a sealed vessel, the ability to separate contaminant removal via biological mechanisms from abiotic removal mechanisms by using sterilized treatments, plus the potential to be applied to a wide variety of hydrologic systems (Chapelle et al., 1996a; Madsen, 1991). However, the results of laboratory evaluations of biodegradation using disturbed, displaced environmental samples are likely to be quantitatively, even qualitatively, different from the results of the same determination performed *in situ*, due to changes in the nutritional, physical, and biological relationships found *in situ* and the physiological status of the associated microorganisms (Madsen, 1991). Therefore, BTEX rate constants estimated from field-scale studies are more appropriate for modeling purposes (Wiedemeier et al., 1995), and the biodegradation rates obtained from microcosms are more appropriate for demonstrating the potential for biodegradation in the field (Madsen, 1991). Microcosm studies may be particularly useful for demonstrating the potential for intrinsic biodegradation of MTBE, because of the difficulties in documenting MTBE biodegradation at the field scale (Anthony et al., 1999). Given the shortcomings discussed above, one practical approach is to use a combination of field and laboratory methods to estimate *in situ* intrinsic biodegradation rates (e.g., Chapelle et al., 1996a).

First order BTEX biodegradation rates obtained in field and laboratory studies under a combination of TEAPS ranged from 0.0002 to 0.0095 day^{-1} for benzene, 0.0017 to 0.2 day^{-1} for toluene, 0.0015 to 0.05 day^{-1} for ethylbenzene, and 0.0009 to 0.21 day^{-1} for the xylenes (Wiedemeier et al., 1999). Most transformation rates for MTBE have been reported as specific degradation activity (i.e., mass/cell mass/T) (as discussion above). However, Wilson et al. (2000) did report first-order removal rates for MTBE in anaerobic microcosms constructed with contaminated aquifer materials. The rate of MTBE removal in microcosms supplemented with alkylbenzenes was 3.01/year (± 0.52/yr at 95% confidence), while removal in corresponding controls was 0.39 ± 0.19/yr at 95% confidence. In comparison, removal in microcosms without added alkylbenzenes was 3.5 ± 0.65/year at 95% confidence (0.3 ± 0.14/year at 95% confidence in controls). Also, Hanson et al. (1999) reported removal rates for the pure culture *Rubrivivax* sp. PM1, which grows on MTBE, of 0.07, 1.17, and 3.56 µg/ml/h for initial MTBE concentrations of 5, 50, and 500 g/ml, respectively.

Although the first-order kinetics assumption is often convenient, it is not necessarily appropriate, particularly under conditions with: (1) high substrate concentrations, (2) toxic substrate concentrations, (3) multiple limiting substrates (e.g., electron donor and acceptor), or (4) expanding or decreasing microbial populations (Chapelle et al., 1996a). Monod kinetics can also be used to describe BTEX biodegradation in the field (e.g., Bekins et al., 1998). However, to our knowledge, no Monod kinetic parameters have been reported for utilization of MTBE and relatively few values are available for the BTEX compounds, especially under anaerobic conditions. Reported half-saturation coefficient (K) values range from 0.31 to 22.16, 0.044 to 56.74, 11.81, and 0.0007 to 67 mg/L for benzene, toluene, ethylbenzene, and the xylenes, respectively. Reported maximum specific growth rate (μ_{max}) values range from 0.784 to 9.3, 0.34 to 10.68, 9, and 3.03 to 11.50 day^{-1} for benzene, toluene, ethylbenzene, and the xylenes, respectively (Wiedemeier et al., 1999). As discussed by Wiedemeier et al., another modeling alternative is the instantaneous reaction model for describing BTEX biodegradation at field sites that are not kinetically-limited, i.e., sites with relatively long hydraulic residence times.

5.1.5 Case Studies

Many case studies demonstrating the natural attenuation of petroleum hydrocarbon releases are available in the literature (e.g., Barker et al., 1987; Chiang et al., 1989; Cho et al., 1997; Davis et al., 1994; Kemblowski et al., 1987; MacIntyre et al., 1993; McAllister and Chiang, 1994; Thierrin et al., 1993). Indeed, the NRC Committee on Intrinsic Remediation (2000) rated the current level of understanding and the likelihood of success of natural attenuation for BTEX compounds as "high". Two very detailed case studies demonstrating natural attenuation of petroleum hydrocarbon releases at Hill Air Force Base, Utah, and Patrick Air Force Base, Florida, are provided in Volume II of Wiedemeier et al. (1995). Perhaps the best documented case study is the natural attenuation of a crude oil spill from a burst pipeline near Bemidji, MN, which contaminated surrounding groundwater and soil, (e.g., Baedecker et al., 1993; Cozzarelli et al., 1990; Franks, 1987; Hult, 1984).

Several extensive compilations of case studies that support the use of monitored natural attenuation at petroleum-release sites are also available, (e.g., Hadley and Armstrong, 1991; Mace et al., 1997; Newell et al., 1990; Rice et al., 1995a; Rice et al., 1995b). These compilations indicate that: (1) a relatively small proportion of petroleum hydrocarbon releases have affected drinking water wells, (2) benzene is destroyed near petroleum release source zones by intrinsic biodegradation, and (3) groundwater plume lengths at these sites generally change slowly and tend to stabilize and/or begin decreasing in concentration and length within relatively short distances (e.g., <76.2 m (250 ft) as defined by the 10 µg/L benzene contour) of the hydrocarbon release area.

Because MTBE has not been routinely measured during remedial investigations at many UST sites, fewer time-series field data sets describe the behavior of MTBE, compared to the BTEX compounds (Landmeyer et al., 1998). Although some detailed cases studies demonstrating natural attenuation of MTBE and BTEX at petroleum release sites are available, the factors controlling MTBE biodegradation in the field are still not well understood. The results of these studies are variable, ranging from observations of low MTBE biodegradation potential, in which reductions in MTBE concentration were primarily attributed to dilution or dispersion (Landmeyer et al., 1998), to observations of limited MTBE biodegradation under mixed aerobic-denitrifying (Borden et al., 1997), and methanogenic conditions (Wilson et al., 2000). In one study, a mass balance analysis and laboratory experiments suggested that intrinsic biodegradation played a major role in MTBE attenuation (Schirmer and Barker, 1998; Schirmer et al., 1999); however, other authors have concluded that there was insufficient field evidence to support intrinsic biodegradation as the mechanism resulting in the reported decrease in MTBE mass at this site (Odencrantz, 1998). Correspondingly, the NRC Committee on Intrinsic Remediation (2000) rated the current level of understanding of natural attenuation of MTBE as "moderate", and the likelihood of success as "low".

Several extensive site data compilations compare MTBE and benzene/BTEX plumes. For example, Happel et al. (1996) analyzed plumes at 30 California leaking underground fuel tank sites and found no significant correlation between benzene and MTBE concentrations. Further, on a site-by-site basis, benzene and MTBE plumes were not well correlated. MTBE plumes (20 µg/L contour) ranged in length from approximately 0.5 to 3.5 times the length of the respective benzene plume (1 µg/L contour). In another study, Reid et al. (1999) analyzed monitoring data from 80 retail gasoline marketing outlets in Florida. The MTBE plumes were generally longer and larger in area than the corresponding benzene plumes, with mean plume length and area, respectively, of 140 ft. and 11,985 ft^2 for MTBE and 115 ft. and 7,919 ft^2 for benzene, using the 10 µg/L contour as the limit. The data analyzed also indicated that benzene, as well as MTBE, plumes do eventually stabilize: 4.4% of the plumes were expanding, 6.6% of the plumes were stable, and 89% were decreasing. Surprisingly, estimated first-order attenuation rates and attenuation rate half-lives for benzene and MTBE were found to be similar.

5.1.6 Summary and Conclusions

Based on this review of the natural attenuation of petroleum releases, in particular, BTEX and MTBE, the following observations can be made:
1. Intrinsic biodegradation is the most important attenuation mechanism contributing to decreases in contaminant concentration and mass for almost all petroleum hydrocarbon plumes. Transformation of BTEX compounds or MTBE by naturally-occurring chemical reactions is not expected under typical groundwater conditions. However, physical attenuation mechanisms may affect the concentration and distribution of BTEX compounds and MTBE, which may in turn affect intrinsic biodegradation rates.
2. The BTEX compounds are readily biodegraded under aerobic conditions. However, anaerobic processes can be a significant component of natural attenuation of BTEX compounds, with the possible exception of benzene at some sites, due to the abundance of anaerobic electron acceptors (e.g., Fe(III), nitrate, sulfate, carbon dioxide) compared to the availability of dissolved oxygen.
3. Until fairly recently MTBE was considered to be recalcitrant; however, increasingly, laboratory and field results indicate that intrinsic biodegradation of MTBE is feasible under aerobic and anaerobic conditions at some sites.
4. Analysis of the "footprints" of natural attenuation mechanisms, coupled with mass balance analyses and solute fate-and-transport models can provide the basis for evaluating natural attenuation. Key footprints providing evidence for natural attenuation of petroleum hydrocarbons and/or MTBE include: electron acceptors and/or their reduced products, DIC, alkalinity, organic acids, TBA, microbial numbers and activity, carbon stable isotopes, and biomarkers.

5. Natural attenuation processes have been observed and documented at a large number of petroleum-hydrocarbon release sites. At these sites, natural attenuation controls the extent of migration of BTEX contaminants away from the source in the dissolved plume.
6. The presence of MTBE in the gasoline-impacted groundwater may compromise the use of natural attenuation at some sites due to the greater mobility and persistence of MTBE compared to BTEX in shallow aerobic aquifers. However, some field data indicate that MTBE plumes do eventually stabilize. Furthermore, field studies indicate that MTBE is biodegradable in shallow aquifers, although the rate of degradation may be slow relative to BTEX compounds. Thus, dispersion and dilution may also be important attenuation mechanisms for MTBE. Factors contributing to MTBE persistence and observations of MTBE plumes extending past the BTEX plume, could include lengthy adaptation periods for MTBE biodegradation under anaerobic conditions and inhibition of MTBE biodegradation by other organic compounds, coupled with the mobility of MTBE in the subsurface.

5.2 Chlorinated Solvents

Chlorinated aliphatic hydrocarbons (CAHs) are ubiquitous pollutants in soil and groundwater (CEQ, 1981; OTA, 1984; Fisher et al., 1987), arising mainly from accidental releases, such as spills or leaks, and inappropriate waste-disposal practices. CAHs are used as solvents, dry cleaning fluids, degreasers for cleaning metal parts, and heat-exchange liquids. They are also commonly used in the manufacturing of electronic components, textiles, and chemicals (Rittmann et al., 1994). Depending on their pre-disposal history of use and handling, CAHs may be released to the environment as aqueous-phase or nonaqueous-phase liquids. Nonaqueous-phase releases of CAHs pose the most important long-term environmental threat, because nonaqueous-phase liquids (NAPLs) can act as continuous, long-term sources of groundwater contamination. Relatively small volumes of NAPL can contaminate huge volumes of groundwater (Mackay and Cherry, 1989). Most CAHs are more dense than water, and therefore, they are classified as dense nonaqueous-phase liquids (DNAPLs). Sometimes, however, they occur in mixtures with petroleum products or other nonchlorinated hydrocarbons in which the density of the mixture is less than that of water, in which case it is classified as a light nonaqueous-phase liquid (LNAPL). The behavior of light and dense NAPLs in the subsurface is very different, both from the perspective of contaminant fate and transport and that of remediation. Although location and recovery of NAPLs is never easy, it is nearly impossible for DNAPLs. Therefore, source removal may not be an option at many sites with CAH contamination.

Both destructive and nondestructive attenuation mechanisms operate on CAHs. The nondestructive processes include dilution by hydrodynamic dispersion and recharge, adsorption to aquifer solids, and loss from the aqueous phase by volatilization. Destructive processes include biological degradation and abiotic dechlorination

168 NATURAL ATTENUATION OF HAZARDOUS WASTES

reactions. Whereas destructive processes are capable of reducing the total mass of contaminants present in a contaminated environment, the nondestructive processes simply reduce the dissolved-phase concentration of the contaminant in groundwater by phase transfer or by dilution with clean water. As such, destructive processes are usually most important in natural attenuation of organic chemicals. This is particularly true for CAHs.

5.2.1 Nondestructive Attenuation Mechanisms for Chlorinated Aliphatic Hydrocarbons

The main nondestructive attenuation mechanisms are dilution, adsorption, and volatilization. The mixing processes are not fundamentally different for chlorinated organics than they are for other dissolved contaminants; so, there is no need to review them here (Sections 3.1.3 and 5.1.3). The contribution of adsorption and volatilization to attenuation depends on the chemical characteristics of the contaminants, however. Therefore, these processes will be described with specific reference to CAH natural attenuation.

5.2.1.1 Adsorption

Adsorption to soil and aquifer solids is primarily due to the interaction of dissolved CAHs with natural organic matter that is present on particle surfaces. As described previously (Section 3.1.4.1), this process is usually modeled as an equilibrium partitioning between the solid and aqueous phases. Table 5-19 shows octanol-water (K_{ow}) and organic-carbon-normalized (K_{oc}) partition coefficients reported for some important chlorinated aliphatic hydrocarbons. Retardation coefficients that could be expected to result from adsorption to sediments with low organic content are also provided to illustrate the relative importance of adsorption in transport of the listed CAHs. In general, these values are relatively low, such that significant retardation of advective transport does not occur for most of these compounds when f_{oc} is less than about 0.01 g C/g sediment. Adsorption of these compounds can be significant in organic-rich sediments, such as peat and silt, however.

5.2.1.2 Volatilization

Most of the environmentally important CAHs are volatile compounds with relatively high vapor pressures and Henry's constants. Vapor pressure describes partitioning between a pure liquid (or solid) phase and the gas phase, and Henry's Law governs partitioning between the aqueous phase and the gas phase. Therefore, vapor pressure is important in controlling evaporation of NAPLs that are present in the vadose zone, and Henry's Law affects volatilization of dissolved contaminants (e.g., in a groundwater plume). A detailed discussion of the partitioning of volatile compounds between aqueous and nonaqueous liquid phases and soil gas are presented in Section 3.1.4.2.

The vapor pressures and Henry's constants for several CAHs are given in Table 5-20. In addition, the gas-phase contaminant concentrations that would result from equilibration of a pure liquid phase (for vapor pressure) and an aqueous solution with

Table 5-19. Reported partition coefficients for some chlorinated organic compounds

Compound	log K_{ow}	log K_{oc}	R^1
tetrachloroethene	2.60	2.56	7.7
trichloroethene	2.38	2.10	3.3
cis-1,2-dichloroethene	0.70	1.90	2.5
trans-1,2-dichloroethene	0.48	1.77	2.1
1,1-dichloroethene	1.84	1.81	2.2
vinyl chloride	1.38	1.76	2.1
1,1,1-trichloroethane	2.5	2.18	3.8
1,1-dichloroethane	1.79	1.48	1.6
carbon tetrachloride	2.64	2.04	3.0
chloroform	1.97	1.49	1.6
methylene chloride	1.30	0.94	1.2

Source: From U.S. EPA (1986) and Verschueren (1983); reported in Rittmann et al. (1994).
[1] Assuming $n = 0.3$ and $\rho_b = 1.86$ g/cm^3 and $f_{oc} = 0.003$ gC/g sediment

the contaminant present at 1% of its saturation concentration (for Henry's constant) are also given. A concentration of 1% of solubility was selected for this illustration because a common rule-of-thumb uses this level as an indication that groundwater is in contact with NAPL. Table 5-20 shows that relatively high equilibrium gas-phase concentrations are achievable for most of these compounds in both situations, especially for partitioning from pure NAPL. In multicomponent NAPLs, such as when CAHs are mixed with petroleum products, the equilibrium gas-phase concentration of the contaminant of interest is reduced in proportion to its mole fraction in the mixture according to Raoult's Law (Eq. 3.1.20). High gas-phase concentrations that can occur at equilibrium but equilibrium concentrations are probably only achieved near the gas-liquid interface. These parameters are useful for establishing boundary conditions for mass-transfer models of volatilization, but cannot predict volatilization rates without additional information. The volatilization rate must also consider the interfacial surface area between the gas and liquid phases and the thickness of the diffusion boundary layer. Since groundwater and the gases in the vadose zone are very poorly mixed, mass transport to and away from gas-liquid interfaces is primarily by diffusion through a relatively large boundary layer. Even though very high interfacial gas-phase concentrations may be achieved, the long diffusion distances make concentration gradients relatively small. Volatilization from the aqueous phase is also retarded by mass transfer resistance in the water.

Table 5-20. Vapor pressure (P^o) and Henry's Law constants (K_H) for volatile CAHs (T = 20°C)

Compound	$P^{o\,1}$ (atm)	$C_g^{\,2}$ (mg/L gas)	$K_H^{\,3}$ (atm-L/mole)	$C_g^{\,4}$ (mg/L gas)
tetrachloroethene	0.032^7	218	13.3	0.831
trichloroethene	0.079	431	7.25	3.31
cis-1,2-dichloroethene	0.26^5	1,060	2.99	0.993
trans-1,2-dichloroethene	0.26^6	1,060	7.32	1.82
1,1-dichloroethene	0.66	2,650	20.9	19.5
vinyl chloride	3.5^5	N/A^8	22.0	10.0
1,1,1-trichloroethane	0.13	730	13.4	24.5
1,1-dichloroethane	0.24	974	4.35	9.94
carbon tetrachloride	0.12	758	23.5	7.82
chloroform	0.21	1,040	2.79	9.28
methylene chloride	0.46	1,620	1.73	14.4

[1] From U.S. EPA (1986) and Verschueren (1983); reported in Rittmann et al. (1994).

[2] Equilibrium gas-phase concentration assuming equilibrium with a pure liquid-phase contaminant: $C_g = (P^o/RT)$ (MW * 1,000 mg/g) where R (0.082054 liter-atm/mole-K) is the ideal gas constant, T (K) is the absolute temperature, and MW (g/mole) is the molecular weight of the CAH of interest.

[3] Estimated using regression coefficients given in Gossett (1987).

[4] Equilibrium gas-phase concentration assuming equilibration with an aqueous solution with contaminant concentration equal to 1% of its solubility limit.

[5] 25°C.

[6] 14°C.

[7] 30°C.

[8] N/A = not applicable; vinyl chloride is a gas at ambient pressure and temperature.

5.2.1.3 Summary of Nondestructive Attenuation Mechanisms for CAHs

From the preceding discussion it should be clear that neither adsorption nor volatilization will normally represent significant long-term attenuation mechanisms for chlorinated solvents. Because of the poor mixing that is characteristic of subsurface environments, the volatilization rate will usually be mass-transfer limited. Adsorption can slow the movement of a contaminant plume, but on its own, adsorption can only be an effective attenuation mechanism for relatively small releases of CAHs. Also, adsorption is only important when the organic content of soils and sediments is relatively high. Therefore, the major attenuation mechanisms for chlorinated solvents will be destructive mechanisms involving chemical and biological dechlorination reactions.

5.2.2 Destructive Mechanisms for Attenuation of Chlorinated Aliphatic Hydrocarbons

Natural attenuation of CAHs occurs primarily via abiotic and biotic dechlorination reactions. Since the toxicity of CAHs is often not proportional to the degree of chlorine substitution (e.g., vinyl chloride and 1,1-DCE are more toxic than the more highly chlorinated parent compounds from which they are commonly derived), complete dechlorination is necessary for effective attenuation.

5.2.2.1 Abiotic Dechlorination

Abiotic dechlorination reactions proceed through either nucleophilic substitution or elimination reactions (Section 3.2). Nucleophilic substitution reactions involve displacement of a chlorine atom, as chloride ion, by a nucleophile, such as hydroxide ion or water. When water or hydroxide is the nucleophile, the reaction is called *hydrolysis*, and the primary products are alcohols. Elimination reactions involve the loss of HCl from adjacent carbon atoms with the formation of carbon-carbon double bonds. Elimination reactions can only occur in CAHs that contain two or more carbon atoms, but hydrolytic dehalogenation can occur for any halogenated aliphatic compound.

For compounds containing multiple halogen substitutions on a single carbon atom, the alcohols that are produced by hydrolysis usually react further by an intramolecular mechanism to eliminate another chloride atom and form aldehydes. Chlorinated aldehydes, which would be produced by hydrolysis of parent CAHs containing carbon atoms with three halogen substituents (e.g., acetyl chloride, which can be formed by hydrolysis of 1,1,1-TCA), are extremely reactive toward further hydrolytic displacement of chloride. Therefore, hydrolytic dehalogenation can lead to complete removal of chlorine from the parent compound. Not surprisingly, hydrolytic dehalogenation is the most important abiotic mechanism for natural attenuation. The rate-limiting step is displacement of the first chloride ion by the nucleophile (either water or hydroxide ion). Elimination reactions of polyhaloalkanes, however, result in formation of haloethenes, which are relatively stable to further abiotic dehalogenation reactions (Cline and Delfino, 1989). For example, 1,1,1-TCA can be completely dechlorinated by hydolysis to form acetic acid, but it also undergoes elimination of HCl to form 1,1-DCE (Vogel and McCarty, 1987; Haag and Mill, 1988; Cline and Delfino, 1989; Wing, 1997). Whereas acetic acid is an environmentally acceptable end product, 1,1-DCE is not and abiotic transformation of 1,1,1-TCA to 1,1-DCE will not substantially reduce the risk of contaminated groundwater, soil, or sediment to human health or the environment. At pHs between about 4 and 10, abiotic transformation of 1,1,1-TCA produces about 20–25% 1,1-DCE by elimination and 75–80% acetic acid by hydrolysis (Haag and Mill, 1988; Cline and Delfino, 1989).

The abiotic dechlorination rate of 1,1,1-TCA is among the fastest reactions for the environmentally important CAHs. The rates of most of these reactions are first

Table 5-21. Abiotic dechlorination rates for CAHs

Compound	k (min^{-1})	$t_{1/2}$ (yrs)
chloroform	7.13×10^{-10}	1,850
carbon tetrachloride	3.26×10^{-8}	40.5
1,1-dichloroethane (1,1-DCA)	2.15×10^{-8}	61.3
1,2-dichloroethane (1,2-DCA)	1.83×10^{-8}	72.0
1,1,2-trichloroethane (1,1,2-TCA)	9.47×10^{-9}	139
1,1,1-trichloroethane (1,1,1-TCA)	1.24×10^{-6}	1.1
1,1,2,2-tetrachloroethane (1,1,2,2-TeCA)	3.03×10^{-6}	0.4
1,1,1,2-tetrachloroethane (1,1,1,2-TeCA)	2.82×10^{-8}	46.8
1,3-dichloropropane	5.87×10^{-7}	2.2
1,1-dichloroethene (1,1-DCE)	1.09×10^{-14}	1.2×10^{8}
1,2-dichloroethene (1,2-DCE)	6.32×10^{-17}	2.1×10^{10}
trichloroethene (TCE)	1.07×10^{-12}	1.3×10^{6}
tetrachloroethene (PCE)	1.37×10^{-15}	9.9×10^{8}

Source: Jeffers et al. (1989).

order with respect to concentration of the halogenated organic compound. The first-order rate constants for some of these compounds are given in Table 5-21, along with the corresponding half-lives (Jeffers et al., 1989). In general, the presence of soil or aquifer solids does not affect the observed dechlorination rates (Haag and Mill, 1988; Miyamoto and Urano, 1996). The half-lives for disappearance of these CAHs by abiotic dechlorination reactions range from 0.4 years (1,1,2,2-tetrachloroethane) to 21 billion years (1,2-dichloroethene). Since about 3.3 half lives are required to reduce the concentration by 90%, the only compounds listed in Table 5-21 that can be expected to be significantly attenuated by abiotic reactions within an average human life time are 1,1,2,2-TeCA, 1,1,1-TCA, and 1,3-DCP. (Several other chlorinated alkanes, such as pentachloroethane and 2,2-dichloropropane, have very short half lives for abiotic dechlorination—on the order of days, but these are so short as to be unlikely to be present at any but very freshly contaminated sites.) Abiotic dechlorination of chlorinated ethenes, such as TCE and PCE, does not occur at rates that are significant from the perspective of natural attenuation.

As stated above, disappearance of the parent compound does not constitute attenuation. Although 1,1,1-TCA appears to be converted primarily to acetate by hydrolysis, elimination to 1,1-DCE is also significant. Since 1,1-DCE persists in the absence of biological degradation processes, abiotic dechlorination of 1,1,1-TCA is probably not sufficient by itself to result in natural attenuation of a contaminated site. The fastest dechlorination rate reported in Table 5-21 is that for 1,1,2,2-TeCA, which reacts primarily by elimination of HCl to form TCE (Haag and Mill, 1988; Cooper et al., 1989), a compound which is recalcitrant to further abiotic dechlorination reactions. Therefore, abiotic dechlorination of 1,1,2,2-TeCA, although fast, does not constitute an attenuation mechanism.

5.2.2.2 Biological Dechlorination

Biological dechlorination reactions are the most important mechanism of natural attenuation for most CAHs. Several biodegradation mechanisms occur: hydrolytic dechlorination converts chlorinated alkanes and acids to the corresponding alcohols, monooxygenase-catalyzed oxidation reactions lead to a variety of fully and partially dechlorinated products, and reductive dechlorination reactions couple cleavage of carbon-chlorine bonds to transfer of electrons to the CAH. Oxidation of chlorinated aliphatics probably occurs exclusively by cometabolism and frequently involves production of toxic intermediates that can harm the organisms that catalyze the reactions. Because a negative selective pressure operates on bacteria that degrade CAHs by this mechanism, oxidative dechlorination does not appear to be a significant process in natural attenuation. Organisms that are able to couple dechlorination of CAHs to growth through hydrolytic and reductive dechlorination mechanisms have been studied, however, and these mechanisms can lead to effective natural attenuation at some sites. Examples of these types of dechlorination reactions are shown in Figure 5-3. Hydrolytic dehalogenation reactions are most commonly observed with compounds that have fewer than two nonhydrogen substituents on the chlorine-substituted carbon atom. Methylene chloride and 1,2-DCA are both known to be degradable by pathways involving hydrolytic dechlorination mechanisms. Reductive dechlorination reactions, on the other hand, are most effective for highly chlorinated compounds, such as PCE, TCE, and carbon tetrachloride.

Although biological degradation reactions have been studied extensively in the laboratory for several decades, the distribution of dechlorinating bacteria in the environment has not been well characterized. Unlike aromatic hydrocarbon degraders, dechlorinators are not always present at CAH-contaminated sites. Therefore, much of the effort during a natural-attenuation demonstration is directed at obtaining evidence that biodegradation of chlorinated aliphatics is occurring.

5.2.3 Demonstration of Natural Attenuation at CAH-Contaminated Sites

Natural attenuation demonstrations are based on a lines-of-evidence approach (Chapter 1), in which historical contaminant concentration data, site geochemical and hydrogeological characteristics, and laboratory experimental results may be used to provide support for the hypothesis that the contaminant mass or concentration will be reduced sufficiently to protect human health and the environment before the contaminants reach a sensitive receptor. Historical data that show a statistically significant trend of decreasing contaminant concentrations or mass over time (e.g., shrinking or stable groundwater plumes) is the most convincing evidence, but it can only be obtained through long-term monitoring. Sites lacking a sufficient historical

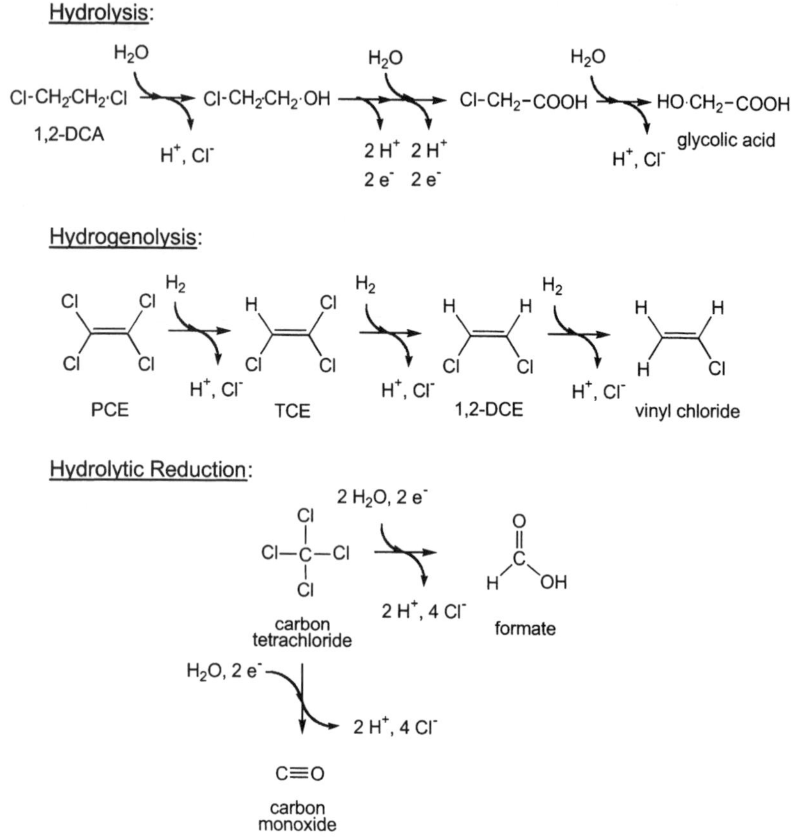

Hydrolysis:

$$Cl\text{-}CH_2CH_2\text{-}Cl \xrightarrow[H^+,\ Cl^-]{H_2O} Cl\text{-}CH_2CH_2\text{-}OH \xrightarrow[2\ e^-\ \ 2\ e^-]{H_2O\quad 2\ H^+\ \ 2\ H^+} Cl\text{-}CH_2\text{-}COOH \xrightarrow[H^+,\ Cl^-]{H_2O} HO\cdot CH_2\text{-}COOH$$

1,2-DCA ... glycolic acid

Hydrogenolysis:

PCE $\xrightarrow[H^+,\ Cl^-]{H_2}$ TCE $\xrightarrow[H^+,\ Cl^-]{H_2}$ 1,2-DCE $\xrightarrow[H^+,\ Cl^-]{H_2}$ vinyl chloride

Hydrolytic Reduction:

carbon tetrachloride CCl_4 $\xrightarrow[2\ H^+,\ 4\ Cl^-]{2\ H_2O,\ 2\ e^-}$ formate (HCOOH)

$\xrightarrow[2\ H^+,\ 4\ Cl^-]{H_2O,\ 2\ e^-}$ C≡O carbon monoxide

Figure 5-3. Dechlorination reactions that may be important in natural attenuation of CAHs.

Source: Janssen *et al.*, 1985; Freedman and Gossett, 1989; Criddle and McCarty, 1991.

Note: Formate, carbon monoxide, and glycolic acid will be rapidly metabolized in most environments. Vinyl chloride may be converted to acceptable products under appropriate conditions.

database will rely more heavily on the latter two lines of evidence: geochemical/ hydrogeological characterization and laboratory experimentation. Since field evidence is generally preferable to evidence from laboratory studies, appropriate geochemical and hydrogeological characterization is critical to making a convincing case in support of natural attenuation. Although there are many advantages to laboratory studies, such as the ability to close mass balances and to limit or quantify losses due to abiotic processes, they are usually very expensive and can never be made truly representative of actual site conditions. Therefore, the case for natural attenuation is usually based primarily on geochemical and hydrogeological characterization of a site. Laboratory studies are used primarily in a supporting role when field data demonstrating the biodegradation potential is inconclusive (EPA, 1999) or when biodegradation rate coefficients cannot be estimated from field data (Wilson et al., 1997; Wiedemeier et al., 1998). Hydrogeological and geochemical site characterization in support of natural attenuation feasibility investigations usually involves three components: (1) evaluation of the redox status of the contaminated environments, (2) identification of the products of contaminant biotransformation, and (3) determination of biotransformation rate coefficients.

5.2.3.1 Evaluation of Redox Conditions

Geochemical characterization is directed at demonstrating that conditions within the contaminated area are suitable for biological degradation. In addition to the general requirement that the pH of the contaminated environment be relatively neutral, efficient biodegradation of CAHs requires highly reducing conditions, because reductive dechlorination is the most common biodegradative mechanism in the absence of engineered manipulation of site conditions. Concentrations of potential electron acceptors such as dissolved oxygen and nitrate must be low so that the redox potential is low. The presence of end products of anaerobic metabolism (e.g., methane, sulfide, or ferrous ion) is a good indication that conditions are appropriate for reductive dechlorination. Redox potential is also measured frequently, because a simple electrochemical method is available. Interpretation of the data is more difficult than its collection, however, and a variety of alternative methods has been used to evaluate the redox status of subsurface environments (e.g., see Section 5.1.4.2).

Among the available alternative methods, measurement of the dissolved hydrogen concentration has been used most frequently and successfully (Chapelle et al., 1996, 1997; Vroblesky et al., 1997). The dissolved hydrogen concentration is particularly important at chlorinated-solvent sites, because H_2 appears to be the most common primary electron donor for bacteria that use reductive dechlorination as a primary electron-accepting process (Holliger et al., 1993; Scholz-Muramatsu et al., 1995; Maymo-Gatell et al., 1995; Wild et al., 1996). Although H_2 is a common electron donor for reductive dechlorination, it is by no means the only substrate that supports this process. Bacteria that couple reductive dechlorination to growth and oxidation of a variety of organic electron donors, including ethanol, lactate, pyruvate, formate, butyrate, glycerol, and succinate, have been isolated (Holliger et al., 1993; Scholz-Muramatsu et al., 1995; Gerritse et al., 1996; and Sharma and McCarty, 1996.) Hydrogen half-saturation concentrations for reductive dechlorinators have been reported to be in the range of 10 to 100 nM (Smatlak et al., 1996; Ballapragada et al., 1997), which is in the range of H_2 concentrations that are expected in methanogenic environments (see Table 5-18); suggesting that the observed reductive dechlorination rates could be very sensitive to the ambient dissolved H_2 concentration, with lower dechlorination rates accompanying lower dissolved hydrogen concentrations.

When reductive dechlorination is thought to be the main destructive attenuation mechanism, it is important to show that, in addition to possessing appropriate redox characteristics, the site also has a sufficient supply of electron-donor substrates present to support complete attenuation of the contaminants of concern. The minimum stoichiometric requirement for reductive dechlorination is 0.45 mg of biochemical oxygen demand (BOD) per mg chloride ion released. Since 0.86 mg Cl^- is released per mg PCE that is converted to ethene, complete reductive dechlorination of 1 mg PCE/L requires an available electron-donor concentration of 0.39 mg BOD/L. Because PCE and other important CAH pollutants exist primarily as NAPLs, an alternative way to express the electron-donor demand is that about 630 grams of available BOD are required to completely dechlorinate 1 liter of PCE. Competition with other electron-acceptor processes (e.g., sulfate reduction or methanogenesis) will increase the stoichiometric requirement for electron-donor substrates. Since the availability of electron donors will usually be determined by chemical means (e.g., COD or TOC analyses), nonbiodegradable or nonbioavailable compounds will further increase the apparent stoichiometric requirement for electron-donor substrates. If the electron-donor concentration is determined by measuring TOC, a relationship between BOD and TOC must be assumed. The ratio can range from 1.3 to 4.6 mg BOD/mg C, with 2.5 to 3 being a reasonable average for many organic compounds.

The U.S. EPA protocol for evaluating natural attenuation (Wiedemeier et al., 1998) identifies three types of behavior for chlorinated solvent plumes. Plume behaviors are distinguished based on the concentration and source of the electron-donor substrate that supports reductive dechlorination. Type 1 behavior is observed when reductive dechlorination is supported by anthropogenic electron donors (e.g., fuel hydrocarbons), whereas reductive dechlorination that is supported by oxidation of natural organic matter is categorized as Type 2 behavior. Type 3 behavior describes oxic conditions (DO > 1 mg/L) where reductive dehalogenation is not expected to occur. Different types of behavior may be observed in different portions of a particular chlorinated solvent plume or at different times during the evolution of a plume.

Since petroleum hydrocarbons are highly reduced (e.g., 3.1 mg BOD/mg toluene), they can support extensive reductive dechlorination: only about 200 g (230 mL) of toluene is required to completely dechlorinate one liter of PCE to ethene. Note, however, that only BTEX are oxidized fast enough under anaerobic conditions to support reductive dechlorination, but a large fraction of the total mass of most common petroleum hydrocarbon mixtures consists of aliphatic hydrocarbons, which are essentially nonbiodegradable under anaerobic conditions. So, the entire mass of the hydrocarbon contamination cannot be considered in the estimate of electron-donor availability when Type 1 behavior is thought to dominate.

Natural organic matter is also relatively reduced. Table 5-22 lists empirical formulas that have been reported for natural organic matter associated with soil or sediment. The corresponding specific theoretical oxygen demands are also given. It is highly unlikely that natural organic matter is completely biodegradable on time scales that are relevant for natural attenuation, however. Also, the degradable fraction probably varies from site to site and probably among geological formations at a specific site. So, where Type 2 behavior is expected, the biodegradability of the natural organic matter should be assessed. Reductive dechlorination supported by

Table 5-22. Theoretical oxygen demand of natural organic matter

Natural organic matter	Empirical formula[1]	ThOD (mg COD/ mg NOM)	ThOD (mg COD/ mg C)
fulvic acid	$C_{100}H_{80}O_{70}N$	1.12	2.25
humic acid (soil)	$C_{17}H_{20}O_{6.7}N$	1.65	2.81
humic acid (lake sediment)	$C_{11}H_{14}O_{5.6}N$	1.42	2.68
humin	$C_{20}H_{38}O_{22}N$	0.88	2.37

[1]Schwarzenbach et al. (1993).

natural organic matter will usually be slower than reductive dechlorination that is supported by anthropogenic organic carbon (Wiedemeier et al., 1998).

Because reductive dehalogenation requires strongly reducing conditions, it is not expected to occur where Type 3 conditions dominate (Jackson, 1998). Some dechlorination products, especially vinyl chloride, are more degradable under aerobic than under anaerobic conditions (Dolan and McCarty, 1995; Freedman and Herz, 1996; Bradley and Chapelle, 1998). Therefore, when oxic conditions are present at the edges of a Type 1 or Type 2 plume, complete biodegradation of the chlorinated organics may be facilitated.

5.2.3.2 Identification of Biotransformation Products

The pH and redox potential of the contaminated environment can provide information on whether conditions are favorable for reductive dechlorination to occur, but they do not show that these reactions are occurring and, if so, at what rate they are occurring. These are critical issues for the evaluation of natural attenuation, however, because they will determine whether it can be protective of human health and the environment. The clearest evidence that biodegradation is occurring is observation of biodegradation products. Organic products of reductive dechlorination reactions are often observed during normal site-characterization activities, because many of these are detectable using the same analytical procedures that are used to monitor the concentrations of the parent contaminants (Wiedemeier et al., 1998). The sequence of reactions leading to complete reductive dechlorination of PCE and TCE is one of the most familiar pathways among remediation professionals (Fig. 5-3). Two of the chlorinated intermediates in this pathway—*cis*-1,2-DCE and vinyl chloride—are rarely released into the environment as primary contaminants. So, when they are observed, they can usually be attributed to biodegradation of PCE or TCE. Note that, unless evidence for the further biodegradation of DCE and vinyl chloride is obtained, this should not be construed as being an indication that natural attenuation will be sufficiently protective of sensitive receptors, because vinyl chloride is a known human carcinogen.

Although reductive dechlorination of PCE and TCE involves formation of products that can unambiguously be attributed to biological activity, reductive dechlorination of other CAHs does not. For example, although biodegradation of carbon tetrachloride and chloroform sometimes proceeds through standard hydrogenolysis pathways in which chloroform and methylene chloride are formed (Egli et al., 1987; Mikesell and Boyd, 1990; deBest et al., 1997), in other cases the main products are nonchlorinated, nonvolatile organic compounds or oxygen-containing compounds such as carbon monoxide or carbon dioxide (Bouwer and McCarty, 1983; Egli et al., 1988; Criddle et al., 1990; Becker and Freedman, 1994; Lewis and Crawford, 1995; Workman et al., 1997). Similar behavior has been observed for 1,1,1-TCA (Vogel and McCarty, 1987; Galli and McCarty, 1989a; Egli et al., 1987; Egli et al., 1988). Formation of these nonchlorinated products reduces the environmental hazard associated with the contaminated medium, and so is an effective mechanism for natural attenuation, but they do not provide a clear indication that the concentration reduction was caused by biological activity. Therefore, evidence in support of natural attenuation may be more difficult to obtain for these compounds. Similar difficulties accompany biodegradation by hydrolytic dechlorination mechanisms (Fig. 5-3): products are either not detected by the standard analytical methods used in site investigations, or the stable products cannot be unambiguously attributed to biotransformation of the chlorinated parent compounds.

One product that is always obtained from dechlorination reactions is chloride ion (Fig. 5-3). Production of 1 mg/L of DCE by reductive dechlorination of PCE should increase the background chloride concentration by about 0.7 mg/L. Similarly, complete reductive dechlorination of 1 mg/L of TCE to ethene should increase the chloride concentration by about 0.8 mg/L. Typical groundwater chloride concentrations range between about 4 to 80 mg/L, with an average of about 10 mg/L (Stumm and Morgan, 1996). So, the expected change in chloride concentration seems small, particularly in light of the relatively low concentrations of chlorinated solvents that are of concern for sensitive receptors. Nevertheless, sites with NAPL contamination should produce easily detectable increases in the chloride ion concentration if biological degradation is occurring (Wiedemeier et al., 1997; Ellis et al., 1997; Biehle et al., 1999). An attempt should be made to reconcile any observed increases in the chloride concentration with either the appearance of dechlorination products or reductions in the concentrations of parent compounds, however, because biotransformation of chlorinated organics is not the only anthropogenic source of chloride in the environment (Norris and Wilson, 1999). This is particularly true for chlorinated solvent plumes emanating from landfills (Christensen et al., 2000).

Because the determination of clearly identifiable biodegradation products may not always be possible, some researchers are attempting to develop other methods for demonstrating the role of biological processes in reducing contaminant concentrations during transport. One promising approach is the use of stable isotope ratios (Section 5.1.4.2). Stable isotope fractionation occurs during biodegradation because covalent bonds formed by heavier isotopes are stronger than the analogous bonds formed by lighter isotopes (March, 1985). As a result, molecules containing heavy isotopes, such as ^{13}C and ^{37}Cl, react more slowly than do molecules composed exclusively of the more abundant light isotopes, and the residual contaminant becomes enriched in the heavier isotope. Similarly, the products are depleted with respect to the heavier isotope. Thus, stable isotope fractionation data involves comparison of the relative isotope ratio ($\delta^{13}C$ or $\delta^{37}Cl$) of a contaminant or the expected degradation products over time or along a flow path, where spatially separated samples represent different extents of reaction. Purely physical processes that can lead to a reduction in contaminant concentration along a flow path, such as dilution by dispersion or recharge, will not affect $\delta^{13}C$ or $\delta^{37}Cl$, but processes that involve chemical reaction may. Volatilization, which is primarily a physical processes but which can be affected by the chemical characteristics of the contaminants, resulted in a slight decrease (−2‰ to −3‰) in the $\delta^{13}C$ values and a relatively large increase in $\delta^{37}Cl$ (8‰ to 9‰) of the residual TCE and methylene chloride after 99% of the solvent was removed by evaporation (Huang et al., 1999). Biodegradation, on the other hand, resulted in enrichment of the residual CAHs with both heavier isotopes. The extent of enrichment was dramatic: aerobic biodegradation of methylene chloride by a methylotrophic bacterium, strain MC8b, resulted in a >150‰ increase in $\delta^{13}C$ of the residual solvent at 95% removal by biodegradation (Heraty et al., 1999), and reductive dechlorination of TCE increased $\delta^{13}C$ by about 15‰ when 90% of the TCE had been degraded (Sherwood Lollar et al., 1999). Despite the clear stable isotope signature that is expected to result from biodegradation, there have been few convincing demonstrations of these effects in the field (Sturchio et al., 1998). Also, care must be used in interpreting these results, because batch-to-batch and manufacturer-to-manufacturer differences in $\delta^{13}C$ and $\delta^{37}Cl$ have been observed in pure PCE, TCE, and 1,1,1-TCA (Beneteau et al., 1999).

5.2.3.3 Biodegradation Rate Constants

A critical aspect of the evaluation of natural attenuation is demonstration that it will be protective of sensitive receptors. This usually involves fate and transport modeling that incorporates all of the hypothesized attenuation processes, including biodegradation. Therefore, it is not sufficient to show that biodegradation is occurring at a particular site. One must also demonstrate that the biodegradation rate, in conjunction with other attenuation processes, is sufficient to reduce the mass or concentration of the contaminant to an acceptable level before it can encounter a sensitive receptor. In order to accomplish this objective, the fate and transport model must incorporate a model for biodegradation kinetics, and kinetic parameters that can be used in the model must be estimated.

Most contaminant fate and transport models use a first-order kinetic model to describe the biodegradation of chlorinated solvents (Suarez and Rifai, 1999; Wiedemeier et al., 1999), although Monod-type kinetics may be more appropriate if a wide range of contaminant concentrations occur at the site (e.g., a contaminant plume). Most kinetic models used in contaminant fate and transport models assume that the size of the contaminant-degrading population is constant over the spatial and temporal domains to which the model is applied. This assumption is probably never true, but our understanding of the factors that limit microbial growth in contaminated environments and our ability to enumerate bacteria with specific metabolic capabilities (especially CAH biodegradation) are too limited to justify the use of more sophisticated models.

Kinetic parameters can be estimated either in laboratory microcosms or from field data. Although parameter estimation is simpler in microcosm studies, where mass balances can be performed and abiotic losses can be prevented or quantified, their use is only recommended when rate constants cannot be obtained from field data (Wilson et al., 1997). Rate constants estimated from laboratory studies are often greater than those estimated from field data (Suarez and Rifai, 1999), in part because the conditions in microcosms can never exactly replicate conditions in the field and in part because field measurements may underestimate the true biodegradation rate by using overly conservative assumptions to account for abiotic attenuation processes or by not considering the effect of preferential flow paths on transport time scales (Wilson et al., 1997; Wiedemeier et al., 1998).

Laboratory estimation of CAH biodegradation kinetic parameters usually involves construction of microcosms containing sediments and water from areas that are representative of conditions at the site (Wilson et al., 1997). The microcosms will usually be spiked with an aqueous or solvent-borne mixture of the contaminants of concern to produce initial concentrations that are similar to the highest concentrations observed at the site, and they should be incubated at the ambient site temperature. Replicate microcosms should be sacrificed at regular intervals over a period of several months to greater than one year and analyzed for the contaminants of concern and their expected degradation products. Kinetic parameters can be estimated by linear or nonlinear regression analysis of the concentration versus time data.

Parameter estimation from field data is more difficult and is based primarily on measurement of concentration versus distance along the centerline of a contaminant plume (Wiedemeier et al., 1998). Alternatively, kinetic parameters can be estimated by analyzing changes in the total mass of CAH present as a function of time. Concentration versus distance methods must account for abiotic attenuation mechanisms, especially adsorption and dilution. This can be accomplished by normalizing observed contaminant concentrations to the concentration of a recalcitrant cocontaminant or some other conservative tracer. If biodegradation kinetics are assumed to be first order in contaminant concentration, the normalized concentration versus distance relationship for a groundwater plume is as follows:

$$\frac{C}{T} = \frac{C_o}{T_o} \exp\left(-\lambda \frac{Rx}{v_x}\right) \tag{5.2.1}$$

where C_o and C ($M_C L^{-3}$) are the contaminant concentrations at the source and at distance "x" from the source, respectively, T_o and T ($M_T L^{-3}$) are the concentrations of the conservative tracer at the same positions, v_x (LT^{-1}) is the groundwater velocity in the direction of the centerline of the plume, λ (T^{-1}) is the first-order biodegradation rate coefficient, and R is the retardation coefficient. With slight modifications, Eq. (5.2.1) can be used in a variety of situations to correct for physical losses of contaminants by mechanisms other than dilution during transport in groundwater (Douglas et al., 1994; Venosa et al., 1996). Trimethylbenzenes, which are usually nonbiodegradable under anaerobic conditions (see Section 5.1.4.2), have been used as conservative tracers in chlorinated solvent plumes that are mixed with petroleum

Table 5-23. First-order biodegradation rate coefficients for reductive dechlorination of CAHs

Compound	λ_{field} (day^{-1})	λ_{lab} (day^{-1})
carbon tetrachloride	0.004–0.490	0.023–0.160
TCA	0–0.125	0–2.33
DCA (all isomers)	0–0.011	0.028–0.044
PCE	0–0.080	0–0.410
TCE	0–0.023	0–3.13
cis-1,2-DCE	0–0.130	0.001–0.200
DCE (other isomers)	0.001–0.006	0.010–0.270
vinyl chloride	0–0.007	0–5.20

Source: Suarez and Rifai (1999).

hydrocarbons (Wiedemeier et al., 1997). Total (i.e., inorganic plus organic) chlorine has also been used (Wiedemeier et al., 1997) and is more generally useful for chlorinated solvent plumes, because it does not rely on co-contamination with petroleum hydrocarbons. High background chloride concentrations may affect the sensitivity of this method, however.

For plumes that have achieved steady state (i.e., plumes for which the rate of all destructive attenuation mechanisms exactly balances the rate of input of contaminants by dissolution from a nonaqueous phase in the source area), first-order biodegradation kinetic parameters can be estimated by determining the concentration versus distance relationship for the contaminants of concern along the centerline of the steady-state plume (Wiedemeier et al., 1998) and applying Eq. (3.5.10). The rate coefficient determined by this method is sensitive to the assumed dispersivity, however, and is only valid for steady-state plumes. Therefore, it is probably less accurate than biodegradation coefficients determined by other methods.

A summary of reported biodegradation kinetic parameters for CAHs has recently been compiled (Suarez and Rifai, 1999). Wide variations were reported among rate coefficients determined for specific compounds in different studies, with the range usually spanning several orders of magnitude. This suggests that rate coefficients for biodegradation of CAHs are highly site-specific, and therefore, they should be estimated independently at each site for which natural attenuation is being evaluated. The range of values reported for first-order rate coefficients for reductive dechlorination reactions are presented in Table 5-23. These rate coefficients represent a wide range of anaerobic conditions, including nitrate-reducing systems. In general, the rates observed under strongly reducing conditions (i.e., sulfate-reducing or methanogenic) were near the high end of the reported range, whereas much lower rates were observed under nitrate- and iron-reducing conditions.

Although first-order kinetics are often assumed to adequately describe biodegradation of CAHs at contaminated sites, the half-saturation concentrations that have been estimated suggest that this may not always be appropriate, especially for highly contaminated sites. Half-saturation concentrations ranging from about 200 µg/L (Ballapragada et al., 1997) to 6.5 mg/L (Barrio-Lage et al., 1987) have been estimated for reductive dechlorination of TCE, and values from about 4.1 mg/L (Galli and McCarty, 1989b) to 25 mg/L (Wrenn and Rittmann, 1996) have been observed for anaerobic biodegradation of 1,1,1-TCA. Half-saturation concentrations of about 300 µg/L and 460 µg/L were estimated for reductive dechlorination of cis-1,2-DCE and PCE, respectively (Ballapragada et al., 1997). Lower half-saturation concentrations for 1,1,1-TCA biodegradation were observed in sulfate-reducing than in methanogenic systems, and the values in both systems decreased with decreasing electron donor concentrations (Wrenn and Rittmann, 1996). Since the aqueous concentration should be less than about 10% of the half-saturation concentration for first-order kinetics to apply, use of a first-order model may not be appropriate for all contaminants or for all portions of some plumes. For example, some CAH concentrations reported by Wiedemeier et al. (1997), Imbrigiotta et al. (1997), Ellis et al (1997), DuPont et al. (1997), Spexet et al. (1999), and Mahaffey et al. (1999) are near or above the reported half-saturation concentrations for at least one of the contaminants of concern. So, although use of first-order biodegradation kinetics in fate and transport models can be justified on the basis of simplicity, caution should be used when attempting to predict plume evolution by extrapolating with parameter estimates of this type.

5.2.4 Case Studies

The natural attenuation of chlorinated solvents has been investigated at numerous sites, especially at military bases, Department of Energy facilities, and some large manufacturing sites. Brief descriptions of these investigations have been published primarily in the proceedings of conferences on natural attenuation and bioremediation (e.g., U.S. EPA, 1997; Alleman and Leeson, 1999). In addition, several case studies are compiled in Wiedemeier et al. (1999).

5.3 Polycyclic Aromatic Hydrocarbons (PAHs)

5.3.1 PAH-Contaminated Sites

PAH-contaminated sites have been associated with, but are not limited to, wood preserving, petroleum coking, and transportation industries as well as former manufactured gas plants (FMGPs), lignite pyrolysis sites, airports, military installations, and municipal and hazardous waste landfills. PAHs at these sites have been found in conjunction with other contaminants including phenols, heterocyclic compounds, monoaromatic compounds, cyanides, chlorinated solvents, pesticides, polychlorinated biphenyls, pentachlorophenol, and arsenic-based wood preservatives. Of the 1,226 sites currently registered on the Superfund National Priority List (NPL),

598 are contaminated with PAH compounds (48.8%). The number of Superfund sites with PAHs as contaminants is only surpassed by sites containing volatile organic compounds, which include benzene, toluene, ethylbenzene and xylene (BTEX) and chlorinated solvents (849 sites (69.2%)), and metals (794 sites (64.8%)).

Although many PAH-contaminated sites exist, much remedial work has been focused on sites originating from creosote works, wood treatment industries, coking industries, asphalt factories, and FMGPs which comprise the largest fraction of PAH-contaminated sites. Burton et al. (1988) reported that there were approximately 700 sites in the United States where wood preservation is currently conducted or has been conducted in the past. Mueller et al. (1989) estimated that the number of creosote-contaminated sites is close to 700 with most of these sites plagued with leaking tanks, drippings from treated lumber, spills, and leachate from unlined holding ponds. The Edison Electric Institute (1984) estimated that approximately 1,500 FMGP sites pose an environmental threat. According to Hatheway (1997), this estimate is conservative as it did not include potentially contaminated FMGP sites. Hatheway argues that the number of contaminated FMGP sites in North America may be greater than 32,000, considering that there were sites operated in rail yards, military posts, arsenals, institutes, and large residential estates that were not reported. Creosote and coal tar compounds have been cited as a widespread problem in nearly all industrialized countries (Broholm et al., 1999). According to Delorme and Carlier (1998), the French National Gas Company owns 467 FMGP plants. In Germany, the total number of FMGP sites is estimated at about 1,000 (Knopp et al., 2000), and in Denmark, Arvin and Flyvbjerg (1992) state that there are 35–45 creosote waste sites per million people.

Contamination at these sites has resulted from leaking tanks and pipe networks, incomplete separation of tar from aqueous liquids, drippings from treated lumber (wood treatment facilities), spills, decommissioning activities, and leachate from unlined storage ponds or shallow wells. In any case, the nonaqueous phase liquids (NAPLs) released from these sites tend to be denser than water and usually migrate downward and laterally into the subsurface by gravity, dispersive, and capillary forces. In many of these sites, the NAPLs pool on the confining layer of the aquifer and move with the water. Given the physical-chemical properties of PAH compounds, they can be very challenging to remediate.

PAH compounds may originate from both anthropogenic processes and organic matter diagenesis. The most common anthropogenic source is incomplete combustion of organic materials. These compounds can be released in the gas or particulate phases or found enriched in the burn residues of fossil fuels including oil, gasoline, diesel fuel, coal, wood, cigarettes, tires, and many other organic-compound containing items (Mueller et al., 1989; Garrett et al., 2000; Fernandez et al., 2000; Fullana et al., 2000; Haefliger et al., 2000; Liu et al., 2000).

5.3.2 PAH Compounds of Concern

PAHs are defined as compounds consisting of carbon and hydrogen in the form of two or more fused aromatic rings forming planar structures with *pi* bonds. They are classified as unsaturated hydrocarbons, yet tend to have low reactivity compared to

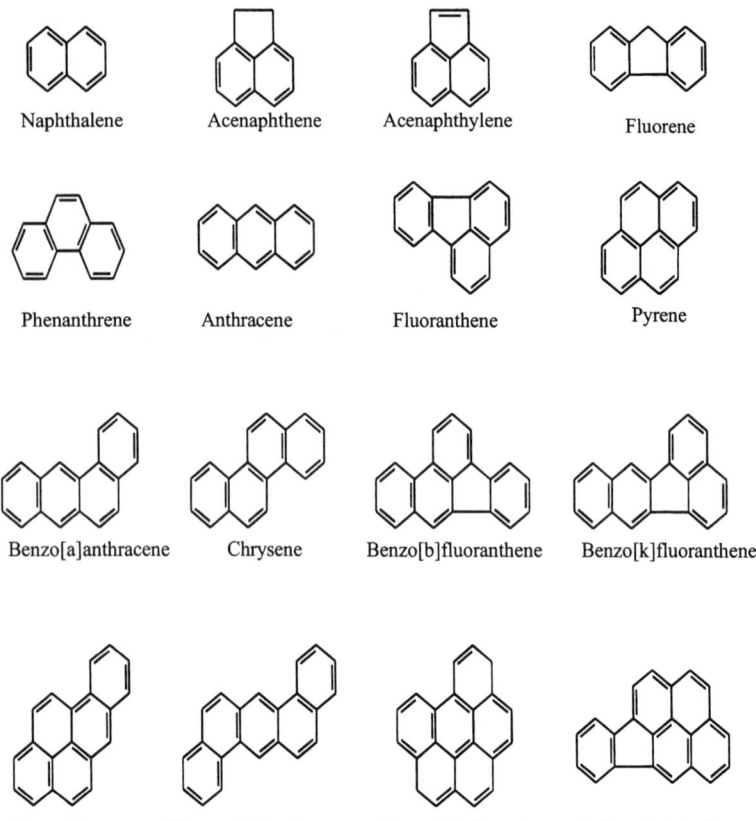

Figure 5-4. Structure of the 16 U.S. EPA priority PAH compounds

other unsaturated hydrocarbons (alkenes and alkynes) as their *pi* electrons are stabilized through delocalization of the *pi* orbitals (Brown et al., 1994).

Of the many PAH compounds, U.S. EPA has identified 16 PAH compounds as priority pollutants that are of environmental concern. Figure 5-4 shows the structures and the names of the 16 PAH compounds. In Table 5-24, a summary of their physical and chemical properties is provided. Some of these compounds are fairly insoluble in water and are solid at most temperatures found in the environment. These compounds tend to have a moderate to low volatility that decreases with increasing molar mass (Brown et al., 1994). Furthermore, PAH compounds adsorbs readily (due to their nonpolar nature), and biodegrade slowly because of the molecular size and shape of the compounds. For a summary of the solid-water partition coefficients of the 16 priority PAH pollutants, see Table 5-24. Because of these properties, PAH compounds tend to be relatively immobile and persistent in soil.

Table 5-24. Physical-chemical properties of polycyclic aromatic hydrocarbons

PAH compound	Chemical formula	CAS RN	Pure solid aqueous solubility ($\mu g/L$)[1]	Molecular weight (g/mole)	Melting temperature (°C)	Vapor pressure (Pa)	Henry's Law constant ($kPa\ m^3\ mole^{-1}$)	n-Octanol water partition coefficient ($\log K_{ow}$)	Organic carbon partition coefficient[1] ($\log K_{oc}$)
Naphthalene	$C_{10}H_8$	91-20-3	31,690	128.18	81.0	10.4	4.89E-2	3.37	3.11
Acenaphthene	$C_{12}H_{10}$	83-32-9	3,420	154.21	95.0	2.9E-1	1.48E-2	4.00	3.65
Acenaphthylene	$C_{12}H_{10}$	208-96-8	3,930	152.20	93.0	8.9E-1	1.14E-3	3.70	3.40
Anthracene	$C_{14}H_{10}$	120-12-7	45	178.24	216.4	8.0E-4	7.30E-2	4.45	4.15
Fluorene	$C_{13}H_{10}$	86-73-7	1,690	166.22	116.0	8.0E-2	1.01E-2	4.18	3.86
Phenanthrene	$C_{14}H_{10}$	85-01-8	1,000	178.24	100.5	1.6E2	3.98E-3	4.46	4.15
Fluoranthene	$C_{16}H_{10}$	206-44-0	206	202.26	108.8	1.2E-3	6.5E-4*	4.90	4.58
Pyrene	$C_{16}H_{10}$	129-00-0	130	202.26	150.4	6.0E-4	1.1E-3	4.88	4.58
Chrysene[2]	$C_{18}H_{12}$	218-01-9	1.8	228.30	253.8	8.4E-5*	NA	5.61	5.30
Benz(a)anthracene[2]	$C_{18}H_{12}$	56-55-3	5.7	228.30	160.7	2.8E-5	NA	5.60	6.14
Benzo(b)fluoranthene[2]	$C_{20}H_{12}$	205-99-2	14	252.32	168.3	6.7E-5*	5.1E-5	6.06	5.74
Benzo(k)fluoranthene[2]	$C_{20}H_{12}$	207-08-9	4.3	252.32	215.7	1.3E-8*	4.4E-5*	6.06	5.74
Benzo(a)pyrene[2]	$C_{20}H_{12}$	50-32-8	3.8	252.32	178.1	7.3E-7	3.4E-5*	6.06	6.74
Dibenz(a,h)anthracene[2]	$C_{22}H_{14}$	53-70-3	0.5	278.36	266.6	1.3E-8*	7.0E-6	6.80	6.52
Indeno(1,2,3-cd)pyrene[2]	$C_{22}H_{12}$	193-39-5	0.53	276.34	163.6	1.3E-8*	2.9E-5*	6.50	6.20
Benzo(g,h,i)perylene	$C_{22}H_{12}$	191-24-2	0.26	276.34	278.3	1.4E-8	2.7E-5*	6.51	6.20

Source: LaGrega et al., 1994; World Health Organization, 1998.
Note: NA = data not available; NR = not reported. [1]From LaGrega et al., 1994. [2]Carcinogenic PAHs.

The hydrophobic nature of these compounds lends to rapid partitioning onto particulate matter or tissues, suggesting bioaccumulation potential (LaGrega et al., 1994). These compounds have been shown to be acutely toxic to aquatic organisms at concentrations ranging from 0.2 mg/L to 10 mg/L, with acute toxicity increasing with increasing molecular weight to a point at which compound solubilities become too low to elicit a response (Neff, 1985). In humans, acute toxic responses may include liver damage or dermatitis. Metabolites of PAH compounds, which are more water soluble and reactive, can bind to protein, DNA, and other macromolecules, leading to cell damage, mutagenesis, or possible cancers of the stomach, lung, or skin (LaGrega et al., 1994). Chrysene, benz(a)anthracene, benzo(b)fluoranthene, benzo(k)fluoranthene, benzo(a)pyrene, dibenz(a,h)anthracene, indeno(1,2,3-cd)pyrene, and benzo-(g,h,i)perylene have been identified by the U.S. EPA as carcinogens.

PAH compounds at contaminated sites are likely to be found in complex mixtures that vary in composition with depth and distance from the source region. Novotny et al. (1981) studied the composition of several coal tar samples derived from coals of different geographical origin and determined that the major constituents were similar. Priddle and MacQuarrie (1994), however, compared four creosote samples and one coal tar sample for chemical composition and found that they all varied significantly. Barbé et al. (1998) studied PAH concentration profiles with depth at a former coke plant and determined that lighter PAH compounds were typically present in shallower depths with heavier compounds (for example, benzo(a)pyrene) becoming increasingly predominant at greater depths in the unsaturated zone and into the saturated region of the subsurface. Delorme and Carlier (1998) studied several former manufactured gas plant sites in France and determined that in the majority of cases, silts and clays tend to be more polluted at these sites than the coarser sands. Furthermore, they determined that the presence of coal tars tended to decrease soil permeability, and that many PAH components in residues showed little mobility and were naturally stabilized in soil.

In complex mixtures, PAH compounds may exhibit properties that differ from their pure phase properties. For example, Bayard et al. (1998) studied the influence of PAH in aqueous and NAPL phases on naphthalene sorption to organic matter on soils and determined that aqueous phase PAH has no effect on naphthalene sorption due to their low solubilities and thereby insignificant competition. However, when coal tar was added to the soil system, soil sorption of naphthalene dropped significantly as the naphthalene preferably partitioned onto the coal tar NAPL rather than the natural organic matter.

In addition to changes in their sorptive behavior, it is expected that the solubility of PAHs in liquid mixtures of organic compounds will be much different from the solubilities expressed in their natural solid form. Raoult's law (Eq. 3.1.20) has been shown to adequately predict equilibrium PAH solubilities (within a factor of two to four) in complex organic mixtures such as diesel fuel, gasoline, coal tar, and creosote (Lee et al., 1992a; 1992b; Cline et al., 1991; King and Barker, 1999). Table 5-25 shows a comparison of the pure aqueous solubility of the 16 priority PAH compounds and the effective solubilities calculated with Raoult's law for two coal tar mixtures and a coal tar creosote. It is interesting to note that the effective solubility of a compound expressed in a mixed liquid state may not necessarily be lower than its

Table 5-25. Composition of several PAH mixtures and predicted effective solubilities of the PAH compounds.

			Source and estimated source molecular weight					
			Coal tar Peters and Luthy (1993) 210 g/mol		Coal tar Ghoshal et al., 1996 226 g/mol		Coal tar creosote Mueller et al., 1989 158 g/mol	
PAH compound	Pure solid aqueous solubility[1] (μg/L)	Fugacity ratio[2] (f^L/f^S)	Mass fraction (% by wt.)	Estimated solubility[3] (μg/L)	Mass fraction (% by wt.)	Estimated solubility[3] (μg/L)	Mass fraction (% by wt.)	Estimated solubility[3] (μg/L)
Naphthalene	31,690	3.57	2.16	4010	10	2000	11.05	15400
Acenapthene	3,420	5.0	1.52	410	1.3	370	3.4	690
Acenaphthylene	3,930	4.55	0.68	150	0.37	86		
Anthracene	45	100	0.59	31	2	114	11.05	440
Fluorene	1,690	7.69	1.4	230			6.8	840
Phenanthrene	1,000	5.56	2.12	140	0.16	11.3	11.05	550
Fluoranthene	206	7.14	0.3	4.6	0.55	9	3.4	39
Pyrene	130	20.0	0.5	14			1.7	35
Chrysene	1.8	200	0.27	0.9	0.36	1.3	1.7	4.2
Benz(a)anthracene	5.7	25.0	0.31	0.4				
Benzo(b)fluoranthene	14	26.6			0.4	1.3		
Benzo(k)fluoranthene	4.3	76.9			0.16	0.47		
Benzo(a)pyrene	3.8	33.3	1.8	1.9	0.36	0.41	0.85	0.67
Dibenz(a,h)anthracene	0.5	250			0.04			
Indeno(1,2,3-cd)pyrene	0.53	NA						
Benzo(g,h,i)perylene	0.26	333						

Note: NA = data not available. [1] At 25°C. [2] Adapted from Peters et al., 1996. [3] Based on Raoult's Law.

solubility in pure solid form (as in anthracene and chrysene). Effective solubilities of PAH compounds based upon Raoult's law have been used in numerical models to simulate steady-state PAH dissolution from complex mixtures of nonaqueous phase liquids (NAPLs) over time with some success (King and Barker, 1999). However, Priddle and Macquarrie (1994) studied the efficacy of such models in columns of glass beads and determined that these models predicted the trends in dissolution but over-predicted aqueous concentrations by factors ranging from 1.5 to 8. These researchers suggested using a reduction factor as an extra model fitting parameter to account for the lower observed aqueous concentrations. Because of the uncertainty in PAH solubilities in the presence of complex mixtures and NAPLs, predictions by models on the fate and transport of PAHs in the presence of NAPLs should be carefully evaluated.

5.3.3 Attenuation Processes of PAH Compounds

Attenuation processes that determine the fate of PAHs in the subsurface are volatilization, hydrolysis, sorption, and biodegradation. Volatilization losses, especially in the vadose zone and at the capillary fringe of the subsurface environment, is one of the attenuation processes for PAH compounds. Bioremediation studies conducted by Bossert and Bartha (1986) and Park et al. (1990) indicated that 2- and 3-ring PAH compounds from soil samples may be lost through volatilization. No significant volatilization losses were found for PAH compounds containing more than three benzene rings. An inverse correlation between the number of rings in PAHs and their volatilization losses is generally assumed. However, quantification of losses of PAHs by volatilization from the subsurface environment under field conditions is not available.

PAHs are chemically stable and are not hydrolyzed by reactive groups under subsurface environmental conditions. Therefore, hydrolysis does not contribute to the abiotic change in the PAHs (Radding et al., 1976; Howard et al., 1991). PAHs can be photodegraded but this effect is minimal in a subsurface environment (Sims and Overcash, 1983). As indicated earlier in Table 5-24, the hydrophobic nature of the PAHs will result in high sorption onto the organic matter and mineral surfaces of the soils. Sorption is therefore one of the major attenuation processes for PAH compounds in the subsurface.

Two- and 3-ring PAHs are easily biodegradable while 4-, 5-, and 6-ring PAHs tend to be recalcitrant. Two- and 3-ring compounds can be utilized as a sole source of carbon and energy for microorganisms (Davis and Evans, 1964; Dean-Raymond and Bartha, 1975). In laboratory studies, Park et al. (1990) showed that the degradation of 2-ring PAHs in sandy soils was extensive with half-lives of approximately two days. In comparison, the half-lives for the 3-ring PAHs such as anthracene and phenanthrene were 16 and 134 days, respectively. The 4-, 5-, and 6-ring PAHs generally exhibited half-lives that were over 200 days. The work of Heitkamp and Cerniglia (1987) showed similar results for PAH degradation in sediment/water microcosms. McGinnis et al. (1988) conducted laboratory treatability studies on creosote-contaminated soils from wood treatment sites and found that PAHs with two rings generally exhibited half-lives of less than 10 days and PAHs with three rings had

Table 5-26. Half-lives and First-order decay rate of 16 PAHs in soil

PAH compounds	Half-life (days)	First order decay rate, k (d^{-1})	Experimental condition	Reference
Naphthalene	0.277	2.5	Soil slurry	Simpkin and Griesbrecht (1994)
Acenaphthene	134	0.0052	Soil[L]	Park et al. (1990)
Acenaphthylene	42 – 60	0.012 – 0.017	Soil[L]	Kincannon and Lin (1985)
Anthracene	55	0.0125	Soil[F]	Kasterner et al. (1999)
Fluorene	7 – 8	0.087 – 0.099	Soil slurry	Durate et al. (1997)
Phenanthrene		0.019	Sediment	Michel et al. (1995)
Fluoranthene	7 – 8	0.087 – 0.099	Soil slurry	Durate et al. (1997)
Pyrene	4.5	0.155	Soil[F]	Schwab et al. (1995)
Chrysene	33	0.021	Sediment	Michel et al. (1995)
Benzo(a)anthracene	75 – 80	0.0087 – 0.0092	Soil[F]	Pott and Henrysson (1995)
Benzo(b)fluoranthene	39	0.018	Sediment	Michel et al. (1995)
Benzo(k)fluoranthene	46	0.015	Sediment	Michel et al. (1995)
Benzo(a)pyrene	25	0.028	Soil[F]	Kanaly et al. (1997)
Dibenzo(a,h)anthracene	420	0.0017	Soil[L]	Park et al. (1990)
Indeno(1,2,3-cd)pyrene	232	0.003	Soil[L]	Park et al. (1990)
Benzo(g,h,i)perylene	590 – 650	0.0011 – 0.0012	Soil[L]	Coover and Sims (1987)

Note: [F] indicates field study; [L] indicates laboratory study.

half-lives of less than 100 days. However, 4- and 5-ring PAHs tend to exhibit half-lives that were over 100 days. The half-lives and the degradation rates of 16 PAHs for various conditions in the soil are summarized in Table 5-26. The half-lives and first-order degradation rates presented are only representative and should be used with care because the degradation rates of PAHs are very site specific and are dependent on the environmental conditions (e.g., electron acceptors, nutrients) present at the site.

Many bacterial, fungal and algal strains have been shown to degrade a wide variety of PAHs containing from 2- to 5-ring compounds. Several researchers have demonstrated that lower molecular weight (LMW) PAHs, 2- to 3-ring compounds, can be removed completely by microorganisms (Heitkamp and Cerniglia 1989; Mueller et al., 1989; Weissenfels et al., 1990). Recently, in laboratory studies, some researchers have isolated microorganisms that have the ability to mineralize 4-ring PAHs as their sole carbon and energy sources from contaminated soils (Mueller et al., 1990; Walter et al., 1991; Weissennfels et al., 1991). However, microbial mineralization of PAHs with four or more rings has generally been reported to occur via cometabolism (Bouchez et al., 1995; Ye et al., 1996; Aitken et al., 1998). A partial list of PAH-degrading microorganisms that have been isolated from contaminated soils is presented in Table 5-27.

Table 5-27. Representative PAHs metabolized by different microorganisms

Compounds	Rings	Microorganism(s)	Comments[1]	References
Naphthalene	2	*Rhodococcus* sp.	S	Bouchez et al. (1996)
		Pseudomonas sp.	S	Bouchez et al. (1996); Aitken et al. (1998)
Acenaphthene	3	*Pseudomonas* sp.	S	Komatsu et al. (1993)
		Neptunomonas naphthovorans	C	Hedlund et al. (1999)
Acenaphthylene	3	*Pseudomonas* sp.	S	Komatsu et al. (1993)
Anthracene	3	*Rhodococcus* sp.	S	Bouchez et al. (1996)
		Pseudomonas sp.	S	Bouchez et al. (1996)
		Bjerkandera sp.	S	Field et al. (1995)
Fluorene	3	*Pseudomonas* sp.	S	Foght and Westlake (1988)
		Rhodococcus sp.	S	Bouchez et al. (1996)
		Pseudomonas saccharophila	C	Stringfellow and Aitken (1995)
		Mycobacterium sp.		Boldrin et al. (1993)
Phenanthrene	3	*Rhodococcus* sp.	S	Bouchez et al. (1996)
		Pseudomonas sp.	S	Bouchez et al. (1996)
		Mycobacterium flavescens	S	Dean-Ross and Cerniglia (1996)
		Mycobacterium sp.	S	
		Flavobacterium sp.	S	Boldrin et al. (1993)
		Beijerinckia sp.	S	Stucki and Alexander (1987)
				Stucki and Alexander (1987)
Fluoranthene	4	*Rhodococcus* sp.	S	Bouchez et al. (1996)
		Pseudomonas sp.	S	Bouchez et al. (1996)
		Mycobacterium flavescens	S	Dean-Ross and Cerniglia (1996)
		Mycobacterium sp.	S	
		Pseudomona paucimobilis	S	Boldrin et al. (1993) Mueller et al. (1990)
Pyrene	4	*Rhodococcus* sp.	S	Bouchez et al. (1996)
		Pseudomonas sp.	S	Bouchez et al. (1996)
		Mycobacterium flavescens	S	Dean-Ross and Cerniglia (1996)
		Rhodococcus sp.	S	
		Xanthamonas sp.	S	Walter et al. (1991)
		Mycobacterium sp.	C	Grosser et al. (1991)
		Mycobacterium sp.	S	Ye et al. (1996) Boldrin et al. (1993); Heitkamp et al. (1988); Heitkamp and Cerniglia (1989)

continued on next page

Table 5-27. Representative PAHs metabolized by different microorganisms (continued)

Compounds	Rings	Microorganism(s)	Comments[1]	References
Chrysene	4	Sphingomonas	C	Ye et al. (1996)
		paucimobilis	S	Caldini et al. (1995)
		Pseudomona fluorescens	S	Cutright and Lee (1994)
		Achromobacter sp.		
Benzo(a)-	4	Sphingomonas	C	Ye et al. (1996)
anthracene		paucimobilis	C	Caldini et al. (1995)
		Pseudomona fluorescens	C	Wolter et al. (1997)
		Pleurotus sp. Florida		
Benzo(b)-	5	Sphingomonas	C	Ye et al. (1996)
fluoranthene		paucimobilis		
Benzo(k)-	5	Achromobacter sp.	C	Cutright and Lee (1994)
fluoranthene				
Benzo(a)-pyrene	5	Pseudomona	C	Chen and Aitken (1999)
		saccharophila P15	C	Kotterman et al. (1998)
		Bjerkandera sp. BOS55	C	Ye et al. (1996)
		Sphingomonas	C	Grosser et al. (1991)
		paucimobilis	C	Heitkamp and Cerniglia
		Xanthamonas sp.		(1989)
		Mycobacterium sp.		
Dibenzo(a,h)-	5	Sphingomonas	C	Ye et al. (1996)
anthracene		paucimobilis	C	Cutright and Lee (1994)
		Achromobacter sp.		

[1] Indicates whether the microorganism utilized the PAH compound as sole carbon (S) source or via cometabolism (C).

5.3.4 Natural Attenuation of PAH-Contaminated Sites: Case Studies

The difficulties associated with the removal of residual contamination at PAH-contaminated sites and the relatively limited movement of PAH compounds in the aquifer from the source areas have prompted several researchers to investigate the application of natural attenuation as a viable remedial alternative. Given that the biodegradation of many PAH compounds has been demonstrated in laboratory and field settings and that the PAH compounds have low solubilities and are highly sorbed by soil organic matter, the use of attenuation principles as a remedial alternative may have some merit. Several studies are available in the literature on the natural attenuation of sites contaminated with PAH compounds. The following is a summary of the results from these publications in which critical site parameters and degradation rates were reported. These results are tabulated in Table 5-28 along with results from other studies that have limited data.

Table 5-28. Summary of various case studies

Contamination site location	Site geology; depth to water, DTW; average pore velocity, v_x; hydraulic conductivity, K_h	Method used for rate estimation	Compounds	Contaminant concentration	Electron acceptor conditions	Linear distance coefficient, K_d (L/kg)	1st order decay rate, K (d^{-1})	References
FMGP Tar Burial Site South Glenn Falls, New York	Graded sand over a confining clay layer DTW = 2.7 m v_x = 0.096 m/d K_h = 0.09–8.16 m/d	Whole field modeling	Naphthalene	7 mg/kg	Mixed	0.81	0.00027	EPRI (1996)
			Acenapththylene	NR	Mixed	0.46	0.00027	
			Phenathrene	NR	Mixed	2.43	0.000027	
			Toluene	NR	Mixed	0.19	0.00068	
Emplaced Creosote Source Material CFB Borden, Ontario, Canada	Unconsolidated sand aquifer underlain by silts and clays DTW = 1.5 m v_x = 0.066–0.0947 m/d K_h = 6.05–8.24 m/d	Whole field modeling	Naphthalene	1200 mg/kg	aerobic	0.22	0.00057	King et al. (1999)
			Phenanthrene	1500 mg/kg	aerobic	1.8	0.014–0.063	
			Carbazole	40 mg/kg	aerobic	0.83	0.0063	
			1-methylnaphthalene	240 mg/kg	aerobic	0.24	0.0040	
FMGP Waste, Charleston, South Carolina	Soft organic clay overlain by sand and artificial fill DTW = 0.46 m v_x = NR K_h = 0.0305–3.05 m/d (fill) K_h = 4.88 m/d (sand)	Laboratory batch studies on site soils from core samples	Naphthalene	NR	aerobic	1.37	0.88	Campbell et al. (1996); Landmeyer et al. (1998)
			Naphthalene	NR	anaerobic	1.37	0.000046	
			Toluene	NR	aerobic	0.94	0.84	
			Toluene	NR	anaerobic	0.94	0.0020	
		Whole field modeling	Naphthalene	NR	anaerobic	0.0014	0.00007	
			Toluene	NR	anaerobic	0.62	0.00009	
Pulsed injection of aqueous compounds, MADE Site Columbus AFB Columbus, Mississippi	Fluvial Sedimentation DTW = NR v_x = NR K_h = 0.086–864 m/d	Whole field modeling	Naphthalene	7.23 mg/L	aerobic	0.085	0.0063	MacIntyre et al. (1993)
			O-dichlorobenzene	32.8 mg/L	aerobic	0.065	0.0059	
			p-Xylene	51.5 mg/L	aerobic	0.048	0.0141	
			Benzene	68.1 mg/L	aerobic	0.059	0.0066	
Landfill Leachate, Vejen Landfill, Vejen, Denmark	Clay deposit overlain by sandy alluvial aquifer DTW = NR v_x = 150–200 m/yr K_h = NR	Whole field modeling of steady-state plume	Naphthalene	NR	Iron(III) Reducing	NR	0.013–0.015	Lyngkilde and Christensen (1992)

continued on next page

Table 5-28. Summary of various case studies (continued)

Contamination site location	Site geology; depth to water, DTW; average pore velocity, v_x; hydraulic conductivity, K_h	Method used for rate estimation	Compounds	Contaminant concentration	Electron acceptor conditions	Linear distance coefficient, K_d (L/kg)	1st order decay rate, K (d^{-1})	References
FMGP Waste, Baltimore Gas and Electric Spring Gardens Facility, Baltimore, Maryland	Shallow unconfined aquifer with fill material underlain by sand and gravel interbedded with silt and clay lenses DTW = 1.6–5 m v_x = NR K_h = NR	Laboratory batch studies on site soils from core samples	Naphthalene	1000 µg/L	Aerobic	NR	0.0014 – 0.0069	Durant et al. (1994); MacFarlane et al. (1994)
	Gravel		Phenanthrene	1000 µg/L	Aerobic	NR	0.0050 – 0.0053	
	Sand		Naphthalene	1000 µg/L	Aerobic	NR	0.039 – 0.0084	
			Phenanthrene	1000 µg/L	Aerobic	NR	0.0015 – 0.0092	
	Clayey Silt		Naphthalene	1000 µg/L	Aerobic	NR	0.0046	
			Phenanthrene	1000 µg/L	Aerobic	NR	0.0053	
	Silty Clay		Naphthalene	1000 µg/L	Aerobic	NR	0.010	
Gasoline leakage, Former Gas Station, Perth, Australia	Thick clay aquitard overlain by 7–12 m fine dune sand DTW = 1 to 1.8 m v_x = 0.27–0.47 m/d K_h = 8.6–29 m/d	In-situ tracer test	Naphthalene, tracer test		Sulfate Reducing	NR	0.018 – 0.026	Davis et al. (1999); Thierrin et al. (1993); Thierrin et al. (1995)
		Whole plume modeling on steady state plume	Naphthalene, whole plume	33 mg/kg 1100 µg/L	Sulfate Reducing	NR	0.0039 – 0.0050	
		Partial plume modeling on steady state plume	Naphthalene, Partial plume	33 mg/kg 1100 µg/L	Sulfate Reducing	NR	0.00095 – 0.0027	
		In-situ tracer test	Toluene, tracer test		Sulfate Reducing	NR	0.0050 – 0.012	
		Whole plume modeling on steady state plume	Toluene, whole plume	670 mg/kg 75 mg/L	Sulfate Reducing	NR	0.0048 – 0.0073	
		Partial plume modeling on steady state plume	Toluene, partial plume	670 mg/kg 75 mg/L	Sulfate Reducing	NR	0.0027 – 0.0063	

Case Study 1: EPRI Site No. 24, South Glenn Falls, Saratoga County, New York

This study was conducted to demonstrate and document the effectiveness of the removal of FMGP tar source material and natural attenuation of the contaminant plume as a remediation alternative for FMGP sites (EPRI, 1996). The 4.5-acre site was approximately 1/2 mile northwest of the Hudson River and west of South Glenn Falls, New York. In the 1960's, several 55-gallon drums of gas main sealant along with 4,000 to 16,000 gallons of coal tar were pumped into a shallow, unlined trench at the site. The FMGP waste has sunk into the underlying aquifer resulting in the contamination of 7,200 cubic yards of soil and forming a dissolved phase plume stretching approximately 430 m down-gradient. The geology of the site consists of a shallow (0.5 m) top soil layer underlain by 4.6 m of coarse to medium grained sands. Fines to very fine-grained sands (silty material) become predominant between 4.6–6 m below ground surface (bgs). Soil borings revealed a confining clay to silty clay layer between 6 to 7.3 m bgs. The sands were well sorted and stratified having a characteristic of a glacier outwash deposit (EPRI, 1996). Hydraulic conductivities, measured by several slug tests and pumping tests, were found to range from 0.09 m/d to 8.16 m/d in the various silts and sediments. The average groundwater velocity through the site was estimated to be 0.096 m/d with an average hydraulic gradient of 0.0086 m/m. The average bulk density and porosity of the site soils were determined to be 1997 kg/m^3 and 0.35, respectively.

Remediation activities including source soil removal, groundwater contaminant evaluation, geochemical studies, microorganism sampling, laboratory batch biodegradation studies, and plume-scale modeling efforts with the numerical code MYGRT™ were initiated in 1987. The focus of the monitoring program was the natural attenuation of the dissolved-phase plume following source removal. Transects of the contaminant plume down the centerline and across the plume in several locations showed clearly that dissolved oxygen concentrations were depleted in regions of high PAH contamination in the plume, suggesting that active bioremediation of the contaminant plume was occurring. Laboratory studies of site soils showed increased numbers of PAH-degrading organisms in the contaminated cores and very high and elevated protozoa counts down-gradient of the source supporting the premise that active bioattenuation was occurring in the contaminant plume (EPRI, 1996).

Investigations of the site soils within the dissolved-phase contaminant plume yielded organic carbon contents of the site soils between 0.5% and 2.1%, but in general less than 1%. Based on these results, the retardation coefficients for naphthalene, acenaphthylene, phenanthrene, and toluene were estimated to be 4.0, 2.7, 10, and 1.7, respectively. Using these coefficients with the MYGRT™ transport code and field sampling data, estimates of the first-order decay rate coefficients for naphthalene, acenaphthylene, phenanthrene, and toluene were found to be 2.7×10^{-4} d^{-1}, 2.7×10^{-4} d^{-1}, 2.7×10^{-5} d^{-1}, and 6.8×10^{-4} d^{-1}, respectively. Over the course of three years of monitoring the dissolved-phase plume after source removal, extensive dissipation in the naphthalene, acenaphthylene, and toluene plume sizes and concentrations were observed. No detectable phenanthrene was observed in any groundwater sample from any well three years after source removal. Modeling and monitoring results suggest that no naphthalene will be present at any location in the groundwater above the 10 µg/L detection limit by the year 2030.

Case Study 2: Former Manufacturing Gas Plant site, Charleston, South Carolina, U.S.A.

This study was conducted to assess whether contamination from a FMGP site which operated from 1855 to 1957 in downtown Charleston, South Carolina would impact the adjacent Cooper River. The path of the plume from the FMGP site to the Cooper River transverses an 8-acre National Park Service property (Campbell et al., 1996). According to Landmeyer et al. (1998), the geology of the property consists of two Quaternary lithostratigraphic marine units, the Wando Formation and Holocene deposits, overlain by artificial fill. The artificial fill is composed of sand, silt, wood, sawdust, concrete, bricks, cinders and various other scrap materials. The depth of the fill varies from 3 to 6.1 m deep, and is considered to be an unconfined (fill) aquifer. The Wando Formation consists of soft organic clay overlain by gray sand with lower boundary about 23 m below ground level and the upper boundary varying between 10.6 to 16.8 m below ground level (Campbell et al., 1996). It is considered to be the confined lower aquifer at the site. The Holocene deposits provide a confining layer between the upper unconfined fill aquifer and lower confined sand aquifer. These deposits are composed of clayey to silty sand and soft, leaky, organic-rich clay with a thickness that varies from 1.5 m to as much as 12.2 m (Landmeyer et al., 1998). The hydraulic conductivities were from 0.03 to 3 m/d for the upper unconfined fill aquifer and 4.9 m/d for the lower confined sand aquifer. The depth to water at the site is approximately 0.5 m.

An intrinsic bioremediation study was initiated in 1993 and included sampling of groundwater, analysis of geochemistry and contaminants, laboratory determination of adsorption coefficients and biodegradation rates with aquifer materials, and modeling of the plumes with the numerical code SUTRA (Campbell et al., 1996). The focus of the study was on the unconfined fill aquifer. In 1994 and again in 1997, sampling showed dissolved oxygen was not present in any of the wells. High dissolved ferrous iron concentrations were observed in some wells. However, the presence of hydrogen at concentrations between 0.95 nM and 4.34 nM in these wells suggested that the ferrous iron was produced in earlier times of iron reducing conditions and that the aquifer was experiencing sulfate-reducing conditions (see Table 5-18) (Landmeyer et al.,1998). This was evidenced by the concentrations of hydrogen sulfide and dissolved sulfate of up to 5.11 mg/L and 633 mg/L, respectively. The presence of methane up to 13.4 mg/L indicated that methanogenic bacteria were present within the contaminant plume as well. Campbell et al. (1996) determined through laboratory analyses that the first-order biodegradation rates of toluene using aquifer sediments were 0.84 d^{-1} and 0.0020 d^{-1} for aerobic and anaerobic environments, respectively while the laboratory adsorption coefficient was determined to be 0.94 L/kg. In the case of naphthalene, the first-order laboratory microbial degradation rates were 0.88 d^{-1} and 4.6×10^{-5} d^{-1} for aerobic and anaerobic environments, respectively. The naphthalene linear adsorption coefficient for the sediments was estimated in the laboratory to be equal to or greater than 137 L/kg. Using site modeling techniques and field sampling data, these researchers reported best estimates for the toluene first-order degradation rate constant and linear sorption coefficient to be 9×10^{-5} d^{-1} and 0.62 L/kg, respectively using hydraulic conductivities between 0.12 to 1.2 m/day. Using the same hydraulic conductivities the best-fit model estimates for the first-order degradation rate constant and linear adsorption coefficient for naphthalene were 6.9×10^{-5} d^{-1} and 0.0014 L/kg, respectively. These results demonstrated that field degradation rates for these compounds more closely conformed to anaerobic laboratory degradation rates. The model simulations using the modeled best-estimates indicated that toluene will not impact the Cooper River within 150 years but naphthalene will impact the Cooper River within this time period at a concentration ranging from 0 to 5 mg/L.

Case Study 3: Creosote Source Emplacement, CFB Borden, Ontario, Canada

This study was performed to investigate natural attenuation processes of coal tar creosote compounds following source emplacement below the water table in two 1.5 m long by 5 m wide by 1.5 m deep excavations of an unused sand pit at CFB Borden, Ontario, Canada (Fowler et al., 1994; King and Barker, 1999). The source material consisted of 74 kg of creosote mixed with approximately 5,800 kg of sand and several kilograms of sodium chloride with a resulting residual creosote content of approximately 7% of the source pore volume. The geology of the site consisted of an unconsolidated sand aquifer of medium to fine sand of glacio-lacustrine origin with a hydraulic conductivity of approximately 6 m/d to 8.4 m/d, longitudinal dispersivity between 0.08 m and 0.036 m, and transverse dispersivity of 0.03 to 0.039 m. The sand aquifer grades into silts and clays at a depth of approximately 9 m and the water table fluctuates between the ground surface and a depth of 1.5 m. The groundwater flows predominantly north-northeast with a time-weighted average gradient of 0.0039 at a velocity that was estimated to range between 0.081 m/d to 0.0947 m/d. The background groundwater from the Borden aquifer was hard and in general aerobic with oxygen contents ranging up to 8.5 mg/L and averaging 2.47 mg/L. Nitrate concentrations ranged from 0.6 mg/L to 6 mg/L and sulfate concentrations were between 10 mg/L and 30 mg/L. The groundwater contained low concentrations of dissolved organic carbon (<0.7 mg/L) and the temperature and pH varied between 6×15 °C and 7.1 to 7.9, respectively. The aquifer material was determined to have organic carbon contents ranging from 0.01% to 0.09% with an average of 0.02%. The porosity, bulk density, and solids density of the aquifer material taken as the volume-weighted arithmetic means of 36 samples were estimated to be 0.33, 1.81 g/cm^3, and 2.71 g/cm^3, respectively.

The natural attenuation study at the CFB Borden site was initiated on August 8, 1991. Groundwater sampling was performed over the course of four years, during which time over 7,800 aqueous samples were taken from a large number of multilevel samplers. Retardation coefficients, affected by both sorption and transformation of the compounds, were estimated from field data based on chloride migration for phenol (1.05), m-xylene (2.5 at 5.9 m from the source and 3.7 at 24.05 m from the source), naphthalene (2.6 at 5.9 m from the source and 3.5 at 24.05 m from the source), and dibenzofuran (3.12 at 5.9 m from the source) (King and Barker, 1999;

King et al., 1999). Laboratory batch sorption experiments yielded linear distribution coefficients of 0.22 for naphthalene, 1.80 for phenanthrene, 0.67 for dibenzofuran, 0.83 for carbazole, and 0.24 for 1-methylnaphthalene, resulting in estimates of the linear retardation coefficients based solely on sorption of 2.2, 10.87, 4.67, 5.55, and 2.31, respectively. Batch sorption experiments for phenol and m-xylene yielded linear distribution coefficients equal to zero. Therefore, estimates of the linear distribution coefficients were made based upon the octanol-water partition coefficients and equaled 0.01 for phenol and 0.11 for m-xylene resulting in linear retardation coefficients of 1.05 and 1.6, respectively (King et al., 1999).

Redox parameters were monitored at 1,008 and 1,357 days after source emplacement both inside and outside the creosote plume. Average dissolved oxygen concentrations dropped from 2.47 mg/L outside the plume to 0.13 mg/L in the plume. Nitrate and ammonia concentrations were greatly variable, however the average values decreased from 2.35 mg/L to 1.51 mg/L and 0.62 mg/L to 0.21 mg/L, respectively, from outside to inside the contaminant plume. In addition, anaerobic degradation of the compounds was suggested by increases in reduced iron (0 mg/L to 0.2 mg/L), reduced manganese (<0.5 mg/L to 0.13 mg/L), and methane (0.001 mg/L to 0.036 mg/L) and a decrease in sulfate (14.1 mg/L to 11.6 mg/L) from outside to inside the contaminant plume. Phospholipid fatty acids analysis of aquifer cores indicated a higher concentration of microorganisms inside the contaminant plume than outside the plume. The microorganisms inside the plume were under more environmental stress, presumably due to either increased toxicity from plume compounds or an imbalance between organic carbon sources and nutrients within the plume. The microbial community tended towards stationary growth outside the plume and log growth inside the plume.

Phenol was observed to deplete quickly from the source material and migrate as a discrete slug after 439 days, at which time the peak concentration was only 7% of the 55 day concentration compared to 44% for chloride, suggesting significant transformation given its low sorptivity. In contrast, the plume for dibenzofuran, naphthalene, phenanthrene, carbazole, 1-methylnaphthalene, and m-xylene were found to be attached to the point of application. m-Xylene was observed to increase in extent from zero to 626 days followed by recession back to the source at 1,008 days and 1,357 days due to transformation. The dibenzofuran plume was observed to reach steady state by 1,008 days as the mass flux into the plume was balanced by mass transformation within the plume. Phenanthrene was observed to rapidly expand into the aquifer from 626 to 1,008 days and then receded and decreased in mass at 1,357 days. Like dibenzofuran, carbazole was observed to reach steady state between 1,008 and 1,357 days, but this was attributed to a decrease in source loading. The naphthalene and 1-methylnaphthalene plumes steadily increased in extent and mass over the time course of the sampling while the source fluxes decreased. All compounds were observed to be transformed in the contaminant plume, with half-lives of 78 d, 1,215 d, 11–49 d, 173 d, 99 d, 41d, and 110 d for m-xylene, naphthalene, phenanthrene, 1-methylnaphthalene, phenol, dibenzofuran, and carbazole, respectively. The estimated first order decay coefficients were 8.9×10^{-3} d^{-1}, 5.7×10^{-4} d^{-1}, $0.014 - 0.063$ d^{-1}, 4.0×10^{-3} d^{-1}, 7×10^{-3} d^{-1}, 0.017×10^{-3} d^{-1}, and 6.3×10^{-3} d^{-1}, respectively. However, the phenanthrene half-life may be subject to error due to the assumption of linearity between sampling events and large changes in phenanthrene concentration. Based upon the modeling results, it was expected that the naphthalene plume would continue to advance for at least two more years before reaching steady-state.

Case Study 4: Pulse Injection, Columbus Air Force Base, Macrodispersion Experiment Site (MADE), Columbus, Mississippi

The objective of this study was to measure degradation due to natural attenuation of a pulse injection of tritiated water, benzene, p-xylene, naphthalene, and o-dichlorobenzene in the saturated region of an unconfined aquifer at the macrodispersion experiment (MADE) site of Columbus Air Force Base, Columbus, Mississippi. At approximately 40 m down gradient of the injection source, lower hydraulic conductivity regions prevail (0.86 m/d). From 40 m to approximately 200 m downgradient of the injection source, the upper 3 m of the aquifer has an average hydraulic conductivity of 86 m/d while the hydraulic conductivity underlying this region remains low. 328 sampling wells were on the site, most of which contained multilevel samplers (Boggs et al., 1992).

The natural attenuation study was initiated on June 26, 1990. A mass balance based upon spatial moments analysis and comparison to tritium migration was used to estimate biodegradation rates. A pulse of 9,600 L of dilute tracer and the organic compounds was released over 47.5 hours through 0.6 m screened intervals, 4 m below the phreatic surface, in five injection wells spaced at 1 m intervals forming a line normal to the direction of the hydraulic gradient. Equal, steady flow was maintained in all the wells. Concentrations in the injection fluid were 55.6 µCi/L tritium, 51.5 mg/L p-xylene containing 2.77 µCi/L ^{14}C radiolabeled p-xylene, 68.1 mg/L benzene, 7.23 mg/L naphthalene, and 32.8 mg/L o-dichlorobenzene. Aqueous samples were taken from multilevel samplers at 27, 132, 224, 328, and 440 days after injection, and dissolved oxygen concentrations were monitored 8 days prior to and 48, 111, 161, 264, and 330 days after injection.

Linear distribution coefficients estimated from batch studies on the aquifer material for naphthalene, o-dichlorobenzene, p-xylene, and benzene were 0.085 L/kg, 0.065 L/kg, 0.048 L/kg, and 0.059 L/kg, respectively. However, because of the strong influence of degradation, the effects of sorption on organic solute distributions was considered minimal and therefore was ignored in assessing degradation rates. The temporal average dissolved oxygen concentration in the contaminant plume was determined to be 3.8 mg/L with a minimum individual value of 2.6 mg/L, suggesting aerobic conditions. Degradation in the Columbus aquifer material was observed to be approximately first order with an initial lag period attributed to microbial adaptation, cell growth and substrate limitation. Non-linear least squares regression was used to fit the first order degradation rate constants for benzene (0.0079 d^{-1}), p-xylene (0.0106 d^{-1}), naphthalene (0.0072 d^{-1}), and o-dichlorobenzene (0.0060 d^{-1}). Maximum first-order degradation plus dilution rates were taken directly from plots of contaminant concentration versus time from the field data and were corrected to yield degradation rates for the four compounds by subtracting the first-order rate of migration of tritium out of the region of interest from these maximum values. The resulting approximated first order degradation rate constants obtained from this method were 0.0066 d^{-1}, 0.0141 d^{-1}, 0.0063 d^{-1}, and 0.0059 d^{-1} for benzene, p-xylene, naphthalene, and o-dichlorobenzene, respectively.

Case Study 5: Unlined Municipal and Industrial Landfill, Vejen, Denmark

A study was done to identify specific redox environments controlling the fate of specific xenobiotic organic contaminants in a contaminant plume as a result of leachate leaking from an unlined municipal and industrial landfill in Vejen, Denmark that was operated between 1962 and 1981. The leachate resulting from the landfill contains BTEX, herbicides (mainly mecoprop (MCPP)), phenols, substituted benzenes, and naphthalene (Lyngkilde and Christensen, 1992). The site geology around the landfill comprised a shallow, unconfined, aerobic, sandy glacioalluvial aquifer, confined at the bottom by a clay deposit at a depth of 20 m close to the landfill and rising to a depth of 10 m at a distance of 400 m to 500 m down-gradient of the landfill. Small clay lenses can be found within the aquifer with a single substantial clay lens stretching out into the aquifer from below the landfill. The pore water velocity in the landfill was estimated to be 0.41 m/d to 0.55 m/d.

Monitoring of the site was conducted at 41 well nests of two monitoring wells with 4-inch screens set at different depths each, all within 130 m of the landfill and located on the plume centerline. Prior to initiation of the study, redox sensitive parameters were monitored and it was estimated that the groundwater within the contaminant plume was primarily anaerobic. A methanogenic region stretching to less than 50 m from the landfill was followed by sulfidogenic, ferro-/manganogenic, and nitrate-reducing redox zones. Aerobic conditions were again observed at about 300 m down-gradient of the landfill.

Using chloride as a conservative tracer, the site was monitored over the course of 285 days to observe the leachate plume characteristics. It was observed that the contaminant plume was stationary, and therefore the rate of source influx was balanced by biodegradation and dispersion. Changes in the chloride concentration were used to estimate dispersive losses and thus allowed the estimation of biodegradation losses by subtracting dispersive losses from the overall change in concentration with distance for a specific compound. Utilizing this method, it was observed that the compounds studied completely degraded under anaerobic (ferrogenic) conditions aside from MCPP which ultimately disappeared at 250–350 m down-gradient of the landfill in nitrate reducing to aerobic conditions. First-order degradation rate constants estimated from half lives taken from the plots of corrected compound disappearance versus distance for BTEX and naphthalene were 0.009 d^{-1} to 0.013 d^{-1} and 0.011d^{-1} to 0.015 d^{-1}, respectively (Lyngkilde and Christensen, 1992). Further studies of indigenous microorganisms in aerobic, denitrifying, ferrogenic, and methanogenic conditions showed that naphthalene could only be degraded under aerobic or ferrogenic conditions (Nielsen and Christensen, 1994).

5.3.5 Summary

There are not many studies on the natural attenuation of PAH-contaminated sites. The main processes affecting the attenuation of PAH compounds are adsorption and biodegradation. However, none of the studies cited above quantified the relative contribution of each of the two attenuation processes. In addition, the few studies available tend to focus on the biodegradation of low molecular weight PAHs such as naphthalene and phenanthrene. The fates of higher molecular weight compounds (more than three rings) were unknown although laboratory studies have indicated that these compounds may be biodegraded as carbon and energy sources or via cometabolism. The first-order decay rates of naphthalene and phenanthrene from the various studies varied from 0.00057 to 0.0063 d^{-1}, and 0.000027 to 0.063 d^{-1}, respectively. In most of the studies, the dominant electron acceptor other than oxygen was not thoroughly investigated. Elucidating the dominant electron acceptor conditions in the degradation of PAHs in PAH-contaminated plumes will assist in formulating strategies to simulate and enhance the attenuation process at these sites. Some of the field conditions which were not addressed in these studies include the impact of residual contamination (nonaqueous phase liquids or coal-tar coating on the soils) on the PAH solubilities and transport of the PAHs, and the presence of other compounds such as benzene and cyanide on the degradation of the PAHs. Overall, natural attenuation appears to be promising for these sites. After all, many of these sites, especially FMGP sites, have been contaminated for more than 50 years, but PAh migration has been minor compared to movement of more soluble compounds such as petroleum and chlorinated compounds.

5.4 Metals

5.4.1 Background

The vast majority of the research, publications, and applications of natural attenuation processes have been focused on sites that are contaminated with synthetic organic compounds. While there are many different attenuation mechanisms for organic compounds, the biological and chemical *degradation processes* are generally considered to be the primary attenuation mechanisms at most of the sites where natural attenuation is successfully occurring. Emphasis on *degradation* mechanisms, makes the use of natural attenuation for remediation at sites contaminated with metals seem ill-conceived, upon initial consideration. Metals, being chemical elements, are not subject to the degradation (i.e., bond breaking) that occurs with organic compounds. Thus, the chemical mass reduction monitoring approach that is common for organics is not appropriate for metals. There are, however, other attenuation mechanisms by which heavy metals concentrations in the aqueous phase can be effectively reduced. Natural attenuation may also be effective for the changing the valence states of some metals from more toxic and mobile forms to less toxic, less mobile states.

The most common analytical methods for measuring metals concentrations in environmental samples are element-based methods (e.g., atomic absorption, graphite furnace, and inductively coupled plasma) that provide the total metal concentrations, but do not provide speciation of the metal-based compounds. Most metals are rarely present in elemental form in the environment, existing as metal-based compounds (e.g., sulfides, hydroxides, oxides, phosphates, carbonates, organometallics, etc.). Fate and transport parameters such as solubility, partition coefficients, toxicity, and volatility can vary, for the same metal, depending upon the species and valence. Therefore, knowledge of the specific metal-based compounds plays an important role in evaluating the potential for natural attenuation to be an effective approach.

Evaluation of the potential for natural attenuation to be an acceptable approach for sites with heavy metals in groundwater involves a complex analysis of the contaminant chemistry, groundwater chemistry, soil chemistry, and other hydrogeologic factors. While considerable research remains to be performed on the topic, the following natural attenuation mechanisms are worthy of consideration:
- Precipitation
- Sorption
- Redox Reactions
- Ion Exchange
- Radioactive Decay (Section 5.5)

The need for additional research is demonstrated by the widespread presence of sites with metals contamination. It has been estimated that about 75 percent of Superfund sites have metals contamination (Evanko and Dzombak, 1997). Common metal contaminants include lead, chromium, arsenic, zinc, cadmium, copper, and mercury.

5.4.2 Technical Considerations and Approaches

Metals can exist in various different chemical forms in the soil/groundwater matrix. These include:
- Soluble contaminants in groundwater or within soil moisture in the vadose zone,
- Sorbed contaminants on soil surfaces, and
- Chemically stable solid compounds.

The physical and chemical properties of the soil and groundwater affect which of these forms are present. For example, naturally occurring inorganic anions (carbonate, phosphate, and sulfide) can convert the metal ions to relatively insoluble complexes. The following section will examine the formation, distribution and transport of some of the best known chemical forms of metallic compounds in the natural environment.

The distribution of metals between the solid phase and aqueous phase is described by the distribution coefficient, K_d, which is defined as follows:

$$K_d = \frac{\text{concentration in solids (mg/kg)}}{\text{concentration in water (mg/l)}}$$

While this concept is simple, actual K_d values for any given metal can vary with many parameters, most notably pH and the specific metal-based compound. Adsorption is sometimes non-linear. While values for K_d are available in the published literature, application of these values to a different site must be made with caution, as the chemistry of the source site and the application site may vary.

Some general trends and observations that may be useful when evaluating natural attenuation on a site specific basis are discussed below.

Sorption of metals on soil organic matter or mineral surfaces can reduce the metals concentration in groundwater. Sorption mechanisms can be generally classified into two types, absorption (also commonly referred to as physisorption) and adsorption (also commonly referred to as chemisorption). Absorption is the physical sorption of one substance into another, and is mediated by the physical parameters of the absorbant material (e.g., surface area, particle size, etc.). Absorption is generally weak and can be reversible. Adsorption (chemisorption) is the sorption of one substance onto the surface of another. Adsorption is often stronger than absorption, however, the attractive force is still relatively small, comparable to van der Waal's forces. Adsorption can be reversible under some conditions. The strength and efficiency of an adsorbate is subject to the number of surface active groups (i.e., pendant hydroxyls, siloxane bridges, etc.) per unit area. Chemisorption is an important mechanism for metal attenuation in the natural environment. For example, iron hydroxides in the soil matrix have been identified as a natural adsorbents for several metals. Likewise, layer-lattice silicates (e.g., smectite clays, zeolites, etc.) are known to strongly adsorb metallic ions from aqueous solutions in the natural environment (Iler, 1979). See discussion of cation exchange capacity, below.

Additional discussion of sorption processes is provided in Section 3.1.

Oxidation-reduction (redox) are reactions in which electrons are transferred. The constituent that accepts electrons is reduced; the constituent that donates electrons is oxidized. Redox reactions change the valence states of metals. These transformations can result in significant changes in both solubility and toxicity. For example, under appropriate conditions, hexavalent chromium is reduced to trivalent chromium. The trivalent valance of chromium is considerably less soluble (i.e., less mobile) and less toxic. Under oxidizing conditions, trivalent arsenic can be converted to pentavalent arsenic, which is less toxic, soluble, and mobile. Individual redox reactions must be identified and evaluated on a site-specific basis to assess their effectiveness and stability. Some redox reactions are more reversible than others. The oxidation-reduction potential (E_h) of a soil, in conjunction with the soil pH, may indicate the valence state and mobility of metals in the soil matrix. E_h-pH (ionic/mineral phase) diagrams identify the aqueous species and/or minerals that are predominant or stable at given E_h-pH conditions. Physical states are also identified. However, it must be kept in mind that these conclusions depend on the assumption of redox equilibrium and reactions are commonly not at equilibrium because of kinetic hinderance.

Cation exchange capacity (CEC), a specific form of chemisorption, is a measure of the soil's ability to exchange heavy metals for cations (positively charged ions) such as Ca^{2+}, Mg^{2+}, Na^+, and K^+. The exchange of cations occurs at negatively charged surfaces on the soil. Ion exchange can also occur at active sites within the lattice of "layer-lattice" silicates. Soils with high clay and/or organic matter typically have higher CEC values than sandy soils. The anion (negatively charged ion) exchange capacity (AEC) is a measure of the soil's affinity for anions. AEC is typically lower than CEC in soils. Naturally-occurring inorganic anions (e.g., carbonates, phosphates, and sulfides) can convert some metal ions to relatively insoluble complexes, thereby causing the metals to desorb and/or precipitate.

Table 5-29 identifies potential natural attenuation pathways and considerations for various metals. Table 5-29 also identifies geochemical parameters and data needs for evaluating the natural attenuation potential for various metals.

Chromium is one of the few metals for which significant laboratory and field-scale natural attenuation studies have been completed. While chromium can exist in oxidation states ranging from +6 to –2, the +6 and +3 states are most common in the environment (Palmer and Puls, 1994). Chromium (III) has relatively low toxicity, and low mobility under slightly acidic to moderately alkaline conditions. Chromium (VI) is relatively mobile, and highly toxic. Specifically, Chromium (VI) is mutagenic, teratogenic, and carcinogenic. Therefore, the reduction of chromium (VI) to chromium (III) is highly desired.

Chromium (VI), being a strong oxidant, can be reduced by naturally-occurring electron donors, including aqueous iron (II), ferrous iron minerals, reduced sulfur, and soil organic matter (Palmer and Puls 1994). Active reducing agents in sulfide minerals such as pyrite (FeS_2) include both ferrous iron and sulfide. Soil organic matter, in the forms of humic and fulvic acids, can be derived from decaying plant and animal matter, microoganisms, and organic materials that are synthesized or secreted from plants and microorganisms. These organic materials typically have high surface areas and high cation exchange capacities.

The reduction of chromium by soil organic matter is well established. The reduction rate increases with soil humic concentration and initial chromium (VI) concentration, and decreases with increasing pH. At neutral pH, reduction occurs over a period of several weeks. Chromium can also be reduced by aerobic and anaerobic microorganisms. While the specific mechanisms are not known, the anaerobic pathway appears more common (Palmer and Puls, 1994). Once the chromium is reduced to the trivalent state, it may bind to the soil organic matter, or precipitate as chromium hydroxide.

The reduction of chromium is not irreversible. Assessments of the potential for natural attenuation of chromium (VI) must acknowledge and evaluate the potential for oxidation of chromium (III) back to chromium (VI). There are two known mechanisms for this: oxidation by dissolved oxygen and oxidation by manganese dioxides (MnO_2).

Another potential application of natural attenuation for metals is the assessment of the toxic concentrations of select divalent metals in anoxic sediments and overlying waters. The bioavailability and toxicity of the metals can be evaluated by assessing the ratio of the concentrations of simultaneously extracted metals to acid volatile sulfides (SEM/AVS).

Table 5-29. Natural attenuation pathways for metals and radionuclides

Constituent	Natural attenuation pathways	Other considerations	Data needs
Metals			
Pb^{2+}	Sorption to iron hydroxides, organic matter, carbonate materials, formation of insoluble sulfides	Low pH destabilizes carbonates, iron hydroxides. Comingled organic acids and chelates (e.g., EDTA) may decrease sorption. Low E_H dissolves iron hydroxides, but favors sulfide formation	Iron hydroxide availability; pH, alkalinity, and Ca^{2+} levels to answer if calcium carbonate is stable. E_H, and if E_H is low, sulfide levels. Organic carbon content.
CrO_4^{2-}	Reduction by organic matter, sorption to iron hydroxides, formation of $BaCrO_4$	Low pH destabilizes carbonates, iron hydroxides. Low E_H dissolves iron hydroxides. Are reductants available?	E_H electron donor levels, pH (reduction rates are faster at low pH).
As (III or V)	Sorption to iron hydroxides, formation of sulfides.	Low pH destabilizes carbonates, iron hydroxides. Low E_H dissolves iron hydroxides.	E_H, and if E_H is low, sulfide levels.
Zn^{2+}	Sorption to iron hydroxides, carbonate minerals, formation of sulfides.	Low pH destabilizes carbonates, iron hydroxides. Comingled organic acids and chelates may decrease sorption. Low E_H dissolves iron hydroxides.	Iron hydroxide availability; pH, alkalinity, and Ca^{2+} levels to answer if calcium carbonate is stable. E_H, and if E_H is low, sulfide levels.
Cd^{2+}	Sorption to iron hydroxides, carbonate minerals, formation of insoluble sulfides.	Low pH destabilizes carbonates, iron hydroxides. Comingled organic acids and chelates (e.g., EDTA) may decrease sorption. Low E_H dissolves iron hydroxides, but favors formation of sulfides.	Iron hydroxide availability; pH, alkalinity, and Ca^{2+} levels to answer if calcium carbonate is stable. E_H, and if E_H is low, sulfide levels.
Ba^{2+}	Sorption to iron hydroxides, carbonate minerals, formation of insoluble sulfate minerals.	Low pH destabilizes carbonates, iron hydroxides. Low E_H dissolves iron hydroxides. What are sulfate levels?	Sulfate levels.
Ni^{2+}	Sorption to iron hydroxides, carbonate minerals.	Low pH destabilizes carbonates, iron hydroxides. Comingled organic acids and chelates may decrease sorption. Low E_H dissolves iron hydroxides, but favors sulfide formation.	Iron hydroxide availability; pH, alkalinity, and Ca^{2+} levels to answer if calcium carbonate is stable. E_H, and if E_H is low, sulfide levels.
Hg^{2+}	Formation of insouble sulfides	Is methylated by organisms.	E_H, and if E_H is low, sulfide levels.

continued on next page

Table 5-29. Natural attenuation pathways for metals and radionuclides (continued)

Constituent	Natural attenuation pathways	Other considerations	Data needs
Radionuclides			
UO_2^{+2}	Sorption to iron hydroxides, precipitation of insoluble minerals, reduction to insoluble valence states.	Low pH destabilizes carbonates, iron hydroxides. Comingled organic acids and chelates may decrease sorption. High pH and/or carbonate levels decrease sorption. Low E_H dissolves iron hydroxides.	Iron hydroxide availability; pH, availability of reducing compound.
Pu (V and VI)	Sorption to iron hydroxides, formation of insoluble hydroxides.	May move as a colloid. Low E_H dissolves iron hydroxides.	Iron hydroxide availability; pH, availability of reducing compound.
Sr^{2+}	Sorption to carbonate minerals, formation of insoluble sulfates.	Low pH destabilizes carbonates.	Iron hydroxide availability; pH, alkalinity, and Ca^{2+} levels to answer if calcium carbonate is stable.
Am^{3+}	Sorption to carbonate minerals.	Low pH destabilizes carbonates. High pH increases solubility of Am-carbonate minerals	Iron hydroxide availability; pH, alkalinity, and Ca^{2+} levels to answer if calcium carbonate is stable.
Cs^+	Sorption to clay interlayers.	High NH_4^+ levels may lessen sorption. How abundant are clays?	Clay content; cation exchange capacity
I^-	Sorption to sulfides, organic matter.	Sorbs to very little else.	Metal sulfide mineral content.
TcO_4^-	Possible reductive sorption to reduced minerals (e.g. magnetite), forms insoluble reduces oxides and sulfides.	Sorbs to very little else.	E_H, and if E_H is low, sulfide levels.
Th^{4+}	Sorption to most minerals, formation of insoluble hydroxide.	May move as a colloid.	
Co^{2+}	Sorption to iron hydroxides, carbonate minerals.	Low pH destabilizes carbonates.	Iron hydroxide availability; pH, alkalinity, and Ca^{2+} levels to answer if calcium carbonate is stable.

Source: Brady and Borns, 1997.

This approach is currently being evaluated and developed by EPA for assessment of cadmium, copper, lead, nickel and zinc (USEPA Science Advisory Board, 1995). The approach is named after the analytical procedures employed. Simultaneously extracted metals (SEM) concentrations are measured using a cold acid extraction. The reactive solid-phase sulfide fraction is also measured using a cold hydrochloric acid extraction, or acid volatile sulfide (AVS) extraction.

It has been demonstrated that the naturally-occurring sulfides will form insoluble metal sulfides in the sediments. Metals sequestered by the sulfides are not available in the aqueous phase. The absence of aqueous phase metals in the interstitial sediment pore space minimizes the toxicity of the sediments. The metals will become biologically available only if their concentrations exceed the amount that will bind with the naturally-occurring sulfides. The SEM/AVS approach indicates that metals releases to the surface waters containing sulfide-rich sediments may "naturally attenuate" through sulfide sequestration.

Evaluation of the SEM/AVS approach should consider whether chemical conditions prevail which allow for the degradation of the sulfides into daughter compounds which may be soluble. For example, FeS_2 in its mineral species Marcasite, can oxidize to form Melanterite ($FeSO_4.7H_2O$) and sulfuric acid. This lowers the pH of the surrounding material, which enhances further oxidation of sulfides (Dana, 1996).

5.4.3 Enhanced Natural Attenuation

Natural remediation is conventionally considered to be the reliance upon natural processes to remediate constituents of concern in the subsurface, without the use of engineered enhancements. Human interaction is typically limited to monitoring and data evaluation. The application of engineered natural processes to remediate the environment can also be considered under the "natural attenuation" umbrella. Under such an approach, human interaction may involve the use of engineering to enhance or promote natural remediation processes. Two such processes that may be effective for heavy metals are phytoremediation and constructed wetlands. These are discussed below.

Phytoremediation is a method by which some metals may be physically removed from soil and/or groundwater. With respect to metals, phytoremediation can be defined as the use of plants to extract the metals from the subsurface. Once captured by the plant roots, the metals are incorporated into the aboveground portions of the plant tissue. Some species, known as hyperaccumulators, are effective at removing significant amounts of metals from the subsurface, and incorporating the metals into the plant tissue. Phytoremediation by the root extraction and plant tissue accumulation approach is also referred to as phytoaccumulation and phytoextraction (USEPA, 1998).

While it is unlikely that hyperaccumulator species are naturally present at sites having metals contamination, planting of such species can provide a relatively low-maintenance remediation approach. Once the metals are incorporated into the plants, the plants must be harvested and managed (e.g., composted, incinerated, or disposed at an off-site facility) to prevent the return of the metals to the source soil. There are many limitations to the use of phytoremediation for metals (Miller, 1996). For example, the metals must be accessible to the plant roots. For plants with shallow root systems, the limitation is obvious. High density planting and multiple planting/harvest cycles may also be required. The metals must be present in a soluble form that is readily captured by the roots. The specific hyperaccumulator species must be able to live under the local weather and soil conditions.

While much of the phytoremediation work to date has focused on organic compounds, some species have been identified that are promising for metals. Metals such as nickel, zinc, and copper have been identified as good candidates for phytoremediation. Research with lead and chromium is ongoing.

Natural wetlands have demonstrated their ability to serve as filters for many metals. The metals can be sequestered within the wetlands sediment by various methods, including sorption on organic matter, sorption onto iron hydroxides and manganese hydroxides, formation/precipitation of reduced metal sulfides, and precipitation of metal hydroxides (Brady and Borns, 1997). A review of the literature finds both successful and unsuccessful examples of wetlands mitigation of metals-containing water. Metals that have reportedly been successfully treated using wetlands include aluminum, arsenic, cadmium, copper, iron, lead, manganese, nickel, selenium, silver, tantalum, uranium, vanadium, and zinc (Brady and Borns, 1997). The specific metal, metal compound, and concentration impact the wetland's ability to sequester metals. Other factors that can impact the success include:

- Flow rate.
- Alkalinity and pH.
- Sediment composition, including organic fraction, and specific organic constituents. These factors can impact both the sequestering capability and the sequestering capacity of the sediment. Some studies have shown good initial sequestering capability, but poor long term capability as the sequestering capacity is exceeded.
- Climate.
- Type and quantity of vegetation. Cattail-based (*Typha* sp.) wetlands have reportedly been successful for treatment of mine water.
- Aerobic/anaerobic conditions.

Some sites have been so successful in sequestering metals that the wetlands sediments have been considered for mining. The recognized success of natural wetlands in the capture of metals has given rise to the use of engineered, man-made wetlands to passively treat metal-containing run-off from sites such as abandoned mines.

5.4.4 Applications

The use of monitored natural attenuation for metals has been approved at many sites, including several Superfund sites. Natural attenuation of groundwater, combined with source control (e.g., source removal, soil treatment, etc.) is a relatively common remediation approach for sites with multi-media contamination. A few examples include:

- **The Geiger (C&M Oil) Site, Charleston County, South Carolina:** During operation between 1969 and 1974, the site included eight unlined lagoons for the storage of waste oil, which was later incinerated. The initial remediation plans for the site, as presented in the June 1987 Record of Decision (ROD), included soil treatment and a groundwater pump-and-treat system. However, subsequent groundwater monitoring has detected lower concentrations of contaminants, including lead, than were detected during the Remedial Investigation. In September 1998, the ROD was amended to eliminate the pump-and-treat approach and replace it with monitored natural attenuation (USEPA, 1999c).
- **Reeves Southeast Galvanizing Corp.:** Galvanizing activities at the site resulted in the contamination of groundwater (the Northern Surficial Aquifer), surface water, soil and sediments. The November 1993 ROD included monitored natural attenuation for the Northern Surficial Aquifer. Groundwater cleanup goals were established for arsenic, cadmium, chromium, lead, nickel, and zinc. The use of groundwater pump-and-treat was included as a contingency remedy, to be implemented in the event that natural attenuation is ineffective (USEPA, 1999b).
- **Preferred Plating Corporation, Suffolk County, New York:** Wastewater produced by plating operations was disposed in unlined leaching pits during operations from 1951 to 1976. Releases from the leaching pits resulted in contamination of soil and groundwater. A 1989 ROD specified pump-and-treat as the groundwater remedy. A September 1992 ROD required excavation and off-site treatment of contaminated soils. Construction of the groundwater remedy was delayed while the soil remedy was implemented. Groundwater monitoring performed after the source area remediation was completed in May 1994 indicated that groundwater contaminant levels had declined significantly. Two additional groundwater monitoring events in 1995 and 1996 indicated a continued decrease in contaminant levels. Based on these findings, a September 1997 ROD amendment eliminated the pump-and-treat requirement in favor of natural attenuation. Subsequent groundwater monitoring performed in January 1998 continued to indicate a decline in cadmium concentrations (USEPA, 1999a).

5.4.5 Conclusions and Recommendations

Metals can and are remediated by natural environmental processes at some sites. At other sites, relatively minor changes in environmental conditions can promote attenuation. There are many different mechanisms by which metals attenuation can occur, with the effectiveness of each mechanism affected by both the physical and chemical properties of the environment. Each of the mechanisms is specific to a handful of metals and/or valence states.

The science of metals attenuation is still developing. Some processes are well understood, while others are still in their infancy. As such, the remediation engineer or scientist attempting to address metals contamination is faced with a significant challenge. First, a thorough understanding of the contaminants of concern and the environmental conditions of the subject site is required. This may require investigation of parameters that are outside the scope of conventional environmental characterization studies. Second, the designer must keep abreast of the latest studies and research in the field, as new concepts and approaches are introduced frequently.

5.5 Radioactive Contaminants

5.5.1 Overview

Abundant evidence indicates the ability of soils and soil constituent surfaces to naturally sequester metals and radionuclides, rendering them non-bioavailable (Ainsworth et al., 1994; Backes et al., 1995; McLaren et al., 1998; Eick et al., 1999). Despite this, the technical basis for implementing monitored natural attenuation (MNA) at radionuclide-impacted sites has only recently received much attention (Brady and Borns, 1997). Case studies outlining natural controls on radionuclide plume attenuation are accumulating (Brady et al. in press). Unlike MNA of organic contaminants, which typically involves contaminant decomposition by microorganisms as discussed in Sections 4.1–4.3, MNA of radionuclides and metals requires sequestering, or transformation of contaminants by the organic and inorganic components of the soil matrix. Because the radionuclide or metal will still remain in the soil matrix (though in a non-bioavailable form), regulators may be hesitant to rely on MNA unless site specific data are presented demonstrating the specific processes responsible for sequestering the contaminant (DOE, 1999). Many of the mechanisms responsible for metal natural attenuation (discussed in Section 4.4) contribute to radionuclide attenuation over and above the extent of attenuation attained through radioactive decay over time.

The rate of radioactive decay or disintegration follows first order kinetics, that is the rate of nuclei decay is directly proportional to the number of nuclei present:

$$-\frac{dA}{dt} = k_d A \quad (5.5.1)$$

where A is the number of radioactive nuclei present, and k_d is the radioactive decay or rate constant (1/s). The half-life ($t_{1/2}$) which is the time required for half of the radioactive nuclei to disintegrate, is often used instead of k_d to characterize radioactive decay:

$$t_{1/2} = \frac{0.693}{k_d} \quad (5.5.2)$$

Since radioactive decay follows first order kinetics the half-life is related only to k_d and is independent of the concentration of the radionuclei present. Within a given time period, the amount of specific radioactivity released will be greater for a radionuclide with a shorter half-life, such as iodine-131 ($t_{1/2} = 8$ days), then one with a longer half-life, such as plutonium-239 ($t_{1/2} = 24,300$ years). In the case of iodine-

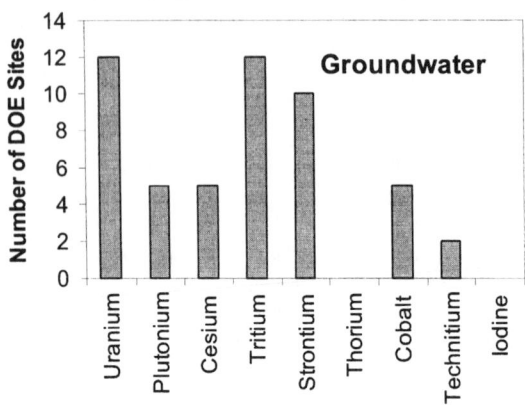

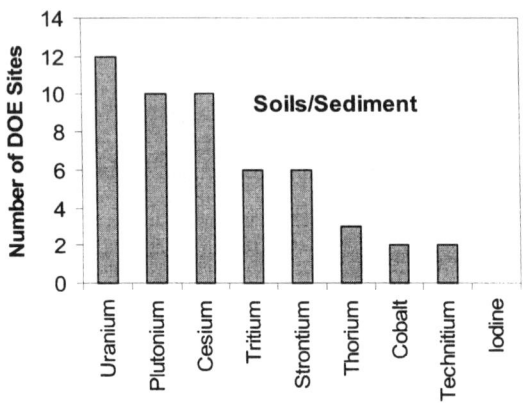

Figure 5-5. Frequency of occurrence of selected radionuclides in groundwater and soils/sediments at DOE facilities

131, its shorter half-life is a critical component of MNA, because it will more rapidly decay to stable forms, while plutonium-239 will persist in the environment for thousands of years. Moreover, many radionuclides decay to other radionuclides, which also produce radiation as they decay to stable forms (Brady et al., 1999). Figure 5-5 indicates the more common radionuclide contaminants seen at U.S. Dept. of Energy Sites.

5.5.2 Radionuclide Attenuation

Radionuclides, and inorganic contaminants in general, can be removed from soils and groundwaters by: (1) sorption to mineral surfaces and/or soil organic matter; (2) coprecipitation or precipitation of sparingly soluble solids; and (3) uptake by plants and organisms. The latter removal mechanism (plant uptake) is not viewed as a viable sink because it provides a concentration mechanism and, as noted above, most radionuclides half-lives are longer than the life of the organism. These processes depend on radionuclide speciation, which depends primarily on the ambient biological and geochemical conditions of the soil or groundwater. Redox state (electron availability), pH, alkalinity, and the presence of chelating agents (e.g., EDTA, natural organic acids) or solid-forming (e.g., phosphate in a number of cases) ligands are critically important (ionic strength to a lesser extent) to defining these conditions (Brady et al., 1999).

As stated previously, the processes controlling radionuclide natural attenuation (i.e., sorption, formation of sparingly soluble carbonates and phosphates, reduction, etc.) have been thoroughly discussed in Section 5.4 on Metals. Table 5-30 provides a summary of the various attenuation mechanisms for some of the more common radionuclides found in soil and groundwater.

An attenuation pathway for tritium could be its short half-life, 12.3 years. Very large amounts of tritium were produced by underground testing of nuclear weapons at the Nevada Test Site (on the order of millions of Ci), some of which were released to the ground water on the site. Although the hydrology of the site is poorly understood because of the geologic complexity of the site, the relatively short half-life (12.3 years) of the tritium, the inert decay product (He), and the remoteness of the site may make MNA the preferred approach. A more thorough discussion of radionuclide-specific natural attenuation mechanisms can be found in the Proceedings from *Natural Attenuation of Metals and Radionuclides: Report from a Workshop held by Sandia National Laboratories* (Brady and Borns, 1997) or the *Screening and Technical Guidance for Monitored Natural Attenuation at DOE Sites* (Brady et al., 1999).

5.5.3 Screening Radionuclide-Impacted Sites for MNA

A necessary "first step" in the MNA process for all impacted sites is a preliminary site-screening (Chapter 2). Site screening relies on an evaluation of existing site characterization data to ascertain whether MNA is applicable for the site. This screening is performed to determine whether site conditions are favorable for natural attenuation to provide protection of human health and the environment. For organic constituents, this is a relatively simple process and can be semi-quantitatively performed by evaluating degradation trends from well data or evaluating redox conditions to ascertain whether favorable conditions are prevalent for biodegradation. A more advanced organic constituent screening could be performed using the Air Force Center for Environmental Excellence (AFCEE) scorecard or one of the other numerous evaluation packages.

Table 5-30. Natural attenuation pathways for selected radionuclides

Radionuclide	Natural attenuation pathway	Mitigating conditions
U(VI)	Sorption to iron hydroxides; precipitation of sparingly soluble hydroxides and phosphates; reduction to sparingly soluble valence states.	Low pH destabilizes carbonates and iron hydroxides. Commingled organic acids and chelates may decrease sorption. High pH and/or carbonate concentrations decrease sorption. Low E_H dissolves iron hydroxides but favors reduction.
Pu(V and VI)	Sorption to iron hydroxides; formation of sparingly soluble hydroxides and carbonates.	May move as a colloid. Low E_H dissolves iron hydroxides.
Sr	Sorption to carbonate minerals and clays; formation of sparingly soluble carbonates and phosphates	Low pH destabilizes carbonates. High dissolved solids favor leaching of exchange sites.
Am(III)	Sorption to carbonate minerals; formation of carbonate minerals.	Low pH destabilizes carbonates. High pH increases solubility of Am-carbonate minerals.
Cs	Sorption to clay interlayers.	High NH_4^+ concentrations may lessen sorption. Low K^+ concentrations may increase plant uptake.
I	Sorption to sulfides and organic matter.	Sorbs to very little else in oxidized state.
Tc(VII) as TcO_4^-	Possible reductive sorption to reduced minerals (e.g., magnetite); forms sparingly soluble reduced oxides and sulfides.	Sorption to other phases extremely limited.
Th	Sorption to most minerals; formation of sparingly soluble hydroxide.	May move as a colloid.
^3H	None	
Co	Sorption to iron hydroxides, organic matter, and carbonate minerals.	Low pH destabilizes carbonates. Low E_H dissolves iron hydroxides. Stable complexes form with chelators.

Source: after Brady et al., 1997.

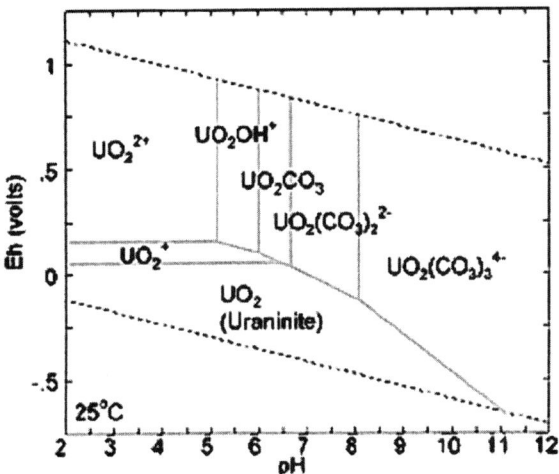

Figure 5-6. Uranium phase diagram
Source: from MNAtoolbox (Brady et al. 1999).

Inorganic species (metals and radionuclides) can be screened rapidly using the *MNAtoolbox*, which is maintained by DOE at www.sandia.gov/eesector/gs/gc/na/mnahome.html. *MNAtoolbox* provides a template for development of the natural attenuation conceptual model. Important environmental conditions that must be considered for successful MNA are pH and electrochemical potential (E_h). The form of aqueous species and solid phases for an inorganic element often depend on these variables. For example, uranium is found in aerobic natural environments as U(VI), and as the uranyl ion (UO_2^{2+}). Uranyl ion sorbs strongly to iron oxide surfaces (Table 5-30). However, at high pH the dominant aqueous species are uranyl carbonates whose ionic charges become more negative with increasing pH (Figure 5-6). In the alkaline pH range where uranyl carbonates dominate, U(VI) is mobile because these uranium carbonate species do not sorb strongly to clay and oxide surfaces. Thus, liming a site to decrease the mobility of a contaminant that is insoluble at high pH (e.g. Pb), might mobilize any comingled U contamination. The phase diagram in Figure 5-6 illustrates the effects of pH and redox on uranium speciation.

MNAtoolbox is useful for identifying those attenuation factors that might contribute to contaminant destruction or immobilization. Additionally, *MNAtoolbox* identifies the most relevant site characterization data needed to develop a natural attenuation argument for radionuclides.

The *MNAtoolbox* provides several pieces of standard biogeochemical data to help the site manager or analyst determine whether MNA may be a promising remediation alternative. The "MNA pathways" section of the *toolbox* indicates the likely biogeochemical process(es) providing significant attenuation of mobile/

bioavailable contaminant. Additional information is provided in a hyperlinked screen associated with each MNA pathway.

A critical component of the *MNAtoolbox* is the Site Screening Scorecard. The Site Screening Scorecard is associated with each contaminant to determine whether natural attenuation may be possible under site-specific conditions. The Scorecard is contaminant-specific and is subdivided into hydrologic and geochemical sections. Credit is given, or taken away, depending on favorable or unfavorable conditions. This methodology is similar to the characterization for potential degradation in the AFCEE approach. However, weight is also placed on physical processes, while the AFCEE protocol only considers conditions that favor biological factors. A total score near 100 indicates that MNA is possible and should be investigated. A low score does not necessarily disqualify MNA at a particular site; but may simply indicate that a greater level of characterization is required to support MNA for the particular metal or radionuclide at the site.

5.5.4 Demonstrating MNA of Radionuclides

Data needs depend primarily on whether the likely fate of the compound is as a component of an insoluble solid, a sorbed contaminant, or, possibly, a species occluded on an iron hydroxide or carbonate mineral surface, or irreversibly sorbed to organic matter or an interlayer clay site. The MNA pathways for the various metal contaminants depend on the radionuclide as well as the soil and/or groundwater type. Moreover, the long-term efficacy of NA of radionuclides often will depend on the permanence of the irreversible sorption mechanism. An effective test of MNA for radionuclides in soil must, in addition to identifying any solubility-controlling phases: 1) Quantify the amount of radionuclide that is irreversibly sorbed (i.e., non-exchangeable), and 2) Identify the sorption host. This process allows for identification of the metal host and thus the attenuation mechanism. One means for roughly constraining the primary MNA sorption pathway(s) in a given sample is through sequential digestion procedures (e.g., Kennedy et al., 1997; Schultz et al., 1996; Tessier et al., 1979; Wasay et al., 1998; Yong et al., 1993). Selective sequential extraction procedure (SEP) uses a series of chemical reagents to sequentially consume the various binding phases. The different fractions are released sequentially upon the separation of each binding phase. Typically the chemical reagents are weak acids, oxidizing agents, reducing agents, and/or concentrated electrolyte solutions. SEP procedures generally target the following soil fractions.

- Exchangeable phase: Radionuclides that are reversibly sorbed (non-specifically) to soil minerals, amorphous solids, and/or organic material.
- Organic phase: Any radionuclide associated with soil organic matter (e.g., humic acids) will either be in the exchangeable phase, or be irreversibly bound in the organic phase.
- Carbonate phase: Radionuclides that are irreversibly sorbed or otherwise bound up in radionuclide carbonate minerals.
- (Hydr)oxide phase: Radionuclides bound to hydroxides of iron, manganese, and/or aluminum.
- Acid/Sulfide fraction: Radionuclides associated with sulfide minerals

Table 5-31. Data needs for natural attenuation of radionuclides

Chemical	Data needs
UO_2^{+2}	Iron hydroxide availability, pH, availability of reducing compounds
Pu(V and VI)	Iron hydroxide availability, pH, availability of reducing compounds
Sr^{2+}	Iron hydroxide availability; pH, alkalinity, and Ca^{2+} levels to answer if calcium carbonate is stable.
Am^{3+}	Iron hydroxide availability; pH, alkalinity, and Ca^{2+} levels to answer if calcium carbonate is stable.
Cs^+	Clay content, cation exchange capacity.
I^-	Metal sulfide mineral content
TcO_4^-	E_H, and if E_H is low, sulfide levels.
Co^{2+}	Iron hydroxide availability; pH, alkalinity, and Ca^{2+} levels to answer if calcium carbonate is stable.

Source: Adapted from Brady et al., 1999.

- Residual fraction: The remaining radionuclides distributed between silicates, phosphates, and refractory oxides.

For example, cobalt, iron and manganese oxides and to a lesser extent organic matter, are typically important sorbents. Measurement of a K_d using actual soils will provide a reasonably clear picture of potential retardation. An SEP would do the same thing, but provide a clearer picture of where the cobalt resides in the particular soil. By using an SEP, a reasonably clear number can be developed for both the presence of iron hydroxides and the amount of contaminant that is associated with it. Table 5-31 describes the types of geochemical data needed to determine whether the particular natural attenuation pathway for radionuclides are operative. Additional measurements of soil or groundwater chemistry (i.e. redox, pH, etc.) may also be required. Note that uncertainties surrounding the controls on irreversible uptake of metals may make predictions of MNA problematic. Several explanations have been proposed for these observed residence time effects including solid-state diffusion within oxide particles (Eick et al., 1999), diffusion into micropores and intraparticle spaces (Ainsworth et al., 1994), change in the type of surface complex (Bruemmer et al., 1988), incorporation into the mineral structure via recrystallization (McLaren et al., 1986), surface catalyzed oxidation and incorporation into the crystal matrix (Ainsworth et al., 1994 and McBride, 1994), and surface catalyzed hydrolysis and precipitation (Ainsworth et al., 1994).

5.5.5 MNA Monitoring Programs and Future Use Considerations

Monitoring is a significant and essential component of any application of MNA. In order for MNA to be a cost-effective technology for radionuclide-impacted soil or groundwater, a clear monitoring plan must be agreed upon early in the site cleanup process. Because the time frame(s) and mechanisms for natural attenuation of radionuclides may be significantly longer than those associated with, for example, organic contaminants, an ill-conceived monitoring program might easily prove more costly than full-scale remediation if expenses are not controlled during the initial monitoring. The monitoring program should address where monitoring will take place, what should be monitored, how monitoring should be performed, and include a contingency plan. A more thorough discussion of long term monitoring requirements for natural attenuation can be found in *Technical Guidance for the Long-Term Monitoring of Natural Attenuation Remedies at DOE Sites* (DOE, 1999) and in Chapter 5.

The major element chemical composition of soil and groundwater determines the mobility of radionuclides. Large-scale changes in major element chemistry in the distant future can conceivably cause very drastic changes in the transport of radionuclides, as well as metals and organics. A clear understanding of the likely range of groundwater compositions, and their effect on contaminant transport, is required to estimate the ultimate permanence of MNA. This is particularly true for radionuclides because of the longer time frames typically required for MNA. Groundwater compositions are to a large extent determined by: 1) chemical equilibrium with soil CO_2, 2) weathering of soil minerals, 3) surface leachate, 4) dissolved organic carbon, and 5) adsorption and ion exchange reactions.

EPA guidelines (EPA, 1997), require an explicit consideration of future use of a site, including an assessment of the time frame in which an aquifer might be needed for use. For organic contaminants that naturally attenuate primarily through biodegradation and dilution, the time frame of future use is important because it is the target against which the predicted efficacy of MNA must be assessed. This is also true for radionuclides, which don't degrade and generally decay slowly. For radionuclides, an assessment of future use must specifically factor in the effect of future use on previously immobilized contaminants.

A natural attenuation remedy that relies on limited infiltration may be invalidated by irrigation for agricultural development. Uranium sorbed to iron hydroxides in an initially aerated soil might be released if the soil became flooded, then anoxic, allowing the dissolution of the iron hydroxide host. Plutonium associated with iron hydroxides under initially oxidizing conditions at the Rocky Flats Plant in Colorado was subsequently mobilized when high rainfall raised the water table, causing more reducing conditions and the dissolution of the original host (Litaor and Ibrahim, 1996).

Often the composition ranges of soil and groundwater is somewhat limited, primarily because there are a host of biogeochemical processes which tend to control or buffer the pH, redox state, alkalinity, and mineralogic makeup of soils. For example, arid land soils are predominately calcareous, which buffers soil pH near 8.0. The chemical behavior of an inorganic constituent in a calcareous system may be very different for slightly acidic systems. Hence, remediation strategies developed for slightly acidic soil conditions, may have no relevance for the mitigation of inorganics in calcareous systems. To illustrate, iodine mobility is limited in slightly acid soils, due to its enhanced sorption to soil particle surface, whereas in calcareous soils, iodine is highly mobile, essentially moving with the water-front, because at higher pH very little iodine sorbs to soil particle surfaces.

Although drastic changes in the compositions of natural waters are more the exception than the rule, it will probably be impossible for site owners to demonstrate that remobilization will never occur. This is a critical obstacle to the implementation of natural attenuation for radionuclides. If monitoring is required in perpetuity to guard against remobilization, natural attenuation will probably never be the chosen remedy for long-lived radionuclides, though it may be chosen for radionuclides possessing a sufficiently short half-life.

References

Acton, D. W., and Barker, J. F. (1992). "In situ biodegradation potential of aromatic hydrocarbons in anaerobic groundwaters." J. Contam. Hydrol., 9(4), 325-352.

Aggarwal, P. K., and Hinchee, R. E. (1991). "Monitoring in situ biodegradation of hydrocarbons by using stable carbon isotopes." Environ. Sci. Technol., 25(6), 1178-1180.

Ahad, J. M. E., Sherwood Lollar, B., Edwards, E. A., Slater, G. F., and Sleep, B. E. (2000). "Carbon isotope fractionation during anaerobic biodegradation of toluene: Implications for intrinsic bioremediation." Environ. Sci. Technol., 34(5), 892-896.

Aitken, M. D., Stringfellow, W. T., Nagel, R. D., Kazunga, C., and Chen, S–H. (1998). "Characteristics of phenanthrene-degrading bacteria isolated from soils contaminated with polycyclic aromatic hydrocarbons." Canada Journal of Microbiology, 44, 743-752.

Ainsworth, S. (1992). "Oxygenates seen as hot market by industry." Chem. Eng. News, 26-30.

Ainsworth, C.C., Pilon, J.L., Gassman, P.L., and Van Der Sluys, W.G. (1994). "Cobalt, cadmium, and lead sorption to hydrous iron oxide: Residence time effects" Soil Sci. Soc. Am. J. 58: 1615-1623.

Alexander, M. (1999). Biodegradation and Bioremediation, Academic Press, San Diego.

Alleman, B.C. and Leeson, A. (eds.). (1999). Natural Attenuation of Chlorinated Solvents, Petroleum Hydrocarbons, and Other Organic Compounds. Battelle Press, Columbus, OH.

Alvarez, P. J. J., Heathcote, R. C., and Powers, S. E. (1998). "Caution against interpreting gasoline release dates based on BTEX ratios in ground water." Ground Wat. Monitor. Remed., 18(4), 69-76.

Amundson, R. G., Chadwick, O. A., Sowers, J. M., and Doner, H. E. (1988). "Relationship between climate and vegetation and the stable carbon isotope chemistry of soils in the eastern Mojave desert, Nevada." Quat. Res., 29(3), 245-254

Angley, J. T., Brusseau, M. L., Miller, W. L., and Delfino, J. J. (1992). "Nonequilibrium sorption and aerobic biodegradation of dissolved alkylbenzenes during transport in aquifer material: column experiments and evaluation of a coupled-process model." Environ. Sci. Technol., 26(7), 1404-1410.

Anthony, J. W., Henry, B. M., Wiedemeier, T. H., Gordon, E. K., and Bidgood, J. B. (1999). "Methodology to evaluate natural attenuation of methyl tertiary-butyl ether." Natural Attenuation of Chlorinated Solvents, Petroleum Hydrocarbons, and Other Organic Compounds, B. C. Alleman and A. Leeson, eds., Battelle Press, Columbus, OH, 121-133.

API. (1972). "The migration of petroleum products in soil and ground water, principles and countermeasures." Publication No. 4149, American Petroleum Institute, Washington, DC.

Arvin, E., and Flyvberg, J. (1992). "Groundwater pollution arising from the disposal of creosote waste." Journal of the Institution of Water and Environmental Management, 6, 646-652.

ASTM. (1998). "Standard guide for remediation of ground water by natural attenuation at petroleum release sites." ASTM E-1943-98, American Society for Testing and Materials, West Conshohocken, PA.

Backes, C.A., McLaren, R. G., Rate, A. W., and Swift, R. S (1995). "Kinetics of cadmium and cobalt desorption from iron and manganese dioxides" Soil Sci. Soc. Am. J. 59: 778-785.

Baedecker, M. J., Cozzarelli, I. M., and Eganhouse, R. P. (1993). "Crude oil in a shallow sand and gravel aquifer--III. Biogeochemical reactions and mass balance modeling in anoxic groundwater." Appl. Geochem., 8, 569-586.

Bakker, G. (1977). "Anaerobic degradation of aromatic compounds in the presence of nitrate." FEMS Letters, 1, 103-108.

Ball, B. C., Smith, K. A., Klemedtsson, L., Brumme, R., Sitaula, B. K., Hansen, S., Prieme, A., MacDonald, J., and Horgan, G. W. (1997). "The influence of soil gas transport properties on methane oxidation in a selection of northern European soils." J. Geophys. Res.-Atmos., 102(D19), 23309-23317.

Ball, H. A., Johnson, H. A., Reinhard, M., and Spormann, A. M. (1996). "Initial reactions in anaerobic ethylbenzene oxidation by a denitrifying bacterium, strain EB1." J. Bacteriol., 178(19), 5755-5761.

Ballapragada, B.S., Stensel, H.D., Puhakka, J.A., and Ferguson, J.F. (1997). "Effect of hydrogen on reductive dechlorination of chlorinated ethenes." Environ. Sci. Technol., 31, 1728-1734.

Barbé, P., Lecomte, P., and Pazdej, R. (1998). Characteristics of soils polluted by PAH, in Contaminated Soil '98; Proceedings of the Sixth International FZK/TNO Conference on Contaminated Soil, 17-21 May 1998, Edinburgh, UK, Volume 2, Thomas Telford Publishing, London, UK, pp. 825 – 828.

Barbaro, J. R., Barker, J. F., Lemon, L. A., and Mayfield, C. I. (1992). "Biotransformation of BTEX under anaerobic, denitrifying conditions: Field and laboratory observations." J. Contam. Hydrol., 11(3/4), 245-272.

Barcelona, M. J. (1980). "Dissolved organic carbon and volatile fatty acids in marine sediment pore waters." Geochim. Cosmochim. Acta, 44(12), 1977-1984.

Barcelona, M. J., Lu, J., and Tomczak, D. M. (1995). "Organic acid derivatization techniques applied to petroleum hydrocarbon transformations in subsurface environments." Ground Wat. Monitor. Remed., 15(2), 114-124.

Barcelona, M. J., Tomczak, D., Lu, J., and Virkhaus, C. "Fractionation and identification of organic matter in natural and fossil-fuel contaminated aquifer systems." API-NGWA 1993 Petroleum Hydrocarbons and Organic Chemicals in Ground Water: Prevention, Detection, and Restoration, Houston, 163-176.

Barker, J. F., Patrick, G. C., and Major, D. (1987). "Natural attenuation of aromatic hydrocarbons in a shallow sand aquifer." Ground Wat. Monitor. Rev., 7(1), 64-71.

Barlaz, M. A., Hunt, M. J., Kota, S., and Borden, R. C. (1995). "Intrinsic bioremediation of a gasoline plume: comparison of field and laboratory results." Bioremediation of hazardous wastes: research, development, and field evaluations, EPA/540/R-95-532, Environmental Protection Agency, Washington, DC, 17-19.

Barrio-Lage, G., Parsons, F.Z., and Nassar, R.S. (1987). "Kinetics of the depletion of trichloroethene." Environ. Sci. Technol., 21, 366-370.

Bayard, R. M., Barna, L., and Gourdon, R. (1998). Influence of organic pollutants on sorption of naphthalene in contaminated soils, in Contaminated Soil '98; Proceedings of the Sixth International FZK/TNO Conference on Contaminated Soil, 17-21 May 1998, Edinburgh, UK, Volume 2, Thomas Telford Publishing, London, UK, pp. 849 - 850.

Becker, J.G. and Freedman, D.L. (1994). "Use of cyanocobalamin to enhance anaerobic biodegradation of chloroform." Environ. Sci. Technol., 28, 1942-1949.

Bekins, B. A., Warren, E., and Godsy, E. M. (1998). "A comparison of zero-order, first-order, and Monod biotransformation models." Ground Water, 36(2), 261-268.

Beller, H. R. (2000). "Metabolic indicators for detecting in situ anaerobic alkylbenzene degradation." Biodegradation, 11(2-3), 125-139.

Beller, H. R., Ding, W.-H., and Reinhard, M. (1995). "Byproducts of anaerobic alkylbenzene metabolism useful as indicators of in situ bioremediation." Environ. Sci. Technol., 29(11), 2864-2870.

Beller, H. R., and Edwards, E. A. (2000). "Anaerobic toluene activation by benzylsuccinate synthase in a highly enriched methanogenic culture." Appl. Environ. Microbiol., 66(12), 5503-5505.

Beller, H. R., Grbic-Galic, D., and Reinhard, M. (1992). "Microbial degradation of toluene under sulfate-reducing conditions and the influence of iron on the process." Appl. Environ. Microbiol., 58(3), 786-793.

Beller, H. R., and Spormann, A. M. (1997). "Anaerobic activation of toluene and o-xylene by addition to fumarate in denitrifying strain T." J. Bacteriol., 179(3), 670-676.

Beneteau, K.M., Aravena, R. and Frape, S.K. (1999). "Isotopic characterization of chlorinated solvents – laboratory and field results." Org. Geochem., 30, 739-753

Biegert, T., Fuchs, G., and Heider, J. (1996). "Evidence that oxidation of toluene in the denitrifying bacterium Thauera aromatica is initiated by formation of benzylsuccinate from toluene and fumarate." Eur. J. Biochem., 238(3), 661-668.

Biehle, A.A., Odom, J.M., Hall, W.B., Caldwell, B.E., Ellis, D.E., and Gregory, G.E. (1999). "Monitored natural attenuation of a methylene chloride plume." Natural

Attenuation of Chlorinated Solvents, Petroleum Hydrocarbons, and Other Organic Compounds, B.C. Alleman and A. Leeson, eds., Battelle Press, Columbus, OH, 35-40.

Boggs, J.M., Young, S.C., and Beard, L.M. (1992). "Field study of dispersion in a heterogeneous aquifer, 1. Overview and site description." Water Resources Research, 28(12), 3281-3291.

Boldrin, B., Tiehm, A., and Fritzsche, C. (1993). "Degradation of phenanthrene, fluorene, fluoranthene, and pyrene by a Mycobacterium sp." Applied and Environmental Microbiology, 59(6),1927-1930.

Bossert, I. D. and Bartha, R. (1986). "Structure-biodegradability relationships of polycyclic aromatic hydrocarbons in soil." Bulletin of Environmental Contamination and Toxicology, 37, 490-495

Bouchez, M., Blanchet, D., and Vandecasteele, J. -P. (1995). "Degradation of polycyclic aromatic hydrocarbons by pure strains and by defined strain associations: inhibition phenomena and cometabolism." Applied Microbiology and Biotechnology, 43, 156-164.

Bouchez, M., Blanchet, D., and Vandecasteele, J. -P. (1996). "The microbiological fate of polycyclic aromatic hydrocarbons: carbon and oxygen balances for bacterial degradation of model compounds." Applied Microbiology and Biotechnology, 45, 556-561.

Bradley, P. M., Chapelle, F. H., and Landmeyer, J. E. (2001a). "Effect of redox conditions on MTBE biodegradation in surface water sediments." *Environ. Sci. Technol.*, 35(23), 4643-4647.

Bradley, P. M., Landmeyer, J. E., Chapelle, F. H. (2001c). "Widespread potential for microbial MTBE degradation in surface water sediments." *Environ. Sci. Technol.*, 35(4), 658-662.

Brady, P. V., and Borns, D.J. (1997). "Natural Attenuation of Metals and Radionuclides: Report from a Workshop held by Sandia National Laboratories." SAND97-2727/UC-800. November 1997. URL: www.prod.sandia.gov/cgi-bin/techlib/access-control.pl/1997/972727.pdf

Brady P. V., Brady M. V., and Borns D. J. (1997). Natural attenuation of hazardous wastes: CERCLA, RBCAs, and the future of environmental remediation. CRC Press.

Brady P.V., Spalding, B.P., Krupka, K.M., Waters, R.D., Zhang, P., Borns, D.J., Lindgren, E.R. and W.D. Brady. (1999). Site Screening and Technical Guidance for Natural Attenuation at DOE Sites. SAND99-0464

Brady P. V., Jove-Colon, C. F. and Huang, F. Y. C. (2002). Soil Radionuclide Plumes, In: Zhang, P. C. and Brady, P. V. (Eds.) Soil Geochemical Processes of Radionuclides. Special Volume of the Soil Science Society of America .

Braddock, J. F., and McCarthy, K. A. (1996). "Hydrologic and microbiological factors affecting persistence and migration of petroleum hydrocarbons spilled in a continuous-permafrost region." Environ. Sci. Technol., 30(8), 2626-2633.

Bradley, P.M. and Chapelle, F.H. (1998). "Microbial mineralization of vinyl chloride and DCE under different terminal electron-accepting conditions." Anaerobe 4, 81-87.

Bradley, P. M., Landmeyer, J. E., and Chapelle, F. H. (1999). "Aerobic mineralization of MTBE and tert-butyl alcohol by stream-bed sediment microorganisms." Environ. Sci. Technol., 33(11), 1877-1879.

Bradley, P. M., Chapelle, F. H., and Landmeyer, J. E. (2001b). "Methyl t-butyl ether mineralization in surface-water sediment microcosms under denitrifying conditions." Appl. Environ. Microbiol., 67(4), 1975-1978.

Bregnard, T. P.-A., Häner, A., Höhener, P., and Zeyer, J. (1997). "Anaerobic degradation of pristane in nitrate-reducing microcosms and enrichment cultures." Appl. Environ. Microbiol., 63(5), 2077-2081.

Broholm, K., Jørgensen, P.R., Hansen, A.B., Arvin, E., and Hansen, M. (1999). "Transport of creosote compounds in a large, intact, macroporous clayey till column." Journal of Contaminant Hydrology, 39, 309 – 329.

Brown, T. L., LeMay Jr., H.E., and Bursten, B. E. (1994). Chemistry: The Central Science. Prentice Hall, Englewood Cliffs, N.J.

Bruemmer, G.W., Gerth, J. and Tiller, K.G.. (1988). "Reaction kinetics of the adsorption and desorption of nickel, zinc, and cadmium by goethite. I. Adsorption and diffusion of metals" J. Soil Sci., 39, 37-52.

Burton, M. B., Martinson, M. M., Barr, K. D. (1988). "Biotech USA." in 5th Annual Industrial Water and Waste Conference, San Francisco, CA, November, Elsevier, New York, NY.

Borden, R. C. (1994). "Natural Bioremediation of hydrocarbon-contaminated groundwater." Handbook of Bioremediation, CRC Press, Boca Raton, FL, 201-225.

Borden, R. C., and Bedient, P. B. (1986). "Transport of dissolved hydrocarbons influenced by oxygen-limited biodegradation. 1. Theoretical development." Water Resour. Res., 22(13), 1973-1982.

Borden, R. C., Bedient, P. B., Lee, M. D., Ward, C. H., and Wilson, J. T. (1986). "Transport of dissolved hydrocarbons influenced by oxygen-limited biodegradation. 2. Field application." Water Resour. Res., 22(13), 1983-1990.

Borden, R. C., Gomez, C. A., and Becker, M. T. (1995). "Geochemical indicators of intrinsic bioremediation." Ground Water, 33(2), 180-189.

Borden, R. C., Daniel, R. A., IV, L. E. L., and Davis, C. W. (1997). "Intrinsic biodegradation of MTBE and BTEX in a gasoline-contaminated aquifer." Water Resour. Res., 33(5), 1105-1115.

Bouwer, E.J. and McCarty, P.L. (1983). "Transformation of 1- and 2-carbon halogenated aliphatic compounds under methanogenic conditions." Appl. Environ. Microbiol., 45, 1286-1294.

Burland, S. M., and Edwards, E. A. (1999). "Anaerobic benzene biodegradation linked to nitrate reduction." Appl. Environ. Microbiol., 65(2), 529-533.

Buscheck, T. E., and Alcantar, C. M. (1995). "Regression techniques and analytical solutions to demonstrate intrinsic bioremediation." Intrinsic Bioremediation, R. E. Hinchee, J. T. Wilson, and D. C. Downey, eds., Battelle Press, Columbus, OH, 109-116.

Caldwell, M. E., and Suflita, J. M. (2000). "Detection of phenol and benzoate as intermediates of anaerobic benzene biodegradation under different terminal electron-accepting conditions." Environ. Sci. Technol., 34, 1216-1220.

Caldini, G., Cenci, G., Manenti, R., and Morozzi, G. (1995). "The ability of an environmental isolate of Pseudomonas fluorescens to utilize chrysene and other four-ring polynuclear aromatic hydrocarbons." Applied Microbiology and Biotechnology, 44,225-229.

Campbell, B.G., Petkewich, M.D., Landmeyer, J.E., and Chapelle, F.H. (1996). Geology, Hydrogeology, and Potential of Intrinsic Bioremediation at the National Park Service Dockside II Site and Adjacent Areas, Charleston, South Carolina, 1993 – 1994, U.S. Geological Services, Report No. 96-4170, Columbia, SC..

Chapelle, F. H. (1999). "Bioremediation of petroleum hydrocarbon-contaminated ground water: The perspectives of history and hydrology." Ground Water, 37(1), 122-132.

Chapelle, F. H., and Bradley, P. M. (1997). "Alteration of aquifer geochemistry by microorganisms." Manual of Environmental Microbiology, C. J. Hurst, G. R. Knudsen, M. J. McInerney, L. D. Stetzenbach, and M. V. Walter, eds., ASM Press, Washington, D. C., 558-564.

Chapelle, F. H., Bradley, P. M., Lovley, D. R., and Vroblesky, D. A. (1996a). "Measuring rates of biodegradation in a contaminated aquifer using field and laboratory methods." Ground Water, 34(4), 691-698.

Chapelle, F. H., Haack, S. K., Adriaens, P., Henry, M. A., and Bradley, P. M. (1996b). "Comparison of E_h and H_2 measurements for delineating redox processes in a contaminated aquifer." Environ. Sci. Technol., 30(12), 3565-3569.

Chapelle, F. H., Vroblesky, D. A., Woodward, J. C., and Lovley, D. R. (1997). "Practical considerations for measuring hydrogen concentrations in groundwater." Environ. Sci. Technol., 31(10), 2873-2877.

Chapelle, F.H., Haack, S.K., Adriens, P., Henry, M.A., and Bradley, P.M. (1996). "Comparison of E_H and H_2 measurements for delineating redox processes in a contaminated aquifer." Environ. Sci. Technol., 30, 3565-3569.

Chapelle, F.H., Vroblesky, D.A., Woodward, J.C., and Lovley, D.R. (1997). "Practical considerations for measuring hydrogen concentrations in groundwater." Environ. Sci. Technol., 31, 2873-2877.

Chen, S. -H. and Aitken, M. D. (1999). "Salicylate stimulates the degradation of high-molecular weight polycyclic aromatic hydrocarbons by Pseudomonas saccharophila P15" Environmental Science and Technology, 33(3), 435-439.

Chen, Y.-M., Abriola, L. M., Alvarez, P. J. J., Anid, P. J., and Vogel, T. M. (1992). "Modeling transport and biodegradation of benzene and toluene in sandy aquifer material: comparisons with experimental measurements." Water Resour. Res., 28(7), 1833-1847.

Chiang, C. Y., Salanitro, J. P., Chai, E. Y., Colthart, J. D., and Klein, C. L. (1989). "Aerobic biodegradation of benzene, toluene, and xylene in a sandy aquifer--data analysis and computer modeling." Ground Water, 27(6), 823-834.

Chiou, C. T., Porter, P. E., and Schmedding, D. W. (1983). "Partition equilibria of nonionic organic compounds between soil organic matter and water." Environ. Sci. Technol., 17(4), 227-231.

Cho, J. S., Wilson, J. T., DiGiulio, D. C., Vardy, J. A., and Choi, W. (1997). "Implementation of natural attenuation at a JP-4 jet fuel release after active remediation." Biodegradation, 8(4), 265-73.

Christensen, T.H., Bjerg, P.L., and Kjeldsen, P. (2000). "Natural attenuation: a feasible approach to remediation of ground water pollution at landfills?" Ground Water Monitor. Rem., 20, 69-77.

Church, C. D., Isabelle, L. M., Pankow, J. F., Rose, D. L., and Tratnyek, P. G. (1997). "Method for determination of methyl tert-butyl ether and its degradation products in water." Environ. Sci. Technol., 31(12), 3723-3726.

Church, C. D., Pankow, J. F., and Tratnyek, P. G. "Assessing the natural attenuation potential of methyl tert-butyl ether." Second Int. Conf. Remediation of Chlorinated and Recalcitrant Compounds (Poster Abstracts), May 22-25, 2000, Monterey, CA.

Cline, P.V. and Delfino, J.J. (1989). "Transformation kinetics of 1,1,1-trichloroethane to the stable product 1,1-dichloroethene." Biohazards of Drinking Water Treatment, R.A. Larson, ed., Lewis Publishers, Inc., Chelsea, MI, 47-56.

Cline, P. V., Delfino, J.J., and Rao, P.S.C. (1991). "Partitioning of aromatic constituents into water from gasoline and other complex solvent mixtures." Environmental Science and Technology, 25, 914 – 920.

Cline, P. V., Delfino, J. J., and Rao, P. S. C. (1991). "Partitioning of aromatic constituents into water from gasoline and other complex solvent mixtures." Environ. Sci. Technol., 25(5), 914-920.

Coates, J. D., et al. (2001). "Anaerobic benzene oxidation coupled to nitrate reduction in pure culture by two strains of *Dechloromonas*." *Nature (London)*, 411(6841), 1039-1043.

Conrad, M. E., Daley, P. F., Fischer, M. L., Buchanan, B. B., Leighton, T., and Kashgarian, M. (1997). "Combined ^{14}C and $d^{13}C$ monitoring of in situ biodegradation of petroleum hydrocarbons." Environ. Sci. Technol., 31(5), 1463-1469.

Conrad, M. E., Templeton, A. S., Daley, P. F., and Alvarez-Cohen, L. (1999). "Isotopic evidence for biological controls on migration of petroleum hydrocarbons." Org. Geochem., 30(8A), 843-859.

Cooney, J. J., Silver, S. A., and Beck, E. A. (1985). "Factors influencing hydrocarbon degradation in three freshwater lakes." Microb. Ecol., 11(2), 127-137.

Cooper, W.J., Slifker, R., Joens, J.A., and El-Shazly, O.A. (1989). "Abiotic transformation of halogenated organic compounds. II. Considerations during water treatment." Biohazards of Drinking Water Treatment, R.A. Larson, ed., Lewis Publishers, Inc., Chelsea, MI, 37-46.

Corcho, D., Watkinson, R., and Lerner, D. "Cometabolic degradation of MTBE by a cyclohexane-oxidising bacteria." The Second International Conference on Remediation of Chlorinated and Recalcitrant Compounds (Poster Abstracts), May 22-25, 2000, Monterey, CA.

Council on Environmental Quality. (1981). Contamination of ground water by toxic organic chemicals, Council on Environmental Quality, Washington, DC.
Cozzarelli, I. M., Baedecker, M. J., Eganhouse, R. P., and Goerlitz, D. F. (1994). "The geochemical evolution of low-molecular-weight organic acids derived from the degradation of petroleum contaminants in groundwater." Geochim. Cosmochim. Acta, 58(2), 863-877.
Cozzarelli, I. M., Eganhouse, R. P., and Baedecker, M. J. (1990). "Transformation of monoaromatic hydrocarbons to organic acids in anoxic groundwater environment." Environ. Geol. Water Sci., 16(2), 135-141.
Council on Environmental Quality (CEQ). (1981). "Contamination of groundwater by toxic organic chemicals." Washington, DC.
Criddle, C.S., DeWitt, J.T., Grbic-Galic, D., and McCarty, P.L. (1990). "Transformation of carbon tetrachloride by Pseudomonas sp. strain KC under denitrification conditions." Appl. Environ. Microbiol., 56, 3240-3246.
Criddle, C.S. and McCarty, P.L. (1991). "Electrolytic model system for reductive dehalogenation in aqueous environments." Environ. Sci. Technol., 25, 973-978.
Coover M. P., and Sims R.C.C. (1987). "The effects of temperature on polycyclic aromatic hydrocarbon persistence in an unacclimated agricultural soil." Haz. Waste. Mat., 4, 181 – 192.
Cutright, T. J. and Lee, S. (1994). "Remediation of PAH-contaminated soil using Achromobacter sp." Energy Source, 16, 279-287.
Dana, J. D. (1966). "Minerology", 17th Edition" (revised by C.S. Hurlbut, Jr.) John Wiley and Sons, New York, NY. 1966.
Davis J.I., and Evans W.C. (1964). "Oxidative metabolism of naphthalene by soil pseudomonas." Biochem. J., 91, 251-261
Davis, G. B., Barber, C., Power, T. R., Theirrin, J., Patterson, B. M., Rayner, J. L., and Wu Q. (1999). "The variability and intrinsic remediation of a BTEX plume in anaerobic sulphate-rich groundwater." Journal of Contaminant Hydrology, 36(3-4), 265-290.
Davis, G. B., Barber, C., Power, T. R., Thierrin, J., Patterson, B. M., Rayner, J. L., and Wu, Q. (1999). "The variability and intrinsic remediation of a BTEX plume in anaerobic sulphate-rich groundwater." J. Contam. Hydrol., 36(3/4), 265-290.
Davis, J. W., Klier, N. J., and Carpenter, C. L. (1994). "Natural biological attenuation of benzene in ground water beneath a manufacturing facility." Ground Water, 32(2), 215-226.
Dean-Ross, D. and Cerniglia, C. E. (1996). "Degradation of pyrene by Mycobacterium flavescens." Applied Microbiology and Biotechnology, 46, 307-312.
Dean-Raymond D., and Bartha R. (1975). "Biodegradation of some polynuclear aromatic petroleum components by marine bacteria." Dev. Ind. Microbiol., 16,97-110.
deBest, J.H., Salminen, E., Doddema, H.J., Janssen, D.B., and Harder, W. (1997). "Transformation of carbon tetrachloride under sulfate reducing conditions." Biodegradation, 8, 429-436.

Deeb, R. A., Hu, H.-Y., Scow, K. M., and Alvarez-Cohen, L. (2001). "Substrate interactions in BTEX and MTBE mixtures by an MTBE-degrading isolate." Environ. Sci. Technol., 35(2), 312-317.

Deeb, R. A., Scow, K. M., and Alvarez-Cohen, L. (2000). "Aerobic MTBE biodegradation: an examination of past studies, current challenges, and future research directions." Biodegradation, 11(2-3), 171-186.

Deines, P. (1980). "The isotopic composition of reduced organic carbon." Handbook of Environmental Isotope Geochemistry, P. Fritz and J. C. H. Fontes, eds., Elsevier Scientific Publishing Co., Amsterdam, 329-406.

Delzer, G. C., Zogorski, J. S., Lopes, T. J., and Bosshart, R. L. (1996). "Occurrence of the gasoline oxygenate MTBE and BTEX compounds in urban stormwater in the United States, 1991-95." Water-Resources Investigations Report 96-4145, United States Geological Survey.

Delorme, P., and Carlier, M. (1998). "Study of transfer mechanisms of the residues in the soils of former manufactured gas plant sites." in Contaminated Soil '98; Proceedings of the Sixth International FZK/TNO Conference on Contaminated Soil, 17-21 May 1998, Edinburgh, UK, Volume 2, Thomas Telford Publishing, London, UK, 831 – 832.

Dolan, M.E. and McCarty, P.L. (1995). "Small column microcosm for assessing methane-stimulated vinyl-chloride transformation in aquifer samples." Environ. Sci. Technol., 29, 1892-1897.

Douglas, G.S., Prince, R.C., Butler, E.L., and Steinauer, W.G. (1994). "The use of internal chemical indicators in petroleum and refined products to evaluate the extent of bioremediation." Hydrocarbon Bioremediation, R.E. Hinchee, B.C. Alleman, R.E. Hoeppel, and R.N. Miller, eds., Lewis Publishers, Boca Raton, FL., 219-236.

DuPont, R.R., Gorder, K., Sorensen, D.L., Kemblowski, M.W., and Haas, P. (1997). "Case study: Eielson Air Force Base, Alaska." Proceedings of the Symposium on Natural Attenuation of Chlorinated Organics in Ground Water, EPA/540/R-97/504, Office of Research and Development, Washington, DC, 106-111.

Durate J. C., David S.D, Eusebio A., and Menaia A.G.F. (1997). "Biotechnology for waste management and site restoration-III Degradation of polycyclic aromatic hydrocarbons by microorganisms from contaminated soil." Kluwer Academic Publishers, pp. 187-192.

Durant, N. D., Wilson, L. P., and Bouwer, E. J. (1994). "Screening for natural subsurface biotransformation of polycyclic aromatic hydrocarbons at a former manufactured gas plant." in Bioremediation of Chlorinated and Polycyclic Aromatic Hydrocarbon Compounds, R. E. Hinchee, A. Leeson, L. Semprini, S. K. Ong (Eds.), Lewis Publishers, Boca Raton, FL, pp. 456 – 461.

Edison Electric Institute. (1984). Handbook on Manufactured Gas Plant Sites, Utility Solid Waste Activities Superfund Committee, Edison Electric Institute, Washington D.C.

EPRI. (1996). Characterization and monitoring before and after source removal at a former manufactured gas plant (FMGP) disposal site. EPRI Report No. TR-105921, Research Projects WO2879-12; WO2879-38.

Edwards, E. A., Edwards, A. M., and Grbic-Galic, D. (1994). "A method for detection of aromatic metabolites at very low concentrations: application to detection of metabolites of anaerobic toluene degradation." Appl. Environ. Microbiol., 60(1), 323-327.
Edwards, E. A., and Grbic-Galic, D. (1994). "Anaerobic degradation of toluene and o-xylene by a methanogenic consortium." Appl. Environ. Microbiol., 60(1), 313-322.
Edwards, E. A., Wills, L. E., Reinhard, M., and Grbic-Galic, C. (1992). "Anaerobic degradation of toluene and xylene by aquifer microorganisms under sulfate-reducing conditions." Appl. Environ. Microbiol., 58(3), 794-800.
Egli, C., Scholtz, R., Cook, A.M., and Leisinger, T. (1987). "Anaerobic dechlorination of tetrachloromethane and 1,2-dichloroethane to degradable products by pure cultures of Desulfobacterium sp. and Methanobacterium sp." FEMS Microbiol. Lett., 43, 257-261.
Egli, C., Tschan, T., Scholz, R., Cook, A.M., and Leisinger, T. (1988). "Transformation of CCl_4 to CH_2Cl_2 and CO_2 by Acetobacterium woodii." Appl. Environ. Microbiol., 54, 2819-2824.
Ellis, D.E. et al. (1997). "Remediation Technology Development Forum intrinsic remediation project at Dover Air Force Base, Delaware." Proceedings of the Symposium on Natural Attenuation of Chlorinated Organics in Ground Water, EPA/540/R-97/504, Office of Research and Development, Washington, DC, 95-99.
Ehrlich, G. G., Schroeder, R. A., and Martin, P. (1985). "Microbial populations in a jet-fuel-contaminated shallow aquifer at Tustin, California." Open-File Report 85-335, U. S. Geological Survey, Sacramento, CA.
Ellis, L. B. M., and Wackett, L. P. (2000). "The University of Minnesota Biocatalysis/Biodegradation Database." , University of Minnesota.
Eick, M.J., Peak, J.D., Brady, P.V., and Pesek, J.D. (1999). "Kinetics of lead adsorption/desorption on goethite: residence time effects" Soil Science, 164, 28-39.
Evanko, C. R. and Dzombak, D.A. (1997). "Remediation of Metals-Contaminated Soils and Groundwater." Technology Evaluation Report TE-97-01, Ground-Water Remediation Technologies Analysis Center, Pittsburgh, PA. October 1997.
Fernández, P., Vilanova, R. M., Martínez, C., Appleby, P., and Grimalt, J. O. (2000). "The historical record of atmospheric pyrolitic pollution over europe registered in the sedimentary PAH from remote mountain lakes." Environmental Science and Technology, 34(10), 1906 – 1913.
Field, J. A., Boelsma, F., Baten, H., and Rulkens, W. H. (1995). "Oxidation of anthracene in water/solvent mixtures by the white-rot fungus, Bjerkandera sp. strain BOS55." Applied Microbiology and Biotechnology, 44, 234-240.
Fisher, A.J., E.A. Rowan, and R.F. Spalding. (1987). "VOCs in ground water influenced by large scale withdrawals." Ground Water, 25, 407-414.
Finneran, K. T., and Lovley, D. R. (2001). "Anaerobic degradation of methyl tert-butyl ether (MTBE) and tert-butyl alcohol (TBA)." Environ. Sci. Technol., 35(9), 1785-1790.

Foght, J. M. and Westlake, D. W. (1988). "Degradation of polycyclic aromatic hydrocarbons and aromatic heterocycles by a Pseudomonas species." Canadian Journal of Microbiology, 34, 1135-1141.

Fowler, M.G., Brooks, P.W., Northcott, M., and King, M.W.G. (1994). "Preliminary results from a field experiment investigating the fate of some creosote compounds in a natural aquifer." Advances in Organic Geochemistry, 22(3-5), 641-649.

Franks, B. J. (1987). "U. S. Geological Survey Program on Toxic Waste--Groundwater contamination: Proc. Third Technical Meeting, Pensacola, Florida, March 23-27, 1987." Open-File Report 87-109, U. S. Geological Survey, Tallahassee, FL.

Freedman, D.L. and Gossett, J.M. (1989). "Biological reductive dechlorination of tetrachloroethylene and trichloroethylene to ethylene under methanogenic conditions." Appl. Environ. Microbiol., 55, 2144-2151.

Freedman, D.L. and Herz, S.D. (1996). "Use of ethylene and ethane as primary substrates for aerobic cometabolism of vinyl chloride." Water Environ. Res., 68, 320-328.

Fullana, A., Font, R., Conesa, J.A., and Blasco, P. (2000). "Evolution of products in the combustion of scrap tires in a horizontal, laboratory scale reactor." Environmental Science and Technology, 34(11), 2092 – 2099.

Galli, R. and McCarty, P.L. (1989a). "Biotransformation of 1,1,1-trichloroethane, trichloromethane, and tetrachloromethane by a Clostridium sp." Appl. Environ. Microbiol., 55, 837-844.

Galli, R. and McCarty, P.L. (1989b). "Kinetics of biotransformation of 1,1,1-trichloroethane by Clostridium sp. strain TCAIIB." Appl. Environ. Microbiol., 55, 845-851.

Garnier, P., Auria, R., Magaña, M., and Revah, S. (1999a). "Cometabolic biodegradation of methyl t-butyl ether by a soil consortium." In Situ Bioremediation of Petroleum Hydrocarbon and Other Organic Compounds, B. Alleman and A. Leeson, eds., Battelle Press, Columbus, OH, 31-35.

Garnier, P. M., Auria, R., Augur, C., and Revah, S. (1999b). "Cometabolic biodegradation of methyl t-butyl ether by Pseudomonas aeruginosa grown on pentane." Appl Microbiol Biotechnol, 51(4), 498-503.

Garrett, R. M., Guénette, C. C., Haith, C. E., and Prince, R. C. (2000). "Pyrogenic polycyclic aromatic hydrocarbons in oil burn residues." Environmental Science and Technology, 34(10), 1934 – 1937.

Gerritse, J., Renard, V., Pedro Gomes, T.M., Lawson, P.A., Collins, M.D. and Gottschal, J.C. (1996). "Desulfitobacterium sp. strain PCE1, an anaerobic bacterium that can grow by reductive dechlorination of tetrachloroethene or ortho-chlorophenols." Arch. Microbiol., 165, 132-140.

Ghoshal, S., Ramaswami, A., and Luthy, R. (1996). "Biodegradation of naphthalene from coal tar and heptamethylnonane in mixed batch systems." Environmental Science and Technology, 30, 1282-1291.

Ghiorse, W. C., and Wilson, J. T. (1988). "Microbial ecology of the terrestrial subsurface." Advances in Applied Microbiology, A. I. Laskin, ed., Academic Press, San Diego, 107-172.

Gibson, D. T., and Subramanian, V. (1984). "Microbial degradation of aromatic hydrocarbons." Microbial Degradation of Organic Compounds, D. T. Gibson, ed., Marcel Dekker, New York, 181-252.

Gieg, L. M., Kolhatkar, R. V., McInerney, M. J., Tanner, R. S., Harris, S. H., Jr., Sublette, K. L., and Suflita, J. M. (1999). "Intrinsic bioremediation of petroleum hydrocarbons in a gas condensate-contaminated aquifer." Environ. Sci. Technol., 33(15), 2550-2560.

Gossett, J.M. (1987). "Measurement of Henry's law constants for C_1 and C_2 chlorinated hydrocarbons." Environ. Sci. Technol., 21(2), 202-208.

Grisham, J. (1999). "Oxygenates do little to reduce ozone, NRC finds." Chem. Eng. News, 77(20), 10-11.

Grosser, R. J., Warshawsky, D., and Vestal, J. R. (1991). "Indigenous and enhanced mineralization of pyrene, benzo[a]pyrene, and carbazole in soils." Applied and Environmental Microbiology, 57(12), 3462-3469.

Hadley, P. W., and Armstrong, R. (1991). ""Where's the benzene?"--Examining California ground-water quality surveys." Ground Water, 29(1), 35-40.

Haefliger, O. P., Bucheli, T. D., and Zenobi, B. (2000). "Laser mass spectrometric analysis of organic atmospheric aerosols. 1. Characterization of emission sources." Environmental Science and Technology, 34(11), 2178 – 2183.

Hagg, W.R and Mill, T. (1988). "Effect of a subsurface sediment on hydrolysis of haloalkanes and epoxides." Environ. Sci. Technol., 22(6), 658-663.

Haggin, J. (1992). "Fuel changes will increase fuel prices." Chem. Eng. News, 22.

Hanson, J. R., Ackerman, C. E., and Scow, K. M. (1999). "Biodegradation of methyl tert-butyl ether by a bacterial pure culture." Appl Environ Microbiol, 65(11), 4788-92.

Happel, A. M., Beckenbach, E., Savalin, L., Temko, H., Rempel, R., Dooher, B., and Rice, D. (1996). "Analysis of dissolved benzene plumes and methyl tertiary butyl ether (MTBE) plumes in ground water at leaking underground fuel tank (LUFT) sites." UCRL-JC-125633, Lawrence Livermore National Laboratory, Livermore, CA.

Hardison, L. K., Curry, S. S., Ciuffetti, L. M., and Hyman, M. R, (1997). "Metabolism of diethyl ether and cometabolism of methyl tert-butyl ether by a filamentous fungus, a Graphium sp." Appl. Environ. Microbiol., 63(8), 3059-3067.

Harkness, M. R., and Bracco, A. A. (1998). "Practical issues in field sampling and analysis for natural attenuation assessments." Natural Attenuation: Chlorinated and Recalcitrant Compounds, G. B. Wickramanayake and R. E. Hinchee, eds., Battelle Press, Columbus, OH, 177-182.

Harms, G., Zengler, K., Rabus, R., Aeckersberg, F., Minz, D., Rosselló-Mora, R., and Widdel, F. (1999). "Anaerobic oxidation of o-xylene, m-xylene, and homologous alkylbenzenes by new types of sulfate-reducing bacteria." Appl. Environ. Microbiol., 65(3), 999-1004.

Harrington, R. R., Poulson, S. R., Drever, J. I., Colberg, P. J. S., and Kelly, E. F. (1999). "Carbon isotope systematics of monoaromatic hydrocarbons: vaporization and adsorption experiments." Org. Geochem., 30(8A), 765-775.

Hartley, W. R., and Englande, A. J., Jr. (1992). "Health risk assessment of the migration of unleaded gasoline-a model for petroleum products." Wat. Sci. Tech., 25(3), 65-72.
Hatheway, A. W. (1997). "Manufactured gas plants: yesterday's pride, today's liability." Civil Engineering, November, 38 – 41.
Hedlund, B.P., Geiselbrecht, A.D., Blair, T.J., and Sraley, J.T. (1999). "Polycyclic aromatic hydrocarbon degradation by a new marine bacterium." Applied and Environmental Microbiology, 65, 251 – 259.
Heermann, S. E., and Powers, S. E. (1998). "Modeling the partitioning of BTEX in water-reformulated gasoline systems containing ethanol." J. Contam. Hydrol., 34(4), 315-341.
Heitkamp, M. A. and Cerniglia, C. E. (1987). "Effects of chemical structure and exposure
on the microbial degradation of polycyclic aromatic hydrocarbons in freshwater and estuarine ecosystems." Environmental Toxicology and Chemistry, 6, 535-546.
Heitkamp, M. A., Freeman, J. P., Miller, D. W., and Cerniglia, C. E. (1988). "Pyrene degradation by a Mycobacterium sp.: identification of ring oxidation and ring fission products." Applied and Environmental Microbiology, 54(10), 2556-2565.
Heitkamp, M. A. and Cerniglia, C. E. (1989). "Polycyclic aromatic hydrocarbon degradation by a Mycobacterium sp. in microcosms containing sediment and water from a pristine ecosystem." Applied and Environmental Microbiology, 55(8), 1968-1973.
Heider, J., and Fuchs, G. (1997). "Anaerobic metabolism of aromatic compounds." Eur. J. Biochem., 243(3), 577-596.
Heider, J., Spormann, A. M., Beller, H. R., and Widdel, F. (1998). "Anaerobic bacterial metabolism of hydrocarbons." FEMS Microbiol. Rev., 22(5), 459-473.
Heraty, L.J., Fuller, M.E., Huang, L., Abrajano, Jr., T., and Sturchio, N.C. (1999). "Isotopic fractionation of carbon and chlorine by microbial degradation of dichloromethane." Org. Geochem., 30, 793-799.
Hernandez-Perez, G., Fayolle, F., and Vandecasteele, J.-P. (2001). "Biodegradation of ethyl t-butyl ether (ETBE), methyl t-butyl ether (MTBE) and t-amyl methyl ether (TAME) by Gordonia terrae." Appl. Microbiol. Biotechnol., 55(1), 117-121.
Hill, J. B., and Moxey, J. G., Jr. (1960). "Gasoline." Petroleum Products Handbook, V. B. Guthrie, ed., McGraw-Hill, New York, 4-1 - 4-37.
Holliger, C., Schraa, G., Stams, A.J.M., and Zehnder, Z.J.B. (1993). "A highly purified enrichment culture couples the reductive dechlorination of tetrachloroethene to growth." Appl. Environ. Microbiol., 59(9), 2991-2997.
Howard, P. H., Boethling, R. S., Jarvis, W. F., Meylan, W. M., and Michalenko, E. M. (1991). Handbook of Environmental Degradation Rates. Lewis Publishers, Chelsea, MI.
Huang, L., Sturchio, N.C., Abrajano, Jr., T., Heraty, L.J. and Holt, B.D. (1999). "Carbon and chlorine isotope fractionation of chlorinated aliphatic hydrocarbons by evaporation." Org. Geochem., 30, 777-785.
Hult, M. F. (1984). "Ground-water contamination by crude oil at the Bemidji, Minnesota, research wite: U. S. Geological Survey Toxic Waste--Ground-Water

Contamination Study." Water-Resources Investigations Report 84-4188, U. S. Geological Survey, St. Paul, MN.
Hunkeler, D., Butler, B. J., Aravena, R., and Barker, J. F. (2001). "Monitoring biodegradation of methyl tert-butyl ether (MTBE) using compound-specific carbon isotope analysis." Environ. Sci. Technol., 35(4), 676-681.
Hutchins, S. R. (1991). "Optimizing BTEX biodegradation under denitrifying conditions." Environ. Toxicol. Chem., 10, 1437-1448.
Hutchins, S. R., Sewell, G. W., Kovacs, D. A., and Smith, G. A. (1991). "Biodegradation of aromatic hydrocarbons by aquifer microorganisms under denitrifying conditions." Environ. Sci. Technol., 25, 68-76.
Hyman, M., Kwon, P., Williamson, K., and O'Reilly, K. (1998). "Cometabolism of MTBE by alkane-utilizing microorganisms." Natural Attenuation, Chlorinated and Recalcitrant Compounds, G. Wickramanayake and R. Hinchee, eds., Battelle Press, Columbus, OH.
Hyman, M., and O'Reilly, K. (1999). "Physiological and enzymatic features of MTBE-degrading bacteria." In Situ Bioremediation of Petroleum Hydrocarbon and Other Organic Compounds, B. Alleman and A. Leeson, eds., Battelle Press, Columbus, OH, 7-12.
Hyman, M., Taylor, C., and O'Reilly, K. "Cometabolic degradation of MTBE by iso-alkane-utilizing bacteria from gasoline-impacted soils." Second Int. Conf. Remediation of Chlorinated and Recalcitrant Compounds (Platform Abstracts), May 22-25, 2000, Monterey, CA.
Iler, R. K. (1979). "The Chemistry of Silica, Solubility, Polymerization, Colloid, Surface Properties, and Biochemistry." John Wiley and Sons, New York, NY. 1979.
Imbrigiotta, T.E., Ehike, T.A., Wilson, B.H., and Wilson, J.T. (1997). "Case study: Natural attenuation of a trichloroethene plume at Picatinny Arsenal, New Jersey." Proceedings of the Symposium on Natural Attenuation of Chlorinated Organics in Ground Water, EPA/540/R-97/504, Office of Research and Development, Washington, DC, 85-91.
Jackson, R.E. (1998). "The migration, dissolution, and fate of chlorinated solvents in the urbanized alluvial valleys of the southwestern USA." Hydrogeol. J. 6, 144-155.
Jamison, V. W., Raymond, R. L., and Hudson, J. O., Jr. (1975). "Biodegradation of high-octane gasoline in groundwater." Developments in Industrial Microbiology, Society for Industrial Microbiology, Arlington, VA, 305-312.
Jamison, V. W., Raymond, R. L., and Hudson, J. O. (1976). "Biodegradation of high-octane gasoline." Third International Biodegradation Symposium, J. M. Sharpley and A. M. Kaplan, eds., Applied Sciences Publishers LTD, 187-196.
Janssen, D.B., Scheper, A., Dijkhuizen, L., and Witholt, B. (1985). "Degradation of halogenated aliphatic compounds by Xanthobacter autotrophicus GJ10." Appl. Environ. Microbiol., 49, 673-677..
Jarvis, W. F., Sage, G. W., Basu, D. K., Gray, D. A., Meylan, W., and Crosbie, E. K. (1989). "Large Production and Priority Pollutants." Handbook of Environmental Fate and Exposure Data for Organic Chemicals, P. H. Howard, ed., Lewis Publishers, Chelsea, MI.

Jeffers, P.M., Ward, L.M., Woytowitch, L.M., and Wolfe, N.L. (1989). "Homogenous hydrolysis rate constants for selected chlorinated methanes, ethanes, ethenes, and propanes." Environ. Sci. Technol., 23(8), 965-969

Johnston, J. J., Borden, R. C., and Barlaz, M. A. (1996). "Anaerobic biodegradation of alkylbenzenes and trichloroethylene in aquifer sediment down gradient of a sanitary landfill." J. Contam. Hydrol., 23(4), 263-283.

Kampbell, D. H., Wiedemeier, T. H., and Hansen, J. E. (1996). "Intrinsic bioremediation of fuel contamination in ground water at a field site." J. Hazard. Mat., 49(2-3), 197-204.

Kanaly R., Bartha R., Fogel S., and Findlay M. (1997). "Biodegradation of benzo[a]pyrene added in crude oil to uncontaminated soil." Applied and Environmental Microbiology, 63, 4511 - 4515

Kasterner M., Streibich S., and Beyer M. (1999). "Formation of bound residue during microbial degradation of [C-14] anthracene in soil." Applied and Environmental Microbiology, 65, 1834 - 1842.

Kazumi, J., Caldwell, M. E., Suflita, J. M., Lovley, D. R., and Young, L. Y. (1997). "Anaerobic degradation of benzene in diverse anoxic environments." Environ. Sci. Technol., 31(3), 813-818.

Kemblowski, M. W., Salanitro, J. P., Deeley, G. M., and Stanely, C. C. "Fate and transport of residual hydrocarbon in groundwater-a case study." NWWA/API Conf. Petroleum Hydrocarbons and Organic Chemicals in Ground Water: Prevention, Detection and Restoration, Houston, 207-231.

Kennedy, V. H., Sanchez, A. L., Oughton, D. H., and Rowland, A. P. (1997) "Use of single and sequential chemical extractants to assess radionuclide and heavy metal availability from soils for root uptake" Analyst, 122, 89R-100R.

Kincannon D. F., and Lin Y. S. (1985). "Microbial degradation of hazardous wastes by land treatment." Proceedings of the Purdue Industrial Waste Conference, Purdue University, IN, 40, 607 – 19.

King, M.W.G., and Barker, J.F. (1999). "Migration and natural fate of a coal tar creosote plume 1. overview and plume development." Journal of Contaminant Hydrology, 39, 249-279.

King, M.W.G., Barker, J.F., Devlin, J.F., and Butler, B.J. (1999). "Migration and natural fate of a coal tar creosote plume 2. mass balance and biodegradation indicators." Journal of Contaminant Hydrology, 39, 281-307.

Kitandis, P. K. (1994). "The concept of the dilution index." Water Resour. Res., 30(7), 2011-2026.

Knopp, D., Siefert, M., Väänänen, V., and Niessner, R. (2000). "Determination of polycyclic aromatic hydrocarbons in contaminated water and soil samples by immunological and chromatographic methods." Environmental Science and Technology, 34(10), 2035 – 2041.

Knox, R. C., Sabatini, D. A., and Canter, L. W. (1993). Subsurface transport and fate processes, CRC Press, Boca Raton, FL.

Komatsu, T., Omori, T., and Kodama, T. (1993). "Microbial degradation of the polycyclic aromatic hydrocarbons acenaphthene and acenaphthylene by a pure bacterial culture." Bioscience Biotechnology and Biochemistry, 57(5), 864-865.

Kotterman, M.J.J., Reiberg, H.J., Huge, A., and Field, J.A. (1998). "Polycyclic aromatic hydrocarbon oxidation by the white-rot fungus, Bjerkandera sp. Bos55 in the presence of nonionic surfactants." Biotechnology and Bioengineering, 57, 220 – 227.

Koenigsberg, S., Sandefur, C., Mahaffey, W., Deshusses, M., and Fortin, N. (1999). "Peroxygen-mediated bioremediation of MTBE." In Situ Bioremediation of Petroleum Hydrocarbon and Other Organic Compounds, B. C. Alleman and A. Leeson, eds., Battelle Press, Columbus, OH, 13-18.

Krieger, C. J., Beller, H. R., Reinhard, M., and Spormann, A. M. (1999). "Initial reactions in anaerobic oxidation of m-xylene by the denitrifying bacterium Azoarcus sp. strain T." J. Bacteriol., 181(20), 6403-6410.

Kuhn, E. P., Colberg, P. J., Schnoor, J. L., Wanner, O., Zehnder, A. J. B., and Schwarzenbach, R. P. (1985). "Microbial transformations of substituted benzenes during infiltration of river water to groundwater: laboratory column studies." Environ. Sci. Technol., 19, 961-968.

Kuhn, E. P., Zeyer, J., Eicher, P., and Schwarzenbach, R. P. (1988). "Anaerobic degradation of alkylated benzenes in denitrifying laboratory aquifer columns." Appl. Environ. Microbiol., 54(2), 490-496.

LaGrega, M. D., Buckingham, P. L., and Evans, J. C. (1994). Hazardous Waste Management. McGraw Hill, Inc., New York, NY.

Landmeyer, J. E., Chapelle, F. H., Petkewich, M. D., and Bradley, P. M. (1998). Assessment of natural attenuation of aromatic hydrocarbons in groundwater near a former manufactured-gas plant, South Carolina, USA, Environmental Geology, 34(4), 79-292.

Landmeyer, J., Chapelle, F., Bradley, P., Pankow, J., Church, C., and Tratnyek, P. (1998). "Fate of MTBE relative to benzene in a gasoline-contaminated aquifer (1993-98)." Ground Water Monit. Rem., 18(4), 93-102.

Landmeyer, J. E., Vroblesky, D. A., and Chapelle, F. H. (1996). "Stable carbon isotope evidence of biodegradation zonation in a shallow jet-fuel contaminated aquifer." Environ. Sci. Technol., 30(4), 1120-1128.

Landmeyer, J. E., Chapelle, F. H., Herlong, H. H., Bradley, P. M. (2001). "Methyl *tert*-butyl ether biodegradation by indigenous aquifer microorganisms under natural and artificial oxic conditions." Environ. Sci. Technol., 35(6), 1118-1126.

Lane, J. C. (1980). "Gasoline and other motor fuels." Kirk-Othmer Encyclopedia of Chemical Technology, M. Grayson, ed., John Wiley & Sons, New York, 652-695.

Larsen, T., Kjeldsen, P., and Christensen, T. H. (1992). "Sorption of hydrophobic hydrocarbons on three aquifer materials in a flow through system." Chemosphere, 24(4), 439-451.

Lee, L.S., Hagwall, M., Delfino, J.J., and Rao, P.S.C. (1992a). "Partitioning of polycyclic aromatic hydrocarbons from diesel fuel into water." Environmental Science and Technology, 26, 104-2110.

Lee, L.S., Rao, P.S. C., and Okuda, I. (1992b). "Equilibrium partitioning of polycyclic aromatic hydrocarbons from coal tar into water." Environmental Science and Technology, 26, 2110-2115.

Lewis, T.A. and Crawford, R.L. (1995). "Transformation of carbon tetrachloride via sulfur and oxygen substitution by Pseudomonas sp. strain KC." J. Bacteriol. 177, 2204-2208.

Litaor, M. I. and Ibrahim, S. A. (1996). "Plutonium association with selected solid phases in soils of Rocky Flats, Colorado, using sequential extraction technique" J. Environ. Qual., 25, 1144-1152.

Lovley, D. R. (1991). "Dissimilatory Fe(III) and Mn(IV) reduction." Microbiol. Rev., 55(2), 259-287.

Lovley, D. R. (1997). "Potential for anaerobic bioremediation of BTEX in petroleum-contaminated aquifers." J. Indust. Microbiol. Biotechnol., 18(2-3), 75-81.

Lovley, D. R., Baedecker, M. J., Lonergan, D. J., Cozzarelli, I. M., Phillips, E. J. P., and Siegel, D. I. (1989). "Oxidation of aromatic contaminants coupled to microbial iron reduction." Nature, 339(6222), 297-299.

Lovley, D. R., Coates, J. D., Woodward, J. C., and Phillips, E. J. P. (1995). "Benzene oxidation coupled to sulfate reduction." Appl. Environ. Microbiol., 61(3), 953-958.

Lovley, D. R., and Goodwin, S. (1988). "Hydrogen concentrations as an indicator of the predominant terminal electron-accepting reactions in aquatic sediments." Geochim. Cosmochim. Acta, 52, 2993-3003.

Lovley, D. R., and Lonergan, D. J. (1990). "Anaerobic oxidation of toluene, phenol, and p-cresol by the dissimilatory iron-reducing organism, GS-15." Appl. Environ. Microbiol., 56(6), 1858-1864.

Lovley, D. R., and Phillips, E. J. P. (1987). "Competitive mechanisms for inhibition of sulfate reduction and methane production in the zone of ferric iron reduction in sediments." Appl. Environ. Microbiol., 53(11), 2636-2641.

Liu, K., Xie, W., Zhao, Z., Pan, W., and Riley, J.T. (2000). "Investigation of polycyclic aromatic hydrocarbons in fly ash from fluidized bed combustion systems." Environmental Science and Technology, 34(11), 2273 – 2279.

Lyman, W. J., Reidy, P. J., and Levy, B. (1992). Mobility and degradation of organic contaminants in subsurface environments, C K Smoley, Chelsea, MI.

Lyngkilde, J., and Christensen, T.H. (1992). " Fate of organic contaminants in the redox zones of a landfill leachate plume (Vejen, Denmark)." Journal of Contaminant Hydrology, 10, 291-307.

Lyngkilde, J., and Christensen, T. H. (1992). "Fate of organic contaminants in the redox zones of a landfill leachate pollution plume (Vejen, Denmark)." J. Contamin. Hydrol., 10(4), 291-307.

Mace, R. E., Fisher, R. S., Welch, D. M., and Parra, S. P. (1997). "Extent, mass, and duration of hydrocarbon plumes from leaking petroleum storage tank sites in Texas." Geological Circular 97-1, Bureau of Economic Geology, Austin, TX.

MacFarlane, D. S., Cherry, J. A., Gillham, R. W., and Sudicky, E. A. (1983). "Migration of contaminants in groundwater at a landfill: A case study, 1. Groundwater flow and plume delineation." J. Hydrol., 63, 1-29.

MacFarlane, I. D., McCleary, G. D., Hoffman, H. L., an Logan, C. M. (1994). "Considering microbial conditions at a former manufactured gas plant." in Bioremediation of Chlorinated and Polycyclic Aromatic Hydrocarbon

Compounds, R. E. Hinchee, A. Leeson, L. Semprini, S. K. Ong (Eds.), Lewis Publishers, Boca Raton, FL, pp. 462 – 468.

MacIntyre, W.G., Boggs, M., Antworth, C.P., Stauffer, T.B. (1993). "Degradation kinetics of aromatic organic solutes introduced into a heterogeneous aquifer." Water Resources Research, 29(12), 4045-4051.

MacIntyre, W. G., Boggs, M., Antworth, C. P., and Stauffer, T. B. (1993). "Degradation kinetics of aromatic organic solutes introduced into a heterogeneous aquifer." Water Resour. Res., 29(12), 4045-4051.

Mackay, D., Shiu, W. Y., and Ma, K. C. (1992). "Monoaromatic Hydrocarbons, Chlorobenzenes, and PCBs." Illustrated Handbook of Physical-Chemical Properties and Environmental Fate for Organic Chemicals, Lewis Publishers, Chelsea, MI.

Mackay, D., Shiu, W. Y., and Ma, K. C. (1993). "Volatile Organic Chemicals." Illustrated Handbook of Physical-Chemical Properties and Environmental Fate for Organic Chemicals, Lewis Publishers, Chelsea, MI.

Mackay, D.M. and Cherry, J.A. (1989). "Groundwater contamination: pump and treat remediation." Environ. Sci. Technol., 23, 630-636.

MacQuarrie, K. T. B., and Sudicky, E. A. (1990). "Simulation of biodegradable organic contaminants in groundwater. 2. Plume behavior in uniform and random flow fields." Water Resour. Res., 26(2), 223-239.

Madsen, E. L. (1991). "Determining in situ biodegradation, facts and challenges." Environ. Sci. Technol., 25(10), 1663-1673.

Madsen, E. L., Sinclair, J. L., and Ghiorse, W. C. (1991). "In situ biodegradation: Microbiological patterns in a contaminated aquifer." Science, 252, 830-833.

Mahaffey, W.R., Gardner, J.C., Mickel, C., Andrews, S., and Santengelo-Dreiling, T. (1999). "Evaluation of in situ biodegradation of chlorinated solvents in a fractured bedrock aquifer." Natural Attenuation of Chlorinated Solvents, Petroleum Hydrocarbons, and Other Organic Compounds, B.C. Alleman and A. Leeson, eds., Battelle Press, Columbus, OH, 71-76.

March, J. (1985). Advanced Organic Chemistry, John Wiley & Sons, Inc., New York, NY.

Maynard, J. B., and Sanders, W. N. (1969). "Determination of the detailed hydrocarbon composition and potential atmospheric reactivity of full-range motor gasolines." J. Air Pollut. Control Assoc., 19(7), 505-510.

Maymo-Gatell, X., Tandoi, V., Gossett, J.M., and Zinder, S.H. (1995). "Characterization of an H_2-utilizing enrichment culture that reductively dechlorinates tetrachloroethene to vinyl chloride and ethene in the absence of methanogenesis and acetogenesis." Appl. Environ. Microbiol., 61(11), 3928-3933.

McAllister, P. M., and Chiang, C. Y. (1994). "A practical approach to evaluating natural attenuation of contaminants in ground water." Ground Wat. Monitor. Remed., 14(2), 161-173.

McBride, M.B. (1994). Environmental Soil Chemistry, Oxford University Press, New York

McCarty, P. L. (1972). "Energetics of organic matter degradation." Water Pollution Microbiology, R. Mitchell, ed., Wiley-Interscience, New York, 91-118.

McKee, J. E., Laverty, F. B., and Hertel, R. M. (1972). "Gasoline in groundwater." J WPCF, 44(2), 293-302.
McLaren, R.G., Backes, C.A., Rate, A.W., and Swift, R.S. (1998). "Cadium and cobalt desorption kinetics from soil clays: Effect of sorption period" Soil Sci. Soc. Am. J., 62, 332-337.
McLaren, R.G., Lawson, D.A. and Swift, R.S. (1986). "Sorption and desorption of cobalt by soils and soil constituents" J. Soil Sci., 37, 413-426.
McNabb, J. F., and Dunlap, W. J. (1975). "Subsurface biological activity in relation to ground-water pollution." Ground Water, 13(1), 33-44.
McGinnis, G. D., Borazjani, H., McFarland, L. K., Pope, D. F., and Strobel, D. A. (1988). Characterization and laboratory testing soil treatability studies for creosote and pentachlorophenol sludges and contaminated soil, USEPA Report No. 600/2-88/055. Robert S. Kerr Environmental Laboratory, Ada, OK.
Michel G. A., Huis, I. V., and Weners, J. (1995). "Biological PAH degradation in dredge sludges." In Bioremediation of Recalcitrant Organics, Hinchee, R.E., Hoppel, R.E., and Anderson, D.B. (eds.), Battelle Press, Columbus, Ohio, pp. 17-22.
Michalenko, E. M., Basu, D. K., Sage, G. W., Meylan, W. M., Beauman, J. A., Jarvis, W. F., and Gray, D. A. (1993). "Solvents 2." Handbook of Environmental Fate and Exposure Data for Organic Chemicals, P. H. Howard, ed., Lewis Publishers, Chelsea, MI.
Miller, R. R. (1996). "Phytoremediation" Technology Overview Report Number TO-96-03, Ground-Water Remediation Technologies Analysis Center, Pittsburgh, PA. October 1996. Homepage: http://www.grtac.org
Mikesell, M.D. and Boyd, S.A. (1990). "Dechlorination of chloroform by Methanosarcina strains." Appl. Environ. Microbiol., 56, 1198-1201.
Miyamoto, K. and Urano, K. (1996). "Reaction rates and intermediates of chlorinated organic compounds in water and soil." Chemosphere 32(12), 2399-2408.
Mo, K., Lora, C. O., Wanken, A. E., Javanmardian, M., Yang, X., and Kulpa, C. F. (1997). "Biodegradation of methyl t-butyl ether by pure bacterial cultures." Appl Microbiol Biotechnol, 47(1), 69-72.
Moldowan, J. M., Dahl, J., McCaffrey, M. A., Smith, W. J., and Fetzer, J. C. (1995). "Application of biological marker technology to bioremediation of refinery by-products." Energy Fuels, 9(1), 155-162.
Molz, F. J., and Widdowson, M. A. (1988). "Internal inconsistencies in dispersion-dominated models that incorporate chemical and microbial kinetics." Water Resour. Res., 24(4), 615-619.
Mormile, M., Liu, S., and Suflita, J. (1994). "Anaerobic biodegradation of gasoline oxygenates: extrapolation of information to multiple sites and redox conditions." Environ. Sci. Technol., 28(9), 1727-1732.
Morrison, R. T., and Boyd, R. N. (1973). Organic chemistry, Allyn and Bacon, Boston.
Mueller, J. G., Chapman, P. J., Blattmann, B. O., and Pritchard, P. H. (1990). "Isolation and characterization of a flouranthene-utilizing strian of Pseudomonas paucimobilis." Applied and Environmental Microbiology, 56(4), 1079-1086.

Mueller, J. G., Chapman, P. J., and Pritchard, P. H. (1989). "Creosote-contaminated sites – their potential for bioremediation." Environmental Science and Technology, 23, 1197-1201.
Neff, J. M. (1985). "Polycyclic aromatic hydrocarbons." in Fundamentals of Aquatic Toxicology, G. M. Rand and S. R. Petrocelli (Eds.), Hemisphere Publishing Corporation, Washington, D.C.
Newell, C. J., Hopkins, L. P., and Bedient, P. B. (1990). "A hydrogeologic database for ground-water modeling." Ground Water, 28(5), 703-714.
Nielsen, P. H., and Christensen, T. H. (1994). "In situ measurement of degradation of specific organic compounds under aerobic, denitrifying, iron(III)-reducing, and methanogenic groundwater conditions." in Bioremediation of Chlorinated and Polycyclic Aromatic Hydrocarbon Compounds, R. E. Hinchee, A. Leeson, L. Semprini, S. K. Ong (Eds.), Lewis Publishers, Boca Raton, FL, pp. 416 – 422.
Novotny, M., Strand, J. W., Smith, S. L., Wiesler, D., and Schwende, F. J. (1981). "Compositional studies of coal tar by capillary gas chromatography/mass spectrometry." Fuel, 60, 213 – 220.
Norris, R.D. and Wilson, D.J. (1999). "Benefits and concerns with application of the U.S. EPA protocol for monitored natural attenuation." Natural Attenuation of Chlorinated Solvents, Petroleum Hydrocarbons, and Other Organic Compounds, B.C. Alleman and A. Leeson, eds., Battelle Press, Columbus, OH, 53-58.
Norris, R. D. (1994). "In-situ bioremediation of soils and groundwater contaminated with petroleum hydrocarbons." Handbook of Bioremediation, R. D. Norris, R. E. Hinchee, R. Brown, P. L. McCarty, L. Semprini, J. T. Wilson, D. H. Kampbell, M. Reinhard, E. J. Bouwer, R. C. Borden, T. M. Vogel, J. M. Thomas, and C. H. Ward, eds., Lewis Publishers, Boca Raton, FL, 17-37.
NRC. (1993). In situ bioremediation: When does it work?, National Research Council, National Academy Press, Washington, DC.
NRC. (2000). Natural attenuation for groundwater remediation, National Research Council, National Academy Press, Washington, DC.
Odencrantz, J. E. (1998). "Implications of MTBE for intrinsic remediation of underground fuel tank sites." Remediation, 9(3), 7-16.
Odermatt, J. R. (1994). "Natural chromatographic separation of benzene, toluene, ethylbenzene and xylenes (BTEX compounds) in a gasoline contaminated ground water aquifer." Org. Geochem., 21(10/11), 1141-1150.
Office of Technology Assessment (OTA). (1984). Protecting the nation's groundwater from contamination, Vol. 1. OTA-0-233, Washington, DC.
Palmer, C. D., and Puls, R.W. (1994). "EPA Ground Water Issue: Natural Attenuation of Hexavalent Chromium in Groundwater and Soils." EPA/540/5-94/505, USEPA Office of Research and Development, USEPA Office of Solid Waste and Emergency Response. October 1994.
Park K.S., Sims, R. C., and Dupont, R. (1990). "Transformation of PAHs in soil systems." J. Envir. Eng., 116, 632 – 640.
Park, K., and Cowan, B. M. (1997). "Biodegradation of gasoline oxygenates." In Situ and On-Site Bioremediation: Volume 1, B. C. Alleman and A. Leeson, eds., Battelle Press, Columbus, OH, 17.
Parkinson, G. (2000). "Gasoline minus MTBE." Chem. Eng., 107(1), 45-46.

Peters, C. and Luthy, R. (1993). "Coal tar dissolution in water-miscible solvents: experimental evaluation." Environmental Science and Technology, 27, 2831-2843.

Peters, C. A., Labieniec, P. A., and Knightes, C. D. (1996). "Multicomponent NAPL composition dynamics and risk,." in Proceedings of the ASCE Annual Convention: Non-aqueous Phase Liquids (NAPLs) in the Subsurface Environment: Assessment and Remediation, L. N. Reddi (Ed.), Washington D.C., November 12-14, pp. 681 – 692.

Phelps, C. D., Kazumi, J., and Young, L. Y. (1996). "Anaerobic degradation of benzene in BTX mixtures dependent on sulfate reduction." FEMS Microbiol. Lett., 145, 433-437.

Phelps, C. D., and Young, L. Y. (1999). "Anaerobic biodegradation of BTEX and gasoline in various aquatic sediments." Biodegradation, 10(1), 15-25.

Pirnik, M. R., Atlas, R. M., and Bartha, R. (1974). "Hydrocarbon metabolism by Brevibacterium erythrogenes: Normal and branched alkanes." J. Bacteriol., 119(3), 868-878.

Piveteau, P., Fayolle, F., Vandecasteele, J.-P., and Monot, F. (2001). "Biodegradation of tert-butyl alcohol and related xenobiotics by a methylotrophic bacterial isolate." Appl. Microbiol. Biotechnol., 55(3), 369-373.

Pott, B., and Henrysson, T. (1995). "Ex situ bioremediation of polycyclic aromatic hydrocarbons in laboratory systems." In Bioremediation of Recalcitrant Organics, Hinchee, R.E., Hoppel, R.E., and Anderson, D.B. (eds.), Battelle Press, Columbus, Ohio, pp. 39 - 44.

Poulsen, M., Lemon, L., and Barker, J. F. (1992). "Dissolution of monoaromatic hydrocarbons into groundwater from gasoline-oxygenate mixtures." Environ. Sci. Technol., 26(12), 2483-2489.

Priddle, M. W., and MacQuarrie, K. T. B. (1994). "Dissolution of creosote in groundwater: an experimental and modeling investigation." Journal of Contaminant Hydrology, 15, 27 – 56.

Prince, R. C., Elmendorf, D. L., Lute, J. R., Hsu, C. S., Halth, C. E., Senius, J. D., Dechert, G. J., Douglas, G. S., and Butler, E. L. (1994). "17a(H), 21b(H)-Hopane as a conserved internal marker for estimating the biodegradation of crude oil." Environ. Sci. Technol., 28(1), 142-145.

Pritchard, P. H., and Costa, C. F. (1991). "EPA's Alaska oil spill bioremediation project." Environ. Sci. Technol., 25(3), 372-379.

Ptacek, C. J., Cherry, J. A., and Gillham, R. W. (1987). "Mobility of dissolved petroleum-derived hydrocarbon in sand aquifers." Oil in Freshwater: Chemistry, Biology, Countermeasure Technology, J. H. Vandermeulen and S. E. Hrudey, eds., Pergamon Press, New York, 195-214.

Radding, S. B., Mill, T., Gould, C. W., Liu, D. H., Johnson, H. L., Bomberger, C. C., and Fojo, C. V. (1976). The Environmental Fate of Selected Polynuclear Aromatic Compounds. U.S. Environmental Protection Agency, EPA-560/5-75-009, Cincinnati, OH.

Reid, J. B., Reisinger, H. J., II, Bartholomae, P. G., Gray, J. C., and Hullman, A. S. (1999). "A comparative assessment of the long-term behavior of MtBE and benzene plumes in Florida, USA." Natural Attenuation of Chlorinated Solvents,

Petroleum Hydrocarbons, and Other Organic Compounds, B. C. Alleman and A. Leeson, eds., Battelle Press, Columbus, OH, 97-102.

Reinhard, M., Goodman, N. L., and Barker, J. F. (1984). "Occurrence and distribution of organic chemicals in two landfill leachate plumes." Environ. Sci. Technol., 18(12), 953-961.

Ribbons, D. W., and Eaton, R. W. (1982). "Chemical transformations of aromatic hydrocarbons that support the growth of microorganisms." Biodegradation and Detoxification of Environmental Pollutants, A. M. Chakrabarty, ed., CRC Press, Boca Raton, FL, 60-84.

Rice, D. W., Dooher, B. P., Cullen, S. J., Everett, L. G., Kastenberg, W. E., Grose, R. D., and Marino, M. A. (1995a). "Recommendations to improve the cleanup process for California's leaking underground fuel tanks (LUFTs)." UCRL-AR-121762, Lawrence Livermore National Laboratory, Livermore, CA.

Rice, D. W., Grose, R. D., Michaelsen, J. C., Dooher, B., MacQueen, D. H., Cullen, S. J., Kastenberg, W. E., Everett, L. G., and Marino, M. A. (1995b). "California leaking underground fuel tank (LUFT) historical case analyses." UCRL-AR-122207, Lawrence Livermore National Laboratory, Livermore, CA.

Riley R. G., Zachara, J. M., and Wobber, F. J. (1992). Chemical contaminants on DOE lands and selection of contaminant mixtures for subsurface science research. US DOE.

Rittmann, B.E., Seagren, E., Wrenn, B.A., Valocchi, A.J., Ray, C., and Raskin, L. (1994). In Situ Bioremediation, 2nd Edition, Noyes Publications, Park Ridge, NJ.

Rittmann, B. E., Seagren, E. A., Wrenn, B. A., Valocchi, A. J., Ray, C., and Raskin, L. (1994). In situ bioremediation, Noyes Publishers, Park Ridge, NJ.

Rivett, M. O. (1995). "Soil-gas signatures from volatile chlorinated solvents: Borden field experiments." Ground Water, 33(1), 84-98.

Robbins, G. A., Wang, S., and Stuart, J. D. (1993). "Using the static headspace method to determine Henry's Law constants." Anal. Chem., 65(21), 3113-3118.

Rogers, R. D., McFarlane, J. C., and Cross, A. J. (1980). "Adsorption and desorption of benzene in two soils and montmorillonite clay." Environ. Sci. Technol., 14(4), 457-460.

Rugge, K., Bjerg, P. L., and Christensen, T. H. (1995) "Distribution of organic compounds from municipal solid waste in the groundwater downgradient of a landfill (Grindsted, Denmark)." Environ. Sci. Technol., 29(5), 1395-1400.

Sage, G. W., Jarvis, W. F., and Gray, D. A. (1990). "Solvents." Handbook of Environmental Fate and Exposure Data for Organic Chemicals, P. H. Howard, ed., Lewis Publishers, Chelsea, MI.

Salanitro, J., Diaz, L., Williams, M., and Wisniewski, H. (1994). "Isolation of a bacterial culture that degrades methyl t-butyl ether." Appl. Environ. Microbiol., 1994(7), 2593-2596.

Salanitro, J. P. (1993). "The role of bioattenuation in the management of aromatic hydrocarbon plumes in aquifers." Ground Wat. Monitor. Remed., 13(4), 150-161.

Schwab A. P., Banks M. K, and Arunachalam M. (1995). "Biodegradation of polycyclic aromatic hydrocarbons in Rhizosphere soil." In Bioremediation of

Recalcitrant Organics, Hinchee, R.E., Hoppel, R.E., and Anderson, D.B. (eds.), Battelle Press, Columbus, Ohio, pp. 23 - 29.
Scholz-Muramatsu, H., Neumann, A., Mem□er, M., Moore, E., and Diekert, G. (1995). "Isolation and characterization of Dehalospirillum multivorans gen. nov., sp. nov., a tetrachloroethene-utilizing, strictly anaerobic bacterium." Arch. Microbiol., 163, 48-56.
Schwarzenbach, R.P., Gschwend, P.M., and Imboden, D.M. (1993). Environmental Organic Chemistry, John Wiley & Sons, Inc., New York, NY.
Schirmer, M., and Barker, J. F. (1998). "A study of long-term MTBE attenuation in the Borden Aquifer, Ontario, Canada." Ground Wat. Monitor. Remed., 18(2), 113-122.
Schirmer, M., Butler, B. J., Barker, J. F., Church, C. D., and Schirmer, K. (1999). "Evaluation of biodegradation and dispersion as natural attenuation processes of MTBE and benzene at the Borden field site." Phys. Chem. Earth (B), 24(6), 557-560.
Schultz M. K., Burnett, W. C., Inn, K. G. W., Thomas, J. W. L., and Lin Z. (1996). Conference Report, NIST speciation workshop. J. Res. NIST, 101, 707-715.
Schmitt, R., Langguth, H.-R., Püttmann, W., Rohns, H. P., Eckert, P., and Schubert, J. (1996). "Biodegradation of aromatic hydrocarbons under anoxic conditions in a shallow sand and gravel aquifer of the Lower Rhine Valley, Germany." Org. Geochem., 25(1/2), 41-50.
Seagren, E. A., and Becker, J. G. (1999). "Organic acids as a bioremediation monitoring tool." Natural Attenuation of Chlorinated Solvents, Petroleum Hydrocarbons, and Other Organic Compounds, B. C. Alleman and A. Leeson, eds., Battelle Press, Columbus, OH, 343-348.
Seagren, E. A., Smets, B. F., Hollander, D. J., Stahl, D. A., and Rittmann, B. E. (1998). "Total alkalinity as a bioremediation monitoring tool." Natural Attenuation: Chlorinated and Recalcitrant Compounds, G. B. Wickramanayake and R. E. Hinchee, eds., Battelle Press, Columbus, OH, 117-122.
Sharma, P.K. and McCarty, P.L. (1996). "Isolation and characterization of a facultatively aerobic bacterium that reductively dehalogenates tetrachloroethene to cis-1,2-dichloroethene." Appl. Environ. Microbiol., 62(3), 761-765.
Sherwood Lollar, B., Slater, G.F., Ahad, J., Sleep, B., Spivack, J., Brennan, M., and MacKenzie, P. (1999). "Contrasting carbon isotope fractionation during biodegradation of trichloroethylene and toluene: Implications for intrinsic bioremediation." Org. Geochem., 30, 813-820.
Sherwood Lollar, B., Slater, G. F., Ahad, J., Sleep, B., Spivack, J., Brennan, M., and MacKenzie, P. (1999). "Contrasting carbon isotope fractionation during biodegradation of trichloroethylene and toluene: Implications for intrinsic bioremediation." Org. Geochem., 30(8A), 813-820.
Simpkin T. J. and Giesbrecht G. (1994). "Slurry bioremediation of polycyclic aromatic hydrocarbons in sediments from industrial complex." In Bioremediation of cChlorinated and Polycyclic Aromatic Hydrocarbon Compounds, Hinchee, R.E., Leeson, A., Semprini, L., and Ong, S.K. (eds.), Lewis Publishers, Boca Raton, FL., pp. 484 - 488.

Sims, R. C. and Overcash, M. R. (1983). "Fate of polynuclear aromatic compounds (PNAs) in soil-plant systems." Residue Reviews, 88, 1-68.
Small, M. C. (1998). "Risk-based corrective action, natural attenuation, and changing regulatory paradigms." Bioremed. J., 2(3&4), 221-225.
Smatlak, C.R., Gosset, J.M., and Zinder, S.H. (1996). "Comparative kinetics of hydrogen utilization for reductive dechlorination of tetrachloroethene and methanogenesis in an anaerobic enrichment culture." Environ. Sci. Technol., 30, 2850-2858.
Smith, R. L., Harvey, R. W., and LeBlanc, D. R. (1991a). "Importance of closely spaced vertical sampling in delineating chemical and microbiological gradients in groundwater studies." J. Contam. Hydrol., 7(3), 285-300.
Smith, R. L., Howes, B. L., and Duff, J. H. (1991b). "Denitrification in nitrate-contaminated groundwater: Occurrence in steep vertical geochemical gradients." Geochim. Cosmochim. Acta, 55(7), 1815-1825.
Somsamak, P., Cowan, R. M., and Häggblom, M. M. (2001). "Anaerobic biotransformation of fuel oxygenates under sulfate-reducing conditions." FEMS Microbiol. Ecol. 37(3), 259-264.
Spexet, A.H., Bedient, P.B., and Marcon, M. (1999). "Biodegradation and DNAPL issues associated with dry cleaning sites." Natural Attenuation of Chlorinated Solvents, Petroleum Hydrocarbons, and Other Organic Compounds, B.C. Alleman and A. Leeson, eds., Battelle Press, Columbus, OH, 7-11
Speight, J. G. (1991). The chemistry and technology of petroleum, Marcel Dekker, New York.
Spormann, A. M., and Widdel, F. (2000). "Metabolism of alkylbenzenes, alkanes, and other hydrocarbons in anaerobic bacteria." Biodegradation, 11(2-3), 85-105.
Squillace, P., Pankow, J., Korte, N., and Zogorski, J. (1998). "Environmental behavior and fate of methyl tert-butyl ether (MTBE)." .
Squillace, P. J., Pankow, J. F., Korte, N. E., and Zogorski, J. S. (1997). "Review of the environmental behavior and fate of methyl tert-butyl ether." Environ. Toxicol. Chem., 16(9), 1836-1844.
Squillace, P. J., Zogorski, J. S., Wilber, W. G., and Price, C. V. (1996). "Preliminary assessment of the occurrence and possible sources of MTBE in groundwater in the United States, 1993-1994." Environ. Sci. Technol., 30(5), 1721-730.
Stringfellow, W.T., and Aitken, M.D. (1995). "Competitive metabolism of naphthalene, methynaphthalene, and fluorene by phenanthrene degrading pseudomonads." Applied and Environmental Microbiology, 61, 357 – 362.
Stucki, G. and Alexander, M. (1987). "Role of dissolution rate and solubility in biodegradation of aromatic compounds." Applied and Environmental Microbiology 53(2), 292-297.
Stumm, W. and Morgan, J.J. (1996). Aquatic Chemistry, Third Ed., John Wiley & Sons, Inc., New York, NY.
Sturchio, N.C., Clausen, J.L., Heraty, L.J., Huang, L., Holt, B.D., and Abrajano, Jr., T.A. (1998). "Chlorine isotope investigation of natural attenuation of trichloroethene in an aerobic aquifer." Environ. Sci. Technol., 32, 3037-3042.
Steffan, R. J., McClay, K., Vainberg, S., Condee, C. W., and Zhang, D. (1997). "Biodegradation of the gasoline oxygenates methyl tert-butyl ether, ethyl tert-

butyl ether, and tert-amyl methyl ether by propane-oxidizing bacteria." Appl. Environ. Microbiol., 63(11), 4216-4222.
Steffan, R. J., Vainberg, S., Condee, C., McClay, K., and Hatzinger, P. "Biotreatment of MTBE with a new bacterial isolate." Second Int. Conf. Remediation of Chlorinated and Recalcitrant Compounds (Platform Abstracts), May 22-25, 2000, Monterey, CA.
Stehmeier, L. G., Francis, M. M., Jack, T. R., Diegor, E., Winsor, L., and Abrajano, T. A., Jr. (1999). "Field and in vitro evidence for in situ bioremediation using compound-specific $^{13}C/^{12}C$ ratio monitoring." Org. Geochem., 30(8A), 821-833.
Stuart, B. J., Bowlen, G. F., and Kosson, D. S. (1991). "Competitive sorption of benzene, toluene and the xylenes onto soil." Environ. Prog., 10(2), 104-109.
Stumm, W., and Morgan, J. J. (1981). Aquatic Chemistry, Wiley-Interscience, New York.
Suarez, M.P. and Rifai, H.S. (1999). "Biodegradation rates for fuel hydrocarbons and chlorinated solvents." Biorem. J., 3(4), 337-362.
Suflita, J., and Mormile, M. (1993). "Anaerobic biodegradation of known and potential gasoline oxygenates in the terrestrial subsurfaces." Environ. Sci. Technol., 27(5), 976-978.
Tangley, L. (1984). "Groundwater contamination, local problems become national issue." BioScience, 34(3), 142-148.
Tessier A., Campbell, P. G. C., and Bisson, M. (1979). "Sequential extraction procedures for the speciation of particulate trace metals" Anal. Chem., 51, 844-851.
Thierrin, J., Davis, G. B., Barber, C., Patterson, B. M., Pribac, F., Power, T. R., and Lambert, M. (1993). "Natural degradation rates of BTEX compounds and naphthalene in a sulfate reducing groundwater environment." Hydrol. Sci. J., 38(4), 309 – 322.
Thierrin, J., Davis, G. B., and Barber, C. (1995). "A groundwater tracer test with deuterated compounds for monitoring in situ biodegradation and retardation of aromatic compounds." Ground Water, 33(3), 469 – 475.
Thierrin, J., Davis, G. B., Barber, C., Patterson, B. M., Pribac, F., Power, T. R., and Lambert, M. (1993). "Natural degradation rates of BTEX compounds and naphthalene in a sulphate reducing groundwater environment." Hydrol. Sci. J., 38(4), 309-323.
U.S. Department of Energy, (1999). Technical Guidance for the Long-Term Monitoring of Natural Attenuation Remedies at Department of Energy Sites, Office of Environmental Restoration, US DOE, Washington, D.C.
U.S. EPA. (1986). Superfund Public Health Evaluation Manual, EPA 540/1-86/060, Office of Solid Waste and Emergency Response, US Environmental Protection Agency Washington, DC.
U.S. EPA. (1997). Proceedings of the Symposium on Natural Attenuation of Chlorinated Organics in Ground Water, EPA/540/R-97/504, Office of Research and Development, US Environmental Protection Agency, Washington, DC.
U.S. EPA. (1999). Use of Monitored Natural Attenuation at Superfund, RCRA Corrective Action, and Underground Storage Tank Sites, Directive 9200.4-17P,

Office of Solid Waste and Emergency Response, US Environmental Protection Agency, Washington, DC.
US EPA, (1998). "A Citizen's Guide to Phytoremediation." EPA 542-F-98-011, Office of Solid Waste and Emergency Response. US Environmental Protection Agency, Washington, DC.
U.S. EPA, (1999a). "National Priority Site Fact Sheet, Preferred Plating Corporation." Office of Solid Waste and Emergency Response. US Environmental Protection Agency, Washington, DC.
URL: http://www.epa.gov/region01/superfund/site_sum/0202245c.htm
U.S. EPA, (1999b). "Record of Decision (ROD) Abstract, Reeves Southeast Galvanizing Corp." Office of Solid Waste and Emergency Response. US Environmental Protection Agency, Washington, DC.
URL:http://www.epa.gov/oerrpage/superfund/sites/query/rods/r0493149.htm
U.S. EPA (1999c). "Site Summary" for the Geiger (C&M Oil) Site. June 1, 1999. URL: http://www.epa.gov/region04/wastepgs/mpl/nplsc/geigersc.htm
U.S. EPA. (2001). "EPA's Decision on California's Request for Waiver from the Reformulated Gasoline Program Oxygen Requirement." EPA 420-F-01-003, Office of Transportation and Air Quality, Office of Solid Waste and Emergency Response. US Environmental Protection Agency, Washington, DC..
U.S. EPA. (1997). Use of Monitored Natural Attenuation at Superfund, RCRA Corrective Action, and Underground Storage Tank Sites. Office of Solid Waste and Emergency Response., US Environmental Protection Agency, Washington, DC.
U.S. Science Advisory Board (SAB) (1995). "Science Advisory Board Review of the Agency's Approach for Developing Sediment Criteria for Five Metals", EPA-SAB-EPEC-95-020. September 20, 1995.
Van de Velde, K. D., Marley, M. C., Studer, J., and Wagner, D. M. (1995). "Stable carbon isotope analysis to verify bioremediation and bioattenuation." Monitoring and Verification of Bioremediation, R. E. Hinchee, G. S. Douglas, and S. K. Ong, eds., Battelle Press, Columbus, OH, 241-257.
Venosa, A.D., Suidan, M.T., Wrenn, B.A., Strohmeier, K.L., Haines, J.R., Eberhart, B.L., King, D., and Holder, E. (1996). "Bioremediation of an experimental oil spill on the shoreline of Delaware Bay." Environ. Sci. Technol., 30, 1764-1775.
Venosa, A. D., Suidan, M. T., King, D., and Wrenn, B. A. (1997). "Use of hopane as a conservative biomarker for monitoring the bioremediation effectiveness of crude oil contaminating a sandy beach." J. Industr. Microbiol. Biotechnol., 18(2/3), 131-139.
Verschueren, K. (1983). Handbook of Environmental Data on Organic Chemicals, 2nd Edition, Van Nostrand Reinhold Co., New York, NY.
Vogel, T.M. and McCarty, P.L. (1987). "Abiotic and biotic transformations of 1,1,1-trichloroethane under methanogenic conditions." Environ. Sci. Technol., 21(12), 1208-1213.
Vroblesky, D.A., Bradley, P.M., and Chapelle, F.H. (1997). "Lack of correlation between organic acid concentrations and predominant electron-accepting processes in a contaminated aquifer." Environ. Sci. Technol. 31, 1416-1418.

Vroblesky, D. A., Bradley, P. M., and Chapelle, F. H. (1997). "Lack of correlation between organic acid concentrations and predominant electron-accepting processes in a contaminated aquifer." Environ. Sci. Technol., 31(5), 1416-1418.

Vroblesky, D. A., and Chapelle, F. H. (1994). "Temporal and spatial changes of terminal electron-accepting processes in a petroleum hydrocarbon-contaminated aquifer and the significance for contaminant biodegradation." Water Resour. Res., 30(5), 1561-1570.

Walter, U., Beyer, M., Klein, J., and Rehm, H. –J. (1991). "Degradation of pyrene by Rhodococcus sp. UW1." Applied Microbiology and Biotechnology, 34, 671-676.

Wasay S. A., Barrington, S., and Tokunaga, S. (1998). "Retention form of heavy metals in three polluted soils" J. Soil Contam., 7, 103-119.

Watts, R. J. (1998). Hazardous wastes: Sources, pathways, receptors, John Wiley & Sons, New York.

Weiner, J. M., and Lovley, D. R. (1998a). "Anaerobic benzene degradation in petroleum-contaminated aquifer sediments after inoculation with benzene-oxidizing enrichment." Appl. Environ. Microbiol., 64(2), 775-778.

Weiner, J. M., and Lovley, D. R. (1998b). "Rapid benzene degradation in methanogenic sediments from a petroleum-contaminated aquifer." Appl. Environ. Microbiol., 64(5), 1937-1939.

Weissenfels, W. D., Beyer, M., and Klein, J. (1990). "Degradation of phenanthrene, fluorene, and fluoranthene by pure bacterial cultures." Applied Microbiology and Biotechnology, 32, 479-484.

Weissenfels, W. D., Beyer, M., Klein, J., and Rehm, H. J. (1991). "Microbial metabolism of fluoranthene: isolation and identification of ring fission products." Applied Microbiology and Biotechnology, 34, 528-535.

Westlake, D. W. S., Jobson, A., Phillippe, R., and Cook, F. D. (1974). "Biodegradability and crude oil composition." Can. J. Microbiol., 20(7), 915-928.

White, G., Russell, N., and Tidswell, E. (1996). "Bacterial scission of ether bonds." Microbiol. Rev., 60(1), 216-232.

Whited, G. M., and Gibson, D. T. (1991). "Separation and partial characterization of the enzymes of the toluene-4-monooxygenase catabolic pathway in Pseudomonas mendocina KR1." J. Bacteriol., 173(9), 3017-3020.

Whiticar, M. J., Faber, E., and Schoell, M. (1986). "Biogenic methane formation in marine and freshwater environments: CO_2 reduction vs. acetate fermentation--isotope evidence." *Geochim. Cosmochim. Acta*, 50(5), 693-709.

Wiedemeier, T.H., Wilson, J.T., and Kampbell. D.H., (1997). "Natural attenuation of chlorinated aliphatic hydrocarbons at Plattsburgh Air Force Base, New York." Proceedings of the Symposium on Natural Attenuation of Chlorinated Organics in Ground Water, EPA/540/R-97/504, Office of Research and Development, Washington, DC., 76-84.

Wiedemeier, T.H., Swanson, M.A., Moutoux, D.E., Gordon, E.K., Wilson, J.T., Wilson, B.H., Kampbell, D.H., Haas, P.E., Miller, R.N., Hansen, J.E. and Chapelle, F.H. (1998). Technical Protocol for Evaluating Natural Attenuation of

Chlorinated Solvents in Ground Water, EPA/600/R-98/128, Office of Research and Development, Washington, DC.
Wiedemeier, T.H., Rifai, H.S., Newell, C.J. and Wilson, J.T. (1999). Natural Attenuation of Fuels and Chlorinated Solvents in the Subsurface, John Wiley & Sons, Inc., New York, NY.
Wild, A., Hermann, R., and Leisinger, T. (1996). "Isolation of an anaerobic bacterium which reductively dechlorinates tetrachloroethene and trichloroethene." Biodegradation 7, 507-511.
Wilson, B.H., Wilson, J.T., and Luce, D. (1997). "Design and interpretation of microcosm studies for chlorinated compounds." Proceedings of the Symposium on Natural Attenuation of Chlorinated Organics in Ground Water, EPA/540/R-97/504, Office of Research and Development, Washington, DC, 23-30.
Wing, M.R. (1997). Apparent first-order kinetics in the transformation of 1,1,1-trichloroethane in groundwater following a transient release. Chemosphere 34(4), 771-781.
Workman, D.J., Woods, S.L., Gorby, Y.A., Fredrickson, J.K., and Truex, M.J. (1997). "Microbial reduction of vitamin B-12 by Shewanella alga strain BrY with subsequent transformation of carbon tetrachloride." Environ. Sci. Technol., 31 2292-2297.
Wolter, M. A., Zadrazil, F., Martens, R., Bahadir, M. (1997). "Degradation of 8 highly condensed polycyclic aromatic hydrocarbons by Pleurotus sp. Florida in solid wheat-straw substrate." Applied Microbiology and Biotechnology, 48, 398 – 404.
World Health Organization. (1998) Selected non-heterocyclic polycyclic aromatic hydrocarbons. Environmental Health Criteria No. 202. World Health Organization, Geneva, Switzerland.
Wrenn, B.A. and Rittmann, B.E. (1996). Evaluation of a model for the effects of substrate interactions on the kinetics of reductive dehalogenation." Biodegradation 7, 49-64.
Wiedemeier, T., Wilson, J. T., Kampbell, D. H., Miller, R. N., and Hansen, J. E. (1995). Technical protocol for implementing intrinsic remediation with long-term monitoring for natural attenuation of fuel contamination dissolved in groundwater, U.S. Air Force Center for Environmental Excellence, San Antonio, TX.
Wiedemeier, T. H., Rifai, H. S., Newell, C. J., and Wilson, J. T. (1999). Natural attenuation of fuels and chlorinated solvents in the subsurface, John Wiley & Sons, New York.
Wiedemeier, T. H., Swanson, M. A., Moutoux, D. E., Gordon, E. K., Wilson, J. T., Wilson, B. H., Kampbell, D. H., Hansen, J. E., Haas, P., and Chapelle, F. H. (1996a). Technical protocol for evaluating natural attenuation of chlorinated solvents in groundwater, Air Force Center for Environmental Excellence, San Antonio, TX.
Wiedemeier, T. H., Swanson, M. A., Wilson, J. T., Kampbell, D. H., Miller, R. N., and Hansen, J. E. (1996b). "Approximation of biodegradation rate constants for monoaromatic hydrocarbons (BTEX) in ground water." Ground Wat. Monitor. Remed., 16(3), 186-194.

Willey, L. M., Kharaka, Y. K., Presser, T. S., Rapp, J. B., and Barnes, I. (1975). "Short chain aliphatic acid anions in oil field waters and their contribution to the measured alkalinity." Geochim. Cosmochim. Acta, 39, 1707-1711.

Williams, D. E., and Wilder, D. G. (1971). "Gasoline pollution of a ground-water reservoir--a case history." Ground Water, 9(6), 50-56.

Wilson, B. H., Smith, G. B., and Rees, J. F. (1986). "Biotransformations of selected alkylbenzenes and halogenated aliphatic hydrocarbons in methanogenic aquifer material: a microcosm study." Environ. Sci. Technol., 20(10), 997-1002.

Wilson, B. H., Wilson, J. T., Kampbell, D. H., Bledsoe, B. E., and Armstrong, J. M. (1990). "Biotransformation of monoaromatic and chlorinated hydrocarbons at an aviation gasoline spill site." Geomicrobiol. J., 8(3/4), 225-250.

Wilson, J. T., Cho, J. S., Wilson, B. H., and Vardy, J. A. (2000). "Natural attenuation of MTBE in the subsurface under methanogenic conditions." EPA/600/R-00/006, Environmental Protection Agency, Cincinnati, OH.

Wilson, J. T., McNabb, J. F., Balkwill, D. L., and Ghiorse, W. C. (1983a). "Enumeration and characterization of bacteria indigenous to a shallow water-table aquifer." Ground Water, 21(2), 134-142.

Wilson, J. T., McNabb, J. F., Wilson, B. H., and Noonan, M. J. (1983b). "Biotransformation of selected organic pollutants in ground water." Developments in Industrial Microbiology, C. H. Nash, III and L. A. Underkofler, eds., Society for Industrial Microbiology, Arlington, VA, 225-233.

Ye, D. -Y., Siddiqi, M. A., Maccubbin, A. E., Kumar, S., and Sikka, H. C. (1996). "Degradation of polynuclear aromatic hydrocarbons by Sphingomonas paucimobilis." Environmental Science and Technology, 30(1), 136-142.

Yeh, C., and Novak, J. (1994). "Anaerobic biodegradation of gasoline oxygenates in soils." Water Environ. Res., 66(5), 744-752.

Yeh, C. K., and Novak, J. T. (1995). "The effect of hydrogen peroxide on the degradation of methyl and ethyl tert-butyl ether in soils." Wat. Environ. Res., 67(5), 828-834.

Yong R. N., Galvez-Cloutier, R., and Phadungchewit, Y. (1993). "Selective sequential extraction analysis of heavy-metal retention in soil" Can. Geotech. J., 30, 834-847.

Zeyer, J., Kuhn, E. P., and Schwarzenbach, R. P. (1986). "Rapid microbial mineralization of toluene and 1,3-dimethylbenzene in the absence of molecular oxygen." Appl. Environ. Microbiol., 52(4), 944-947.

Zytner, R. G. (1994). "Sorption of benzene, toluene, ethylbenzene, and xylenes to various media." J. Hazard. Matl., 38(1), 113-126.